THE ART OF LEGO® MINDSTORMS® EV3 PROGRAMMING

THE ART OF LEGO® MINDSTORMS® EV3 PROGRAMMING

terry **griffin**

Printed in USA
First printing

18 17 16 15 14   1 2 3 4 5 6 7 8 9

ISBN-10: 1-59327-568-4
ISBN-13: 978-1-59327-568-6

Publisher: William Pollock
Production Editor: Laurel Chun
Interior Design: Octopod Studios
Developmental Editors: Seph Kramer and Jennifer Griffith-Delgado
Technical Reviewers: Daniele Benedettelli and Rob Torok
Copyeditor: Gillian McGarvey
Compositors: Lynn L'Heureux and Riley Hoffman
Proofreader: Emelie Burnette

For information on distribution, translations, or bulk sales, please contact No Starch Press, Inc. directly:
No Starch Press, Inc.
245 8th Street, San Francisco, CA 94103
phone: 415.863.9900; info@nostarch.com
www.nostarch.com

The Library of Congress has cataloged the first edition as follows:

Griffin, Terry, 1962-
The art of LEGO Mindstorms NXT-G programming / Terry Griffin.
p. cm.
Includes index.
ISBN-13: 978-1-59327-218-0
ISBN-10: 1-59327-218-9
1. Robots--Design and construction. 2. Robots--Programming. 3. Lego Mindstorms toys. I. Title.
TJ211.G75 2010
629.8'9252--dc22

2010017757

To my family, who make all the work worthwhile.
And to Bella, who gives me a reason to get up in the morning.

about the author

Terry Griffin has been a software engineer for more than 20 years and has spent most of that time creating software to control various types of machines. He earned a master's degree in Computer Science from the University of Massachusetts and has taught programming at the college and adult education levels. A lifelong LEGO enthusiast, he wrote *The Art of LEGO MINDSTORMS NXT Programming* (No Starch Press) to help his wife, a dynamic middle school science and math teacher, learn how to use these incredible robots in her classroom. He works for the Ion Microscopy Innovation Center division of Carl Zeiss, writing software to control charge particle microscopes.

about the technical reviewers

Daniele Benedettelli is known worldwide for his original LEGO robots, including his Rubik's Cube solvers and his humanoid robots. As a LEGO MINDSTORMS Community Partner (MCP), he helped to test and develop new MINDSTORMS products. He earned a master's degree in Robotics and Automation from the University of Siena in Italy. He holds educational presentations and workshops on Robotics and Information and Communications Technology around the world, teaches robotics at the high school level, and designs LEGO models as a freelancer for LEGO educational programs. He's the author of *The LEGO MINDSTORMS EV3 Laboratory* (No Starch Press).

Rob Torok is a teacher in Tasmania, Australia, and has been using LEGO robotics with his students since 2001. He has mentored teams in RoboCup Junior and the FIRST Robotics Competition, and teaches an online robotics class called SmartBots. In 2010, Rob spent six months based at the Tufts Center for Engineering Education and Outreach (CEEO) in Boston and continues to work closely with the center. He is currently the content editor for both *http://LEGOengineering.com/* and *http://LEGOeducation.com.au/*.

acknowledgments

I'd like to thank my family for all their patience while I wrote this book. Special thanks to my wife, Liz, who spent countless hours reviewing the text and tolerated the robots inhabiting the dining room table.

This work would not have been possible without the help and support of Bill Pollock and the staff at No Starch Press. It was a pleasure to work with Seph Kramer, Laurel Chun, and Jennifer Griffith-Delgado. Their knowledge and expertise were instrumental in the completion of this project.

I'd also like to thank my tech reviewers, Daniele Benedettelli and Rob Torok. Their knowledge of the EV3, and robotics in general, was a big help in keeping the material relevant and technically correct.

brief contents

contents in detail

5

6

program flow 69

7

the WallFollower program: navigating a maze 81

8
data wires

9

10

17

data logging ... 209

18

multitasking ... 219

19

a PID-controlled LineFollower program ... 229

introduction

This book is about learning how to write programs for LEGO MINDSTORMS EV3 robots. The EV3 software is a powerful tool, and this book will teach you how to get the most out of it as you acquire the programming skills necessary to create your own programs.

who this book is for

This book is for anyone who wants to learn how to create programs to control their EV3 robot, whether you're a young robotics enthusiast; an adult teaching children about robotics; a parent; a FIRST LEGO League coach; or a teacher using the EV3 in a classroom. One of my goals in writing this book was to make the material accessible to young learners while going into enough depth to help students and teachers understand the hows and whys of EV3 programming.

prerequisites

This book can be used with either the Home or Education Edition, and you'll use a single general purpose robot for testing your programs. There are only a few relevant differences between the programs for each edition, and I point them out as appropriate. Almost all the material presented here applies to either edition.

No previous programming experience is required. The EV3 software is powerful but easy to use, and makes a great introductory tool for first-time programmers.

what to expect from this book

This book focuses on programming EV3 robots rather than on the mechanical aspects of building them. All of the programs in this book are designed to work with one general-purpose robot or with just the EV3 Intelligent Brick. You'll learn how to work with the core parts of the EV3 software, such as blocks, data wires, files, and variables, and how these pieces work together. You'll also learn some good programming practices, bad habits to avoid, and debugging strategies that will help you have fun while programming and keep your frustration level low.

In this book you'll find step-by-step instructions and explanations for many EV3 programs, including small examples designed to help you understand exactly how EV3 programs work, as well as complete, sophisticated programs designed to perform complex behavior. Along the way, you'll also see programming challenges, which will prompt you to explore EV3 programming on your own to practice the concepts you've learned.

The book begins with an introduction to the EV3 set and the software you'll be using to create your programs. This is followed by the building instructions for the test robot. The next few chapters cover the basics of the EV3 software, culminating in a maze-solving program in Chapter 7. That's followed by several chapters covering the more advanced language features, and the book finishes up with a sophisticated line-following program using a PID controller. Here's an overview of what you'll learn in each chapter.

chapter 1: LEGO and robots: a great combination

The first chapter provides a brief introduction to the LEGO MINDSTORMS EV3 software. It also presents some important differences between the Home and Education Editions, and how they impact this book.

chapter 2: the EV3 programming environment

This chapter gives a tour through the features of the EV3 software. Two simple programs demonstrate how to create programs and run them on the EV3. This chapter also covers the basics on changing block parameters, adding comments, and using Port View.

chapter 3: TriBot: the test robot

In this chapter you'll build TriBot, the test robot. You'll use this general-purpose robot to test the programs throughout the rest of the book.

chapter 4: motion

This chapter is about the EV3 motors and the blocks that control them. You'll build several programs designed to show how these blocks are typically used and to point out some common pitfalls.

chapter 5: sensors

This chapter covers the EV3 sensors: the Touch, Color, Ultrasonic, Infrared, Gyro and Rotation Sensors. You'll build an example program for each sensor and learn how to use Port View to monitor a sensor's value while developing or running a program.

chapter 6: program flow

This chapter focuses on the Switch block (which lets a program make decisions) and the Loop block (which makes a program repeat certain actions). You'll use these program flow blocks to create a simple line-following program.

chapter 7: the WallFollower program: navigating a maze

With all the basic features of EV3 programming covered, at this point you'll be ready to start tackling more complex problems. In this chapter you'll learn how to design, create, and debug a large wall-following program to make your robot solve a maze.

chapter 8: data wires

Data wires are one of the most powerful features in EV3 programming. This chapter explains what data wires are and how to use them effectively. Example programs show how to use data wires to get information from a sensor and how to use a sensor to control a motor.

chapter 9: data wires and the switch block

This chapter covers the advanced features of a Switch block that are available when using data wires. You'll also learn how to use data wires to move data into and out of a Switch block.

chapter 10: data wires and the loop block

In this chapter you'll learn how to use data wires with a Loop block. You'll build a program that makes the robot search in a rectangular spiral pattern using new techniques which involve loop counters and controlling the loop exit condition.

chapter 11: variables

This chapter covers the Variable and Constant blocks. You'll learn how to add and manage variables to store and update values.

chapter 12: my blocks

A My Block is a new block created by grouping other blocks together. In this chapter you'll learn how create My Blocks, use them in your programs, and share My Blocks between projects.

chapter 13: math and logic

This chapter covers the blocks that deal with math and logic: the Math, Logic, Range, Round and Random blocks. You'll learn advanced uses of these blocks as you enhance some programs developed in previous chapters.

chapter 14: the EV3 lights, buttons, and display

In this chapter you'll learn how to use the Brick Button block to control a program and how to use the Brick Status Light block to control the colored lights on the EV3 brick. You'll also learn how to use the Display block, which you'll use to design a simple drawing program.

chapter 15: arrays

This chapter covers arrays and how to use them in EV3 programming. You'll develop a program that lets you give the TriBot a list of commands to execute.

chapter 16: files

This chapter covers how to use files to store information on the EV3 brick, how to manage the EV3's memory, and how to transfer files between the EV3 and a computer. You'll create a program that uses a file to save and restore program settings.

chapter 17: data logging

The programs in this chapter show how to use the EV3 as a data logger. I'll cover the basics of collecting and analyzing data, and you'll use data logging to gain a deeper understanding of how the Move Steering block works.

chapter 18: multitasking

The EV3 can multitask by executing multiple groups of blocks in parallel. You'll learn how to use multiple sequences effectively and how to avoid some common problems.

chapter 19: a PID-controlled LineFollower program

The final chapter uses advanced EV3 programming features to create a complex line-following program. You'll learn how to use a proportional-integral-derivative (PID) controller to create a fast and accurate line-following machine.

appendix A: NXT and EV3 compatibility

This appendix discusses how to use the older NXT MINDSTORMS product with the new EV3 set.

appendix B: EV3 websites

This appendix is a list of websites that provide information about EV3 programming.

how best to use this book

To get the most out of this book, you should work through the step-by-step instructions for building the example programs on your computer as you are reading. Programming is a learn-by-doing activity, and you'll learn a lot more by writing and experimenting with the programs than you will by just reading about them.

The programs and accompanying discussions will make the most sense if you read the chapters in order. Several of the example programs are introduced in the early chapters and then expanded in the later chapters as you learn more about the EV3 programming. By the time you get to the end of the book, you'll have the knowledge and skills you need to be an expert EV3 programmer.

LEGO and robots: a great combination

Welcome to the world of robotics. Not long ago, the only place you could find a robot was in a good science fiction story. Today robots are very real and perform a wide variety of important jobs, like exploring other planets, investigating deep-sea volcanoes, assembling automobiles, and performing surgery. Figure 1-1 shows the Mars Exploration Rover *Curiosity*. You can even buy a robot at the hardware store to sweep your floors while you sleep!

Figure 1-1: Mars Exploration Rover (courtesy NASA/JPL-Caltech)

Figure 1-2: Living Room Rover

LEGO MINDSTORMS EV3

With the LEGO MINDSTORMS EV3 set, you can build your own robot. In fact, you can build lots of robots. Figure 1-2 shows a simple robot you could build to explore your living room.

The EV3 set is a lot of fun to play with, but it's more than just a toy. Teachers in middle and high schools use the sets to teach science and engineering. The LEGO Group even has an education division, *LEGO Education*, which provides resources for teachers who use LEGO products in the classroom.

In educational competitions, such as the *FIRST LEGO League (FLL)*, the World Robot Olympiad, and RoboCup Junior, students from all over the world use MINDSTORMS sets to build robots to solve a given challenge.

The EV3 set comes in two editions: Home and Education. The Home Edition, LEGO set #31313, is sold in stores and is targeted at the general public. The Education Edition, LEGO set #45544, is sold through LEGO Education distributors to schools, educators, and FLL teams. Each edition has a slightly different mix of LEGO pieces and sensors. The Education Edition of the software also has some additional features for using the EV3 for science experiments. For the purposes of this book, the difference in building pieces isn't an issue; you can build the robot with either the Home or Education set. Robots built from the two sets won't be exactly the same (for example, tires are different sizes), but these differences won't matter too much.

The EV3 set is the third generation of LEGO MINDSTORMS. Many of the parts from the previous generation, the NXT, can be used with the EV3. See Appendix A for details.

the LEGO MINDSTORMS EV3 set

Your EV3 set includes the EV3 Intelligent Brick, three motors, several sensors, instructions for downloading the MINDSTORMS EV3 software, and LEGO pieces for building your robot. As discussed in the previous section, the exact mix of parts and sensors included depends on your version of the set.

The building pieces include gears, axles, pins, and beams from the LEGO TECHNIC line, as shown in Figure 1-3. These pieces are both strong and lightweight, and you can connect them to create sophisticated moving parts, which makes them ideal for creating robots. You can also easily add your own parts from other TECHNIC, BIONICLE, or even traditional LEGO sets to enhance your robotic creations.

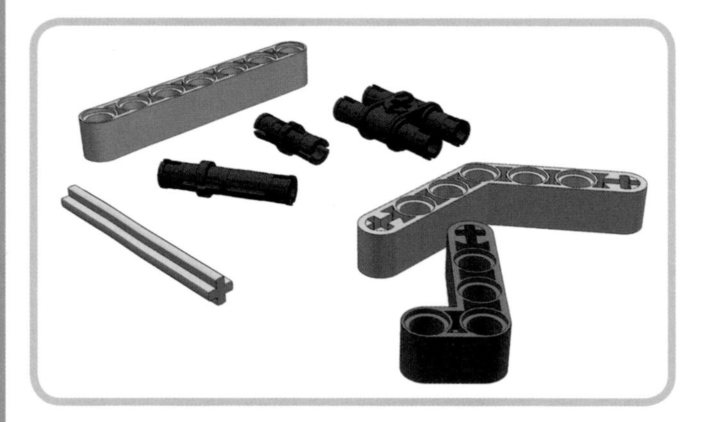

Figure 1-3: Beams and pins

The EV3 Intelligent Brick (often abbreviated as just "the EV3" or "the Brick") is the brains of your robot. The EV3 is really a small computer that you program to make your creations move. Instead of a full-size monitor and keyboard, it contains a small screen, a set of buttons, and connections for the motors and sensors. You can program the EV3 directly using the "on-brick programming," or create a program using the EV3 software for Windows or OS X and then download it to the EV3. When you run the program, the EV3 collects data from the sensors and moves the motors according to the instructions you provided in the program.

Using the EV3 motors, you can turn an ordinary LEGO model into a moving robot. The two large motors make it easy to build mobile robots using either wheels or treads. You can also use these motors or the smaller third motor to create robotic hands, cranes, catapults, and other contraptions. Many robots use two of the motors to move and the third for another function, but some robots use all three motors for other tasks and don't move around at all.

The EV3 sensors allow your robot to react to its environment according to your commands. The EV3 sensors include the Ultrasonic, Infrared, Touch, Color, Gyro, and Rotation Sensors. A Rotation Sensor is built into every EV3 motor; the other sensors are separate. The sensor functions are described here:

Ultrasonic Sensor Measures the distance to an object or obstacle. It can also detect the presence of another Ultrasonic Sensor.

Infrared Sensor Measures the distance to an object or obstacle. It can also measure the distance and direction to the Infrared Remote, and detect buttons pressed on the remote.

Touch Sensor Detects when the button on the front of the sensor is pressed. It can tell when the robot runs into something and whether an object is touching the robot.

Color Sensor Determines the color of objects. This sensor can also measure the brightness of light shining into the front of the sensor. The sensor has a small light on the front, which allows it to measure both reflected and ambient light.

Gyro Sensor Measures rotational motion. It can measure the rate of rotation and the angle at which your robot moves.

Rotation Sensor Measures the distance a motor moves. Each EV3 motor contains a built-in Rotation Sensor.

The Home version includes a Touch Sensor, a Color Sensor, and an Infrared Sensor and Remote. The Education version includes two Touch Sensors, a Color Sensor, an Ultrasonic Sensor, and a Gyro Sensor. This means you will have either an Ultrasonic or an Infrared Sensor, both of which can be used for measuring the robot's distance from an object. Most of the example programs in this book will use the Ultrasonic and Infrared Sensors interchangeably.

The LEGO Group also makes a Temperature Sensor (sold separately), and other companies make additional sensors for the EV3 sets. For example, products from HiTechnic, Vernier,

Dexter Industries, and Mindsensors include a Compass Sensor, Acceleration Sensor, and Barometric Sensor.

the LEGO MINDSTORMS EV3 software

The EV3 software is a graphical programming environment that contains all the tools you need to create a program for an EV3 robot. This type of application is often called an *integrated development environment* (or *IDE* for short). The EV3 IDE is considered a *graphical* programming environment because you use colored icons called *blocks* to create a program. There are blocks for controlling the motors, using the sensors, and doing many other things. You create a program by dragging blocks around the screen, connecting them, and changing their settings.

The EV3 software features a remarkable balance between ease of use and programming power. It's easy to write simple programs yet still possible to create very complex ones. Some advanced features may be difficult to understand at first, but they will become clear with a little practice.

software, firmware, and hardware

Your program is one of three components that work together to control your robot. The program you create is called *software*, which is a set of instructions that a computer can perform. In this case, the computer is the EV3 Intelligent Brick. The *soft* part of the word *software* comes from the ability to make changes easily. This malleability allows you create an endless variety of programs using only the EV3, three motors, and a few sensors.

Arranging blocks on a virtual canvas is a convenient way for humans to create programs, but to run your program, the EV3 needs something a little different. Your program is called the *source* or *source code*, and it needs to be translated into a set of instructions that the EV3 can execute. Then these instructions need to be copied from your computer to the Brick. After your program is translated and downloaded, you can run it.

The program that runs directly on the Brick is *firmware*, which is a program that rarely changes and is effectively part of the device. The EV3 firmware functions in the same way as a computer or smartphone operating system such as Windows, iOS, Linux, or Android. The firmware is the program that makes the sound when you turn on the Brick, controls the display, and responds to the buttons on the EV3. When you connect the EV3 to your computer, the MINDSTORMS environment communicates with the EV3's firmware.

NOTE LEGO occasionally releases updates of the EV3 firmware to add new features or fix problems. If your computer is connected to the Internet, the MINDSTORMS application will check for updates and prompt you to download them when necessary.

The EV3 Brick is the hardware on which your program runs. *Hardware* refers to the physical components of a computer. This includes the Brick, motors, sensors, and LEGO building pieces. The hardware does not change; you can rearrange them, and even use them in different ways, but the capabilities of each piece do not change.

art and engineering

For me, the most fascinating part of creating a robot is writing the program to make it come alive. Computer programming is a combination of art and engineering. We use principles of *engineering* when we follow a set of logical steps to solve a practical problem. As you move through this book—and especially with the longer programs toward the end—you'll learn engineering principles and programming practices to help you write better programs (and avoid some common bad habits).However, the basic process of writing a program to solve a particular problem is often more art than engineering. Creating a program isn't always a step-by-step process, and it usually involves a lot of creativity and ingenuity. In my opinion, this use of creative thinking is what makes programming so much fun.

But programming can also be frustrating when things don't work quite the way you want. When a software program breaks, figuring out why can be a bit of a mystery. Throughout this book, I will show you how to diagnose and fix problems with your program. Just remember, solving a mystery should be fun!

qualities of a good program

Many of the decisions you make when creating your programs will depend on your individual taste, and you will develop your own programming style. There is almost always more than one correct way to solve a problem. However, there are three rules you can use to judge the quality of a program. A program should do the following:

1. Perform the desired function
2. Be easy to modify
3. Be understandable by someone who knows the programming language used to create the program

The first rule seems pretty obvious, but there's more to it. Before you can be sure a program works, you first need to know the program's *requirements*—the complete description of what a program should do. If you are creating a program for a school project or an FLL challenge, you might receive the requirements before starting. If you're just building a robot for fun, you can make up the requirements as you go along. In either case, you need to know what your robot is being built to do before you can judge whether it's a success.

The second rule exists because requirements often change after you start a program. You might find that you can't solve a problem the way you first thought, or you might decide to expand the requirements to solve a harder problem. It's better if you can easily change your program to adapt to the new requirements. A program that's easier to modify is also more likely to be reused to solve similar problems. Reusing existing programs instead of starting new programs from scratch can save a lot of time.

The third rule is about keeping your program as simple and easy to understand as possible. Programs that are more complex than necessary tend to have more errors and are harder to reuse. If you want to make your program easier to understand, you can use comments to explain how the program works. Well-placed comments are a simple way to make a program useful to other programmers.

what you'll learn from this book

The secret to becoming a successful programmer is knowledge and practice. Throughout this book, I will concentrate on three areas of knowledge that are important to becoming a successful EV3 programmer:

The behavior of each block Learning how each block works is the first step to using it in a program. Although there are many blocks, each of which has several options, learning about each block is not difficult. The EV3 help file gives a comprehensive description of each block, and it's pretty easy (and fun) to write little test programs that let you discover exactly what each block can do.

Joining several blocks together into a working program To do this, you need to learn about program flow, data wires, and variables. This is where things get a little more complicated. Learning some details about how an EV3 program works can help you avoid the confusion that many users experience when they move beyond simple programs.

General programming practices The three rules listed earlier are the first examples. As we go along, I'll introduce more concepts that are useful regardless of the programming language you're using or the type of program you're writing.

Programming is one of those learn-by-doing activities, and this is where practice comes in. Many of the concepts you need to understand will only make sense when you see them in action. The more programs you write, the more comfortable you will become.

the LEGO MINDSTORMS online community

LEGO robotics has a thriving online community, including websites that show hundreds of innovative robot designs. One site in particular, mindBOARDS (*http://www.mindboards.net/*), is well known for its message forums where users can exchange ideas and find answers to questions. These are great resources when you can't figure out why your robot isn't working the way you think it should. A quick search of the forums often provides the answer you're looking for. If you don't find a solution already posted, you can ask a question describing your particular problem. See Appendix B for a list of useful MINDSTORMS-related websites.

what's next?

In the next chapter, I'll introduce the EV3 programming environment and then present some simple programming concepts and easy example programs that show how to use the software. The following chapters gradually introduce more blocks and programming concepts using increasingly complex programs.

You can download the source code for all the programs in this book from *http://www.nostarch.com/ev3programming.htm*.

the EV3 programming environment

This chapter explores the EV3 programming environment and presents a few simple programs. We'll start with the basics by looking at sample programs that use the EV3 Intelligent Brick without any motors or sensors. Chapter 4 covers programming motors, and Chapter 5 explains how to use sensors.

a tour through the MINDSTORMS software

When you start the MINDSTORMS EV3 software, the Lobby appears. The Home and Education Editions have different Lobby screens, but they function in the same way. In the Lobby, you can create or open a project, access the user guide and help file, and read instructions for building example robots. The Lobby for the Home Edition is shown in Figure 2-1, and Figure 2-2 shows the Lobby for the Education Edition.

Before writing our first program, let's take a look at the main areas of the MINDSTORMS EV3 environment. Select **File ▸ New Project (File ▸ New Project ▸ Program** for the Education Edition) to create a new project. A screen like the one shown in Figure 2-3 appears.

NOTE The images in this book are taken from the EV3 Home Edition running on Windows 7.

Figure 2-1: The Lobby screen for the EV3 Home Edition software

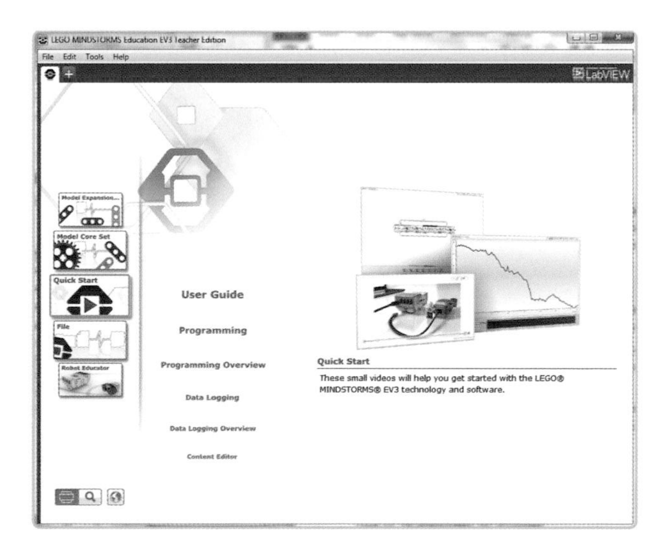

Figure 2-2: The Lobby screen for the EV3 Education Edition software

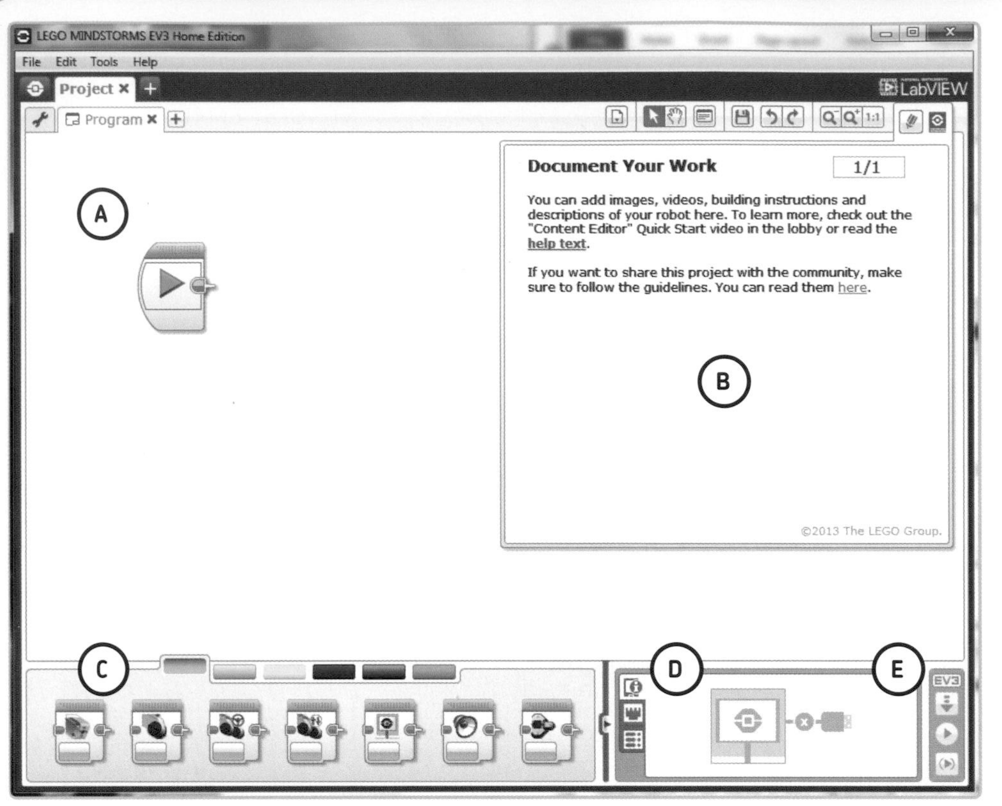

Figure 2-3: The MINDSTORMS EV3 environment

A: the programming canvas

The main part of the screen is the *Programming Canvas*, which is where you build your program. You can use the tabs at the top of the window to navigate among open projects. Figure 2-3 shows one open project, named Project. The tab to the left of the Project tab, which has a small MINDSTORMS icon, takes you back to the Lobby screen.

A project can contain multiple programs. Below the project tab is another group of tabs that you can use to select a program. Figure 2-3 shows one open program, named Program. We'll talk about how to rename programs and projects later in this chapter. The tab with the small wrench icon opens the Project Properties page, which is also described later.

B: the content editor

The *Content Editor* allows you to document your project by creating a presentation that includes text, images, and videos. You could include a description of how the program works, instructions for building the robot, or a video showing the robot in action. The presentation is saved with the project, so you don't need to keep track of a separate file.

When you are not using the Content Editor, you can close it by clicking the small tab with the MINDSTORMS icon in the top right corner. This will give you more space on the Programming Canvas to build your program.

C: the programming palettes

The *Programming Palettes* appear at the bottom of the Programming Canvas. They contain the blocks that you will use to create your programs. To keep things simple, the blocks are organized into six color-coded palettes. You can use the colored tabs to navigate among them. Going from left to right, the six groups of blocks are Action (green), Flow Control (orange), Sensor (yellow), Data Operations (red), Advanced (blue), and My Blocks (cyan).

D: the hardware page

The *Hardware page* displays information about the EV3 Brick. This information is divided into three sections, which you can select using the tabs on the left side (shown in Figure 2-4). The top tab, Brick Information, shows the battery level, firmware version, and how much EV3 memory you've used. The middle tab, Port View, shows which sensors and motors are attached to the Brick. The bottom tab, Available Bricks, connects your brick to the EV3 software.

E: the download and run buttons

The *Download and Run* buttons (shown in Figure 2-5) let you transfer the program from your computer to the EV3 Brick and run it. The process of transferring the program to your

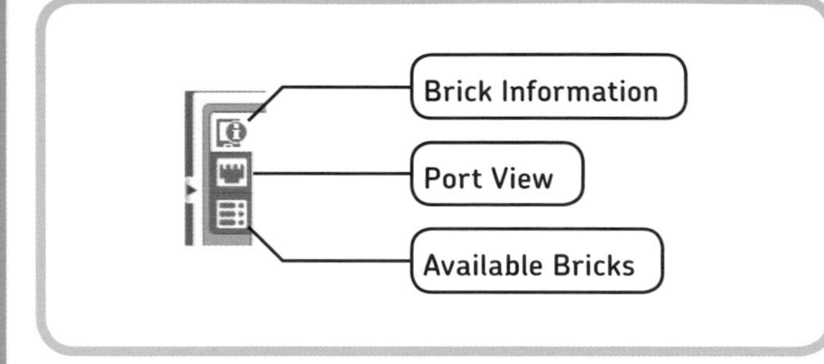

Figure 2-4: The Hardware page tabs

brick is called *downloading*. The middle button, Download and Run, will download and run your program immediately. The top button, Download, will download your program to the Brick but not run it. You can then start your program using the buttons on the Brick itself. This is useful if you need to move or unplug your robot before running the program. The bottom button, Run Selected, will download and run only the blocks you have selected, which can be helpful for finding and fixing problems in your program.

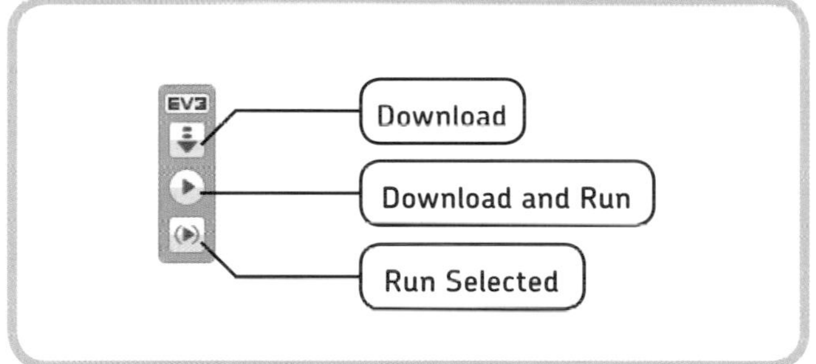

Figure 2-5: The Download and Run buttons

writing an EV3 program

When you create a new project, the Programming Canvas will already contain a Start block (shown in Figure 2-6). Create your program by dragging blocks from the Programming Palettes onto the Programming Canvas and connecting them in a line, beginning with the Start block. Each block has a number of different options, and its behavior will depend on how you configure them. When you run the program, the EV3 executes each block in the arranged order, from left to right. Usually, the blocks run one at a time—meaning that each block must finish its operation before the next block starts—but as you'll see, there are some exceptions to this, where multiple blocks can run simultaneously. The program ends after executing the last block.

Figure 2-6: The Start block

general layout of a block

Before writing our first program, let's look at how to set the options for a block, known as *parameters*. At the bottom of each block, you will see a group of controls that can tell the block how to behave. Each block's parameters are unique, but there is a consistent look and feel to the configuration options, which is one of the features that make EV3 easy to use.

Let's take a look at the controls for the Sound block. This is a fairly typical block, so all the ideas discussed here apply to most other blocks. Figure 2-7 shows how the Sound block's parameters look when you first add the block to a program, before you make any changes.

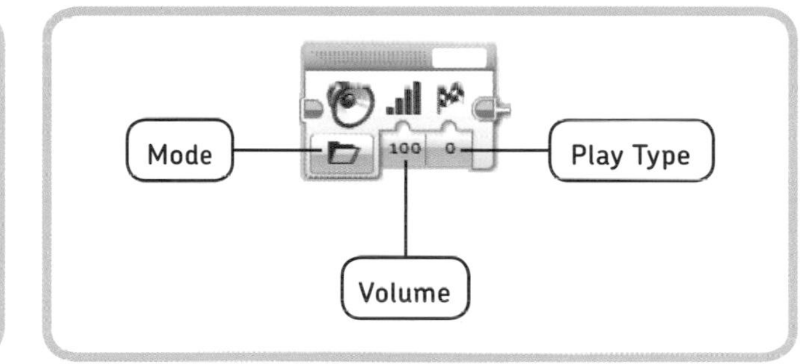

Figure 2-7: The Sound block's parameters

The first control, Mode, selects the *mode* of the block. A block's mode determines the main function of the block. The four modes of the Sound block allow you to play a sound file, play a note, play a tone, and stop a sound that is already playing. When you select the mode for a block, the remaining options change to match the selected mode. Figure 2-7 shows the Sound block with the Play File mode selected. In this mode, you can set the volume and Play Type, which tells the program how to play the sound—it can be played just once while the program waits for it to complete, played once while the program continues to the next block, or played over and over until the program ends or encounters a Sound block in Stop mode.

Other modes have different options. Figure 2-8 shows the Sound block with the Play Tone mode selected. In this mode, you can set the frequency and duration of the tone to be played.

To change a mode or setting, click the current value. The way in which you enter a new value depends on the option type. For some parameters, you just enter a number. For others, you might use a list or a slider to select a value. In the Play Note mode, you use a small keyboard to select the note for the Sound block, as shown in Figure 2-9.

Frequency Duration

Figure 2-8: The Sound block in Play Tone mode

Figure 2-9: Selecting the note for the Sound block

your first program

For your first program, you'll use the Sound block to make the EV3 say "Hello." To begin, start the MINDSTORMS EV3 software and create a new project. You won't be using the Content Editor, so you can close it to make more space for the Programming Canvas. Follow these steps to add a Sound block to your new program:

1. Select the Sound block from the Action Palette, as shown here:

2. Drag the block onto the Programming Canvas. Place the Sound block just to the right of the Start block, which should already be on the canvas.

Your program should look like this:

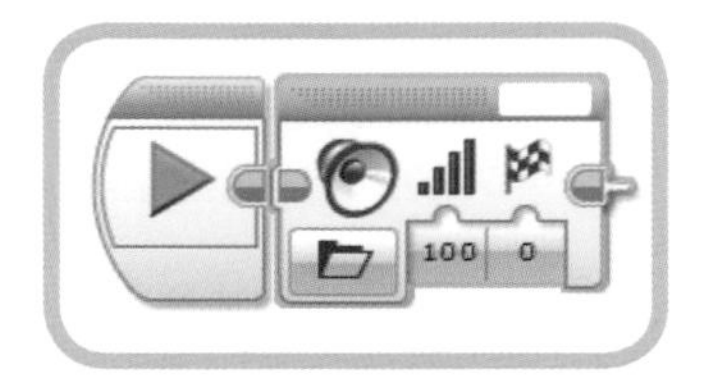

If you accidentally grab the wrong block or drop it in the wrong place, select **Edit ▸ Undo** and start again.

You can leave most of the Sound block settings at their defaults. All you need to do is select the sound file you want the block to play. The EV3 software comes with a large selection of sound files, which are arranged in a folder structure similar to the ones on computer hard drives.

3. Click the white box in the top right corner of the Sound block. A box containing a list of sound files appears.

4. Click the **LEGO Sound Files** folder to open it and then select the **Communications** folder. Scroll down and select the **Hello** file. The white box at the top right of the Sound block should now contain the word `Hello`.

saving your work

Before continuing, name your program and save the project. In this book, I generally use one project for each chapter, named *Chapter2*, *Chapter3*, and so on. Each project contains all the programs developed in that chapter. Let's change the name of this program to *Hello* and save the project as *Chapter2*.

To change the name, double-click the program name tab. Type the new name, *Hello*, over the highlighted text. The tabs should look like this:

Now save the project by selecting **File ▸ Save Project**. When you first save a project, a dialog appears to let you select the location and name for the project file. Type the name *Chapter2* and click **Save** to create the *Chapter2.ev3* file. This file contains all the information about your program, including the configuration and arrangement of the blocks you used. The *.ev3* file format is unique to the MINDSTORMS environment; you won't be able to edit it using other programs.

NOTE Save your work often. Save before downloading your program and certainly before getting up to answer the phone or walk the dog. Having to redo several hours of work because you neglected to save your program is really annoying!

MAKING BACK-UP COPIES

Every now and then when you change your program, instead of making it better, you end up with a horrible mess and can't get back to what you had before. You might not be able to remember how things were arranged when they did work.

Professional software developers use fancy tools called *source code control systems* to save versions of their work in order to avoid this problem, but you can get the same benefits by saving copies of working versions of your program as you go along. Use one of these back-up copies if you run into trouble. It's a good idea to save a copy after getting each new feature working and before making large changes. For example, if you are working on a program with four tasks, you might save it as *Task1* after you get the first part working. When working on the second task, you might save it as *Task1_Task2*. This way, you could always go back a step if necessary.

running your program

After you save your program, it's time for a test run. The first step is to make sure the EV3 is turned on and connected to your computer with either a USB cable or a Bluetooth or Wi-Fi connection. The USB connection is easier to set up (just plug in the cable between the EV3 and your computer) but requires that you keep your EV3 Brick nearby. A Bluetooth or Wi-Fi connection can be more difficult to set up, but it allows your robot to roam free. The EV3 help file, available from **Help ▸ Show EV3 Help**, provides instructions for connecting the Brick to your computer.

Click the center **Download and Run** button to download and run your program. Your EV3 should respond by saying "Hello."

project properties

An EV3 project can contain multiple programs, and each program can use numerous sound or image files. The *Project Properties page* gives you an overview of your project and lets you manage all of the resources it uses. Click the small wrench icon to the left of the program name to open the Project Properties page. Figure 2-10 shows the Project Properties for the Chapter 2 project, with a few of the items filled in.

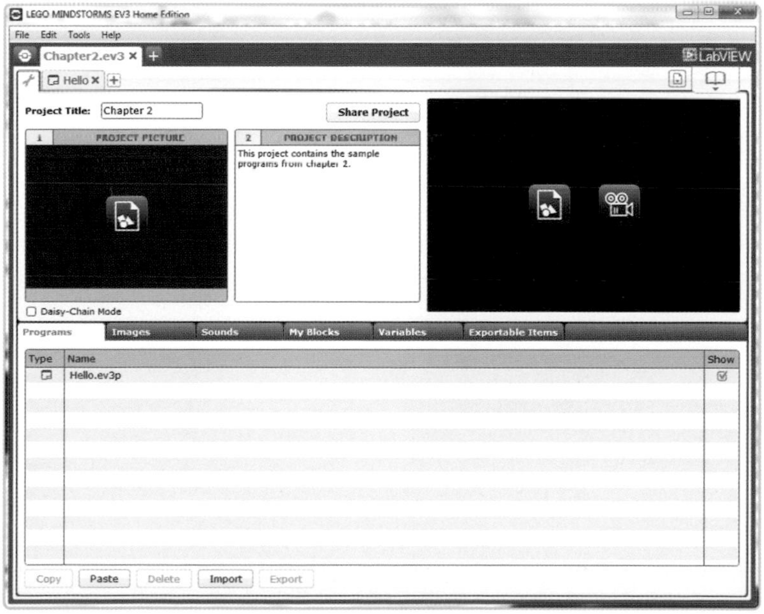

Figure 2-10: The Project Properties page

In the Project Properties page, you can give your project a title and description, and see all of its programs, images, sound files, and My Blocks. Later in this book, I'll show you how to import and export programs and My Blocks using this window so you can share them between projects.

your second program

Your second program, *HelloDisplay*, is similar to the Hello program, except that you'll use the Display block instead of the Sound block to write *Hello* on the EV3 display. Your first attempt at making this program won't work, which gives you a chance to see what happens when a program doesn't run as expected and to learn how to fix it. Use the following steps to create the initial version of the program:

1. Select **File ▸ New Program**. This adds a new program to the project.
2. Change the name of the program to *HelloDisplay*. The tabs at the top of the Programming Canvas area should look like this:

3. Drag the Display block from the Action palette, as shown here:

4. Place the Display block next to the Start block. Your program should look like this:

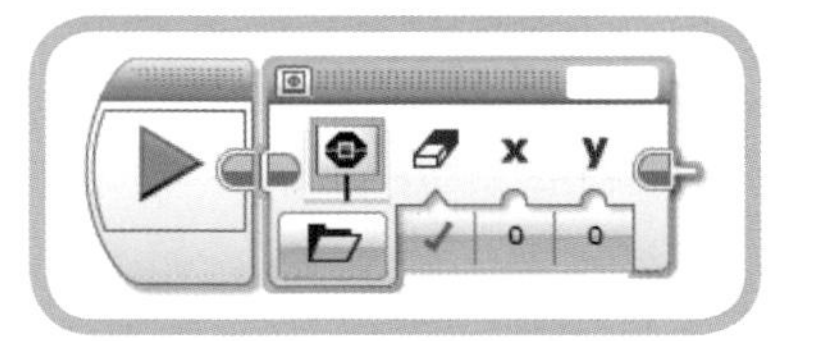

The next step is to set the Display block to print the text "Hello." The Display block has several options, which are described in Chapter 14. By default, the block is set to display an image, so the first thing you need to do is configure it to display text.

5. Click the folder at the bottom left corner of the block. Then select **Text ▸ Grid**, as shown here:

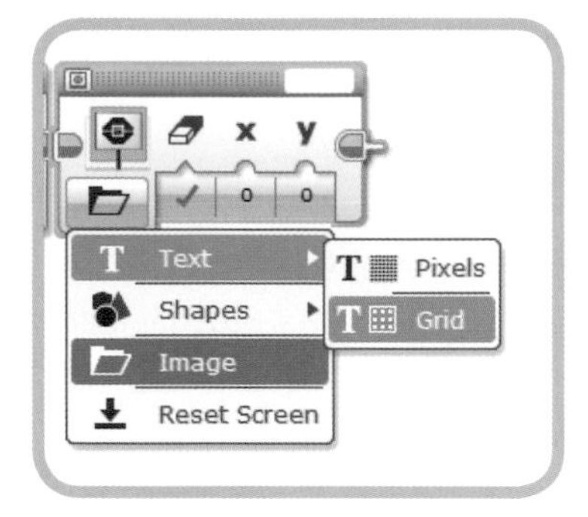

The default display text is "MINDSTORMS," so the next step is to change this to "Hello."

6. Click the text in the white box in the top right corner of the Display block, and change *MINDSTORMS* to ***Hello***.

   The complete program should now look like Figure 2-11.

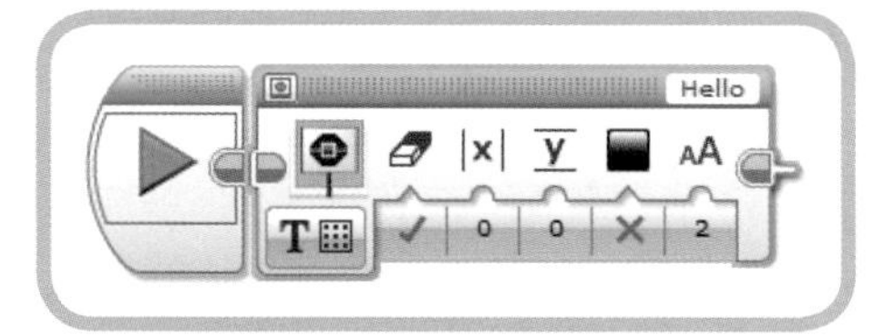

Figure 2-11: The HelloDisplay *program*

Now download and run your program. The EV3 should make a noise to let you know it has downloaded a program, but *Hello* does not appear on the display. What happened?

To put it simply, this program has a bug. A *bug* is a program error. *Debugging* is the process of finding and fixing errors. Like all programmers, you'll spend a lot of your time debugging. In fact, it's very rare to write a program on the first try without any. Running your program, finding out what's wrong, and fixing those problems are all normal parts of the programming process. Fixing a bug can be frustrating, but it can also be incredibly rewarding. Think of it as solving a puzzle, and remember that you should always have fun!

One way to fix this program is by adding a Wait block after the Display block. You can use the Wait block to tell the program to pause for five seconds before ending, which will give you enough time to read the display.

The Wait block is on the Flow Control Palette, which you select by clicking the orange tab at the top of the Programming Palettes area. Figure 2-12 shows the Flow Control blocks, with the Wait block circled.

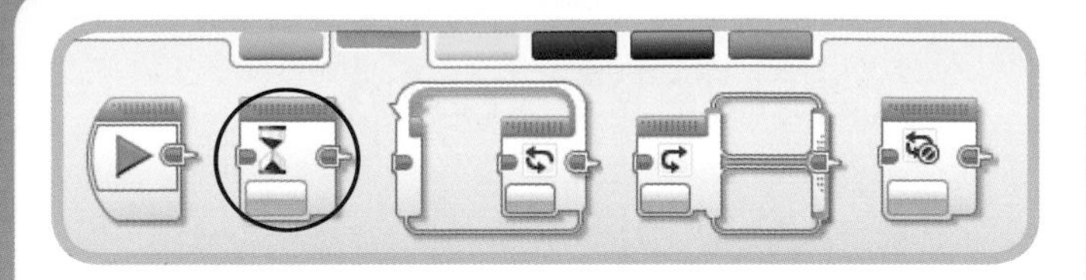

Figure 2-12: The Wait block on the Program Flow palette

Follow these steps to fix the program:

1. Drag a Wait block to the right of the Display block. Your program should look like Figure 2-13.

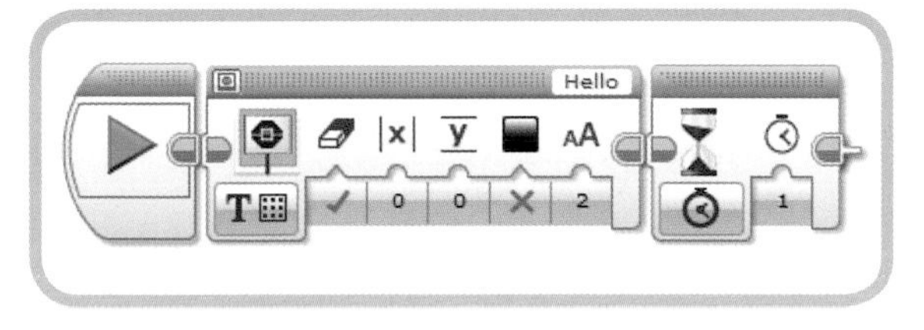

Figure 2-13: The program with the Wait block added

2. By default, the Wait block pauses for one second. To have a little more time to read the display, change the value under the small clock face icon from 1 to **5**. This causes the program to wait five seconds before ending. Figure 2-14 shows the Wait block with the change.

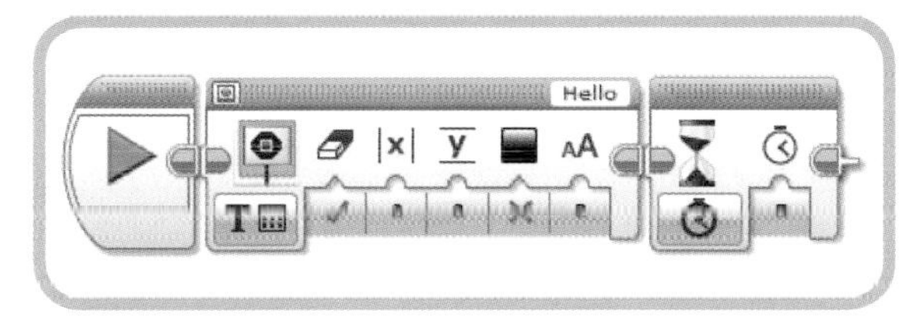

Figure 2-14: The Wait block set for five seconds

Now when you download and run the program, the display should show *Hello* for five seconds before clearing, which is the behavior you want. Adding the Wait block is a successful solution to this bug.

NOTE Why didn't the first program have the same problem? Unlike the Display block, the Sound block enables the *Wait for Completion* option by default (see Figure 2-15). This makes the program wait until the sound plays before continuing. If you select the *Play Once* option instead of the *Wait for Completion* option, the first program fails in the same way that the second one did.

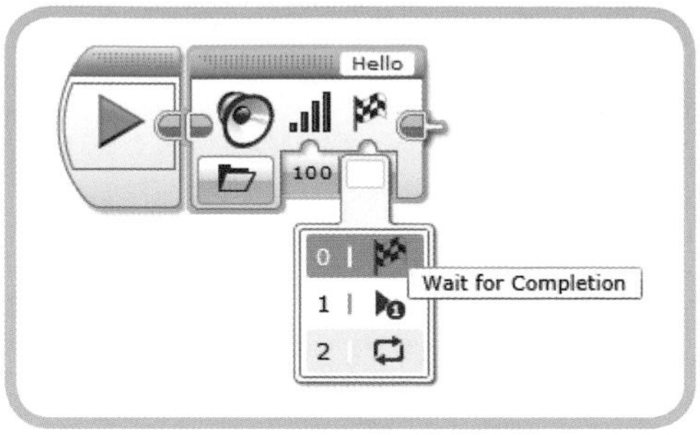

Figure 2-15: The Sound block's Wait for Completion option

comments

Programmers use *comments* to add descriptive text to their programs; these explain how the program works or why the programmers made certain decisions while they were building it. For example, you could add a comment to the previous program that explains why you added the Wait block.

In the previous chapter, I mentioned that a good program should be easy to modify and understandable to other programmers. Good comments are important in achieving both of these goals. It can be very difficult to figure out how a program works just by looking at the settings for each block. A short description in plain English will make your program much easier to understand. Think of how you might describe your program to a friend. You wouldn't just list the blocks you use; instead, you'd describe what the program does as a whole, perhaps explaining the more complicated parts in depth. Comments also help you remember why you wrote a program in a particular way, making it easier to reuse your own programs.

Comments do not affect how a program runs; the EV3 will completely ignore them. It only needs to know about the arrangement and configuration of the blocks making up your program.

adding a comment

Try adding a comment to the *HelloDisplay* program to explain why you added the Wait block. A reasonable comment is *Wait 5 seconds before ending the program to give the user time to read the display.* To add comments to your program, use the Comment tool (▤) on the Toolbar.

Follow these steps to add the comment:

1. Click the Programming Canvas above the Wait block. The new comment will appear here when you select the Comment tool.

2. Click the **Comment** tool on the Toolbar. A small comment box will appear where you clicked in the above step, as shown in Figure 2-16.

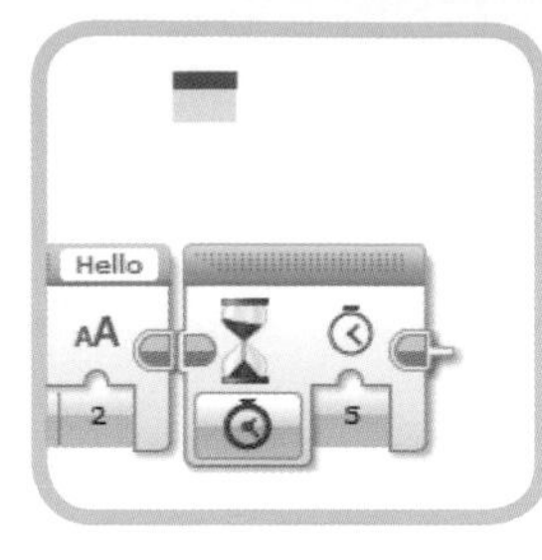

Figure 2-16: A new comment box

3. Click the center of the comment box to select the comment.
4. Start typing the comment **Wait 5 seconds before ending the program to give the user time to read the display**. Press the ENTER key to move to the next line when the text gets wider than the block (it's easier to read if the comment does not extend too far past the block).

Figure 2-17 shows the program with the comment added. Now, anyone who looks at this program will know why the Wait Time block is there.

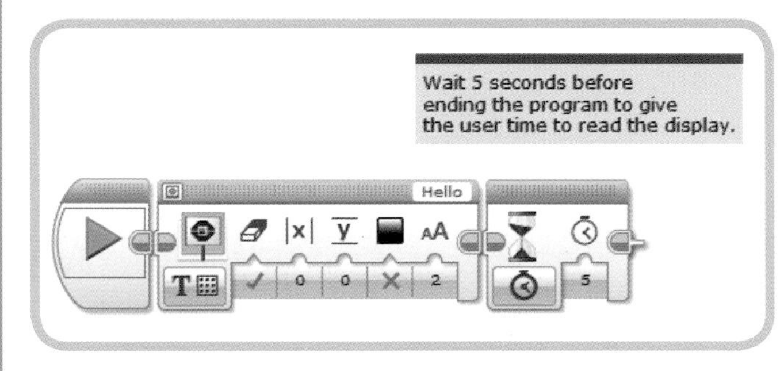

Figure 2-17: Explaining the Wait Time block

tips for working with comments

Keep the following tips in mind when writing comments:

* Pressing ENTER while typing a comment makes it continue on the next line.
* Clicking a comment selects the comment.
* When a comment is selected, clicking the text of a comment lets you change the comment.
* You can delete a selected comment by pressing the DELETE key.
* You can resize a comment box using the handles on the edge of the box. The handles appear when you move your mouse over the comment.
* You can move a selected comment by dragging it with your mouse.

context help

Having so many blocks with so many options gives us the power to build a wide variety of robotic creations. However, learning what all these blocks and options do can be a daunting task. The EV3 software has a useful feature to help with this: *context help*. To access it, select **Help ▸ Show Context Help**. A small window appears and displays brief but useful information about any item that you hover over with your cursor. Every topic contains a link to **More Information** in case you need more details about it. Keeping this little window open is a great way to quickly come up to speed on all the choices you have while creating programs. Figure 2-18 shows the help that is displayed when you select the note for the Sound block in Play Note mode.

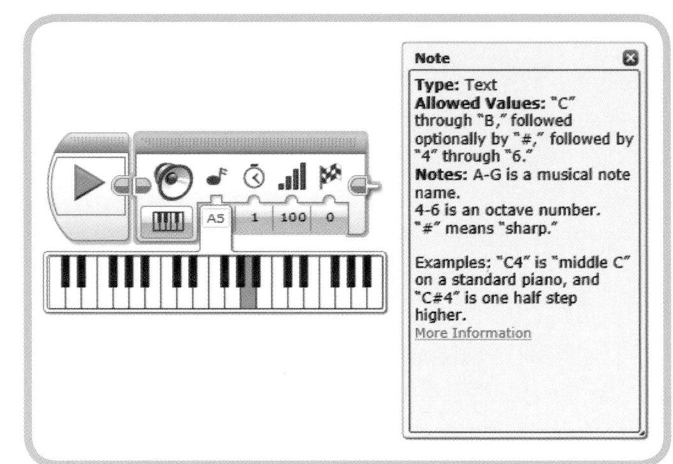

Figure 2-18: Context help for selecting a note

conclusion

This concludes the introductory tour of the MINDSTORMS environment. Next, you'll build a simple robot, called the *TriBot*, to use with the example programs in the following chapters. Then I'll introduce the multitude of blocks available in the EV3 software and show you how to combine them to make the TriBot perform a variety of tasks.

TriBot: the test robot

This chapter shows you how to build a simple, three-wheeled TriBot (Figure 3-1) that uses all of the EV3 sensors. You can use the TriBot to run the example programs in the rest of this book and to test your own programs later on. You'll also build a simple Lift Arm that you'll use to experiment with the EV3 Medium motor.

TriBot components

You can build the TriBot using either the EV3 Education or Home Edition. For the most part, the images in this chapter show Home Edition components. Let's review the differences between the two editions before we proceed. The wheels in the Home Edition are smaller than the ones in the Education Edition (see Figure 3-2).

The Home Edition includes an Infrared Sensor and Remote, while the Education Edition includes an Ultrasonic Sensor and a Gyro Sensor. If you're using the Home Edition, skip the steps involving the Gyro Sensor. If you're using the Education Edition, substitute the Ultrasonic Sensor for the Infrared Sensor.

The components in each set differ in color. For example, some beams are black in the Home Edition but gray in the Education Edition. Color doesn't matter (except where indicated in the instructions); just make sure the parts are the correct size and shape. Figure 3-3 and Figure 3-4 show the parts you'll need when using the Home Edition and the Education Edition, respectively.

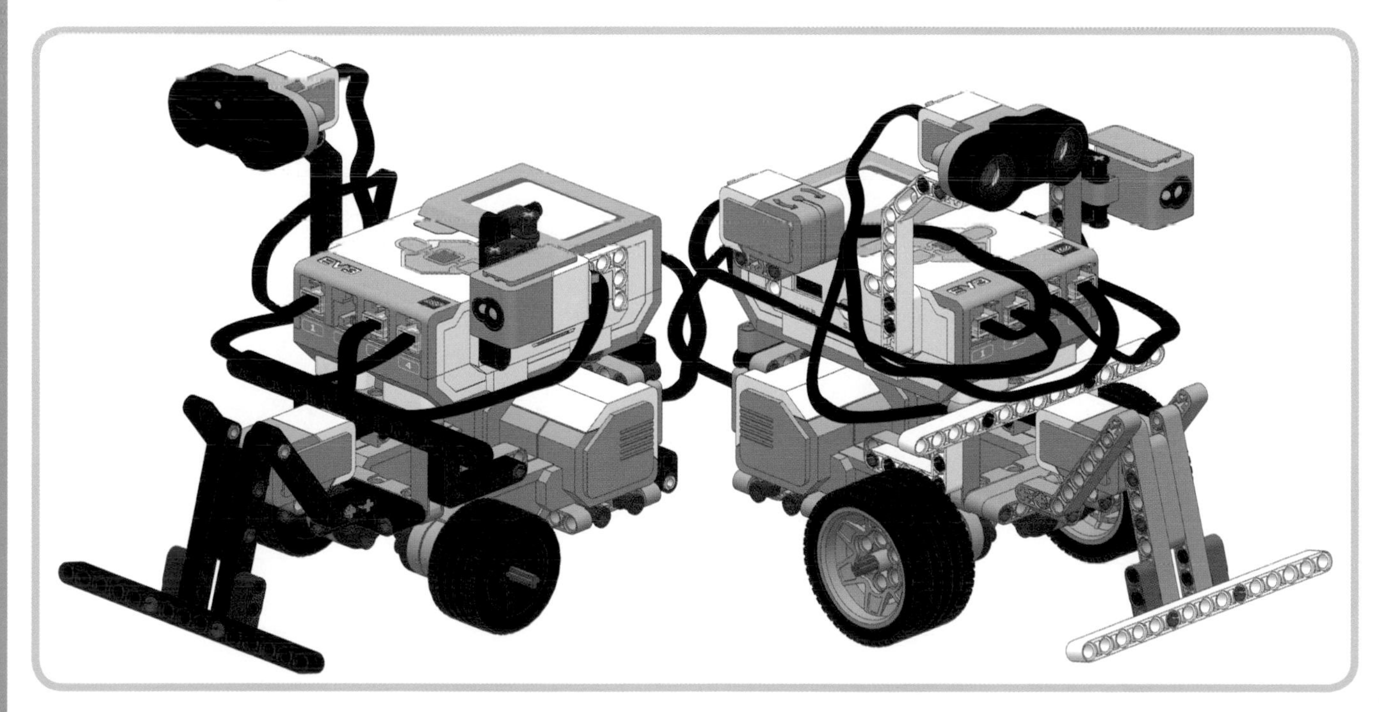

Figure 3-1: On the left, a TriBot built from the Home Edition; on the right, a TriBot built from the Education Edition

Figure 3-2: Wheels from the Education Edition (left) and the Home Edition (right)

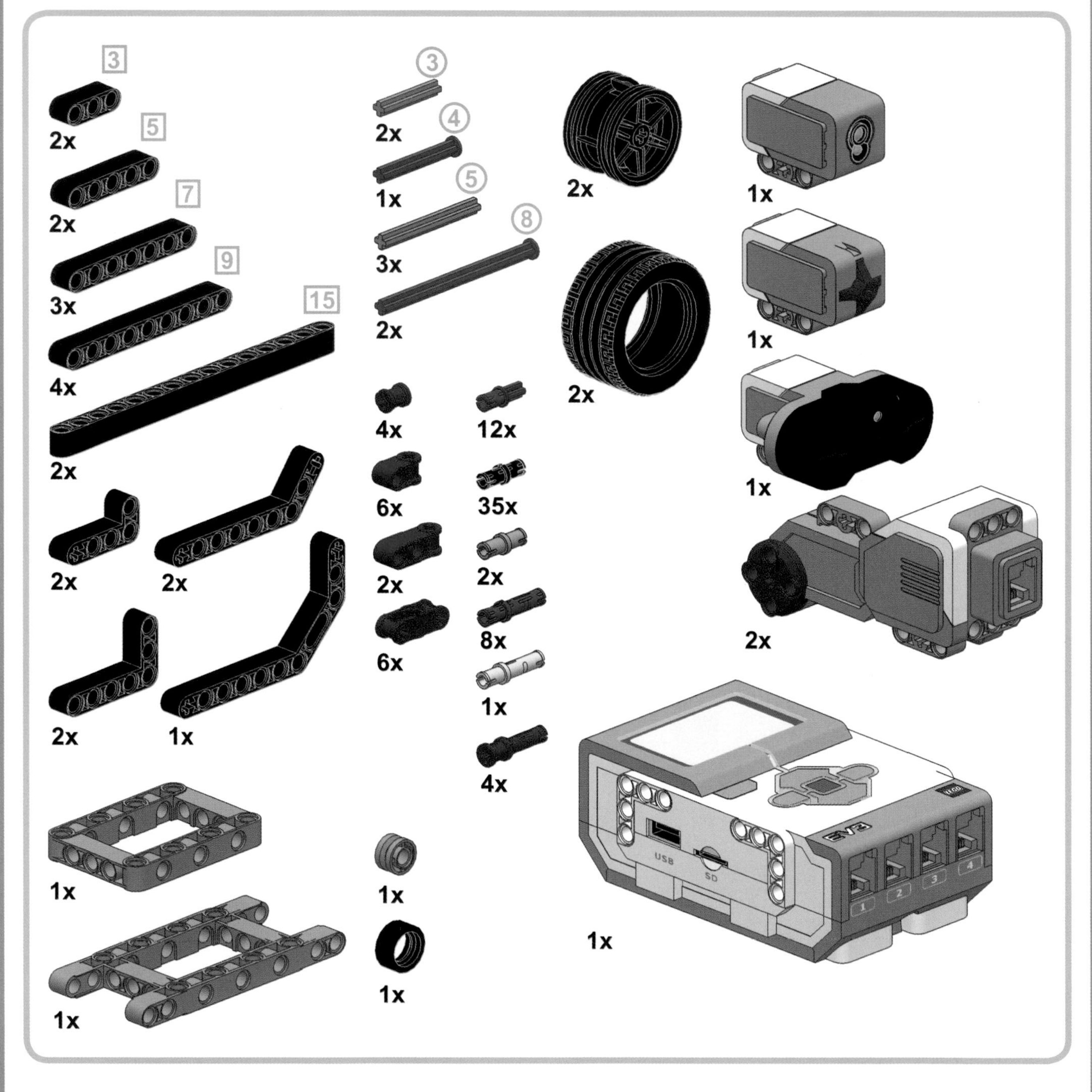

Figure 3-3: Parts required to build the TriBot with the EV3 Home Edition (#31313)

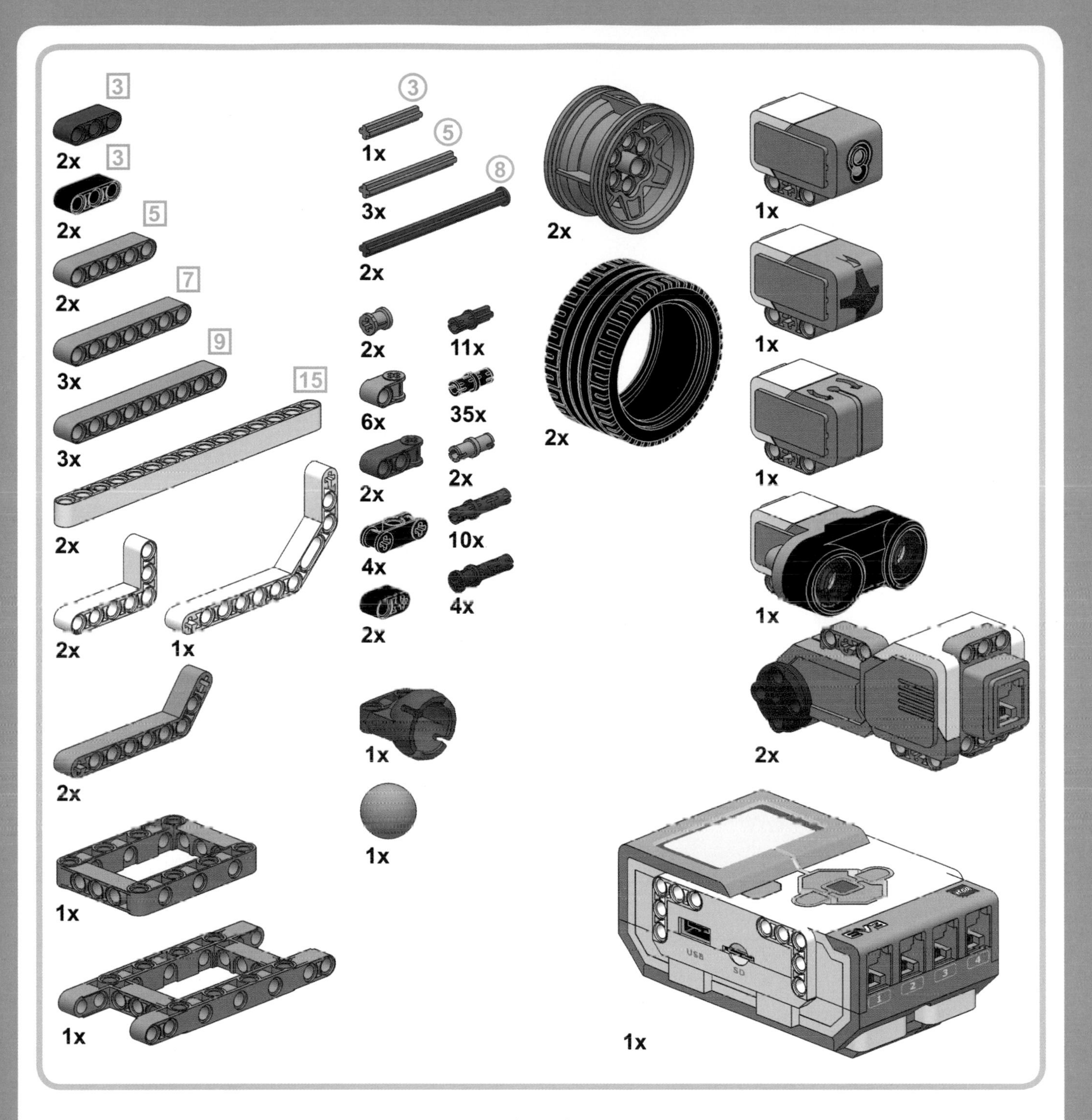

Figure 3-4: Parts required to build the TriBot with the EV3 Education Edition (#45544)

building the motor and wheel assembly

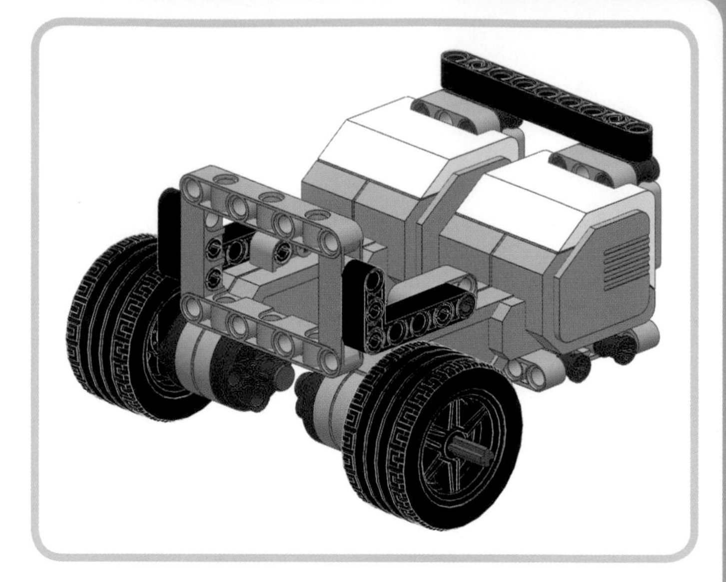

First, build the motor and wheel assembly.

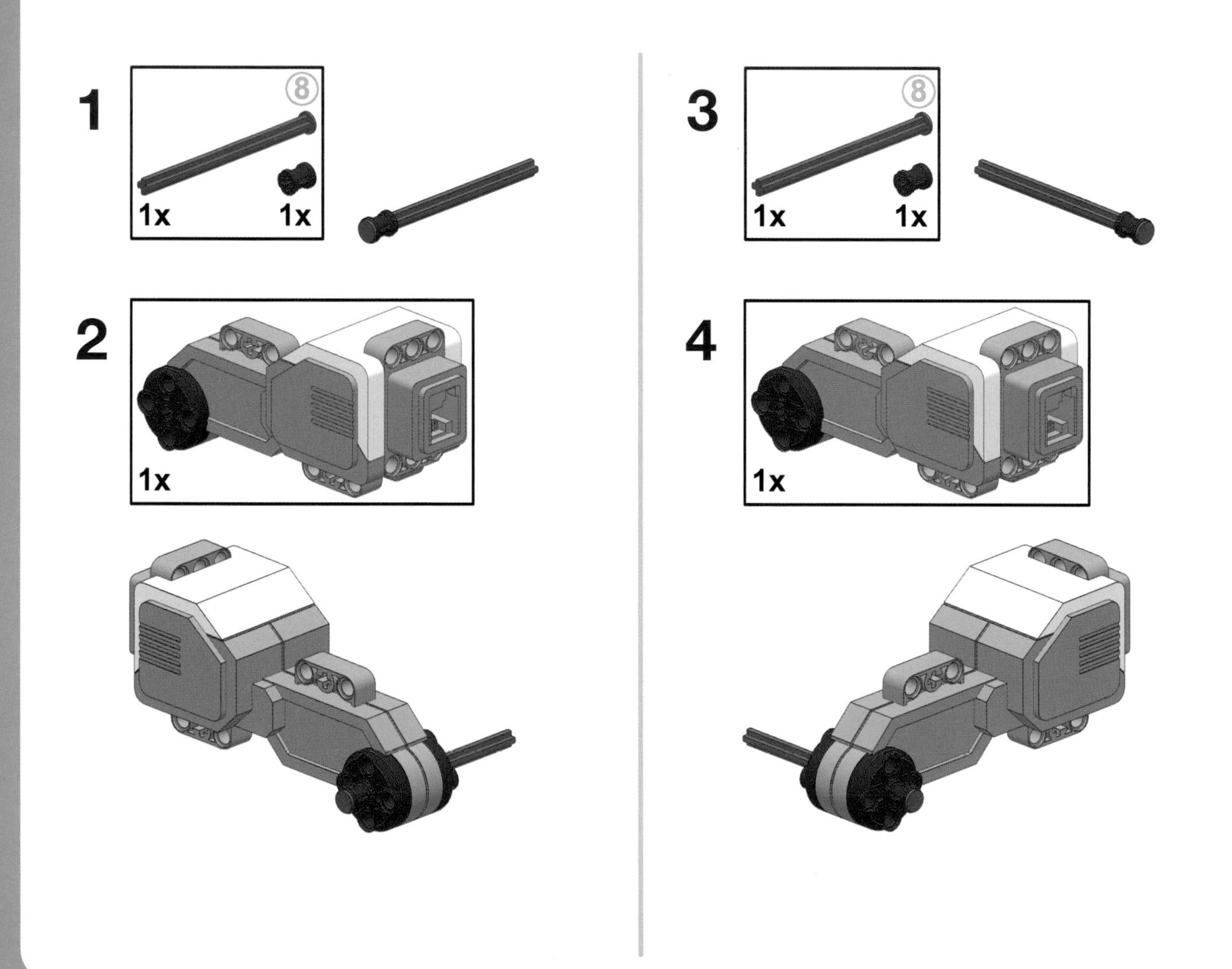

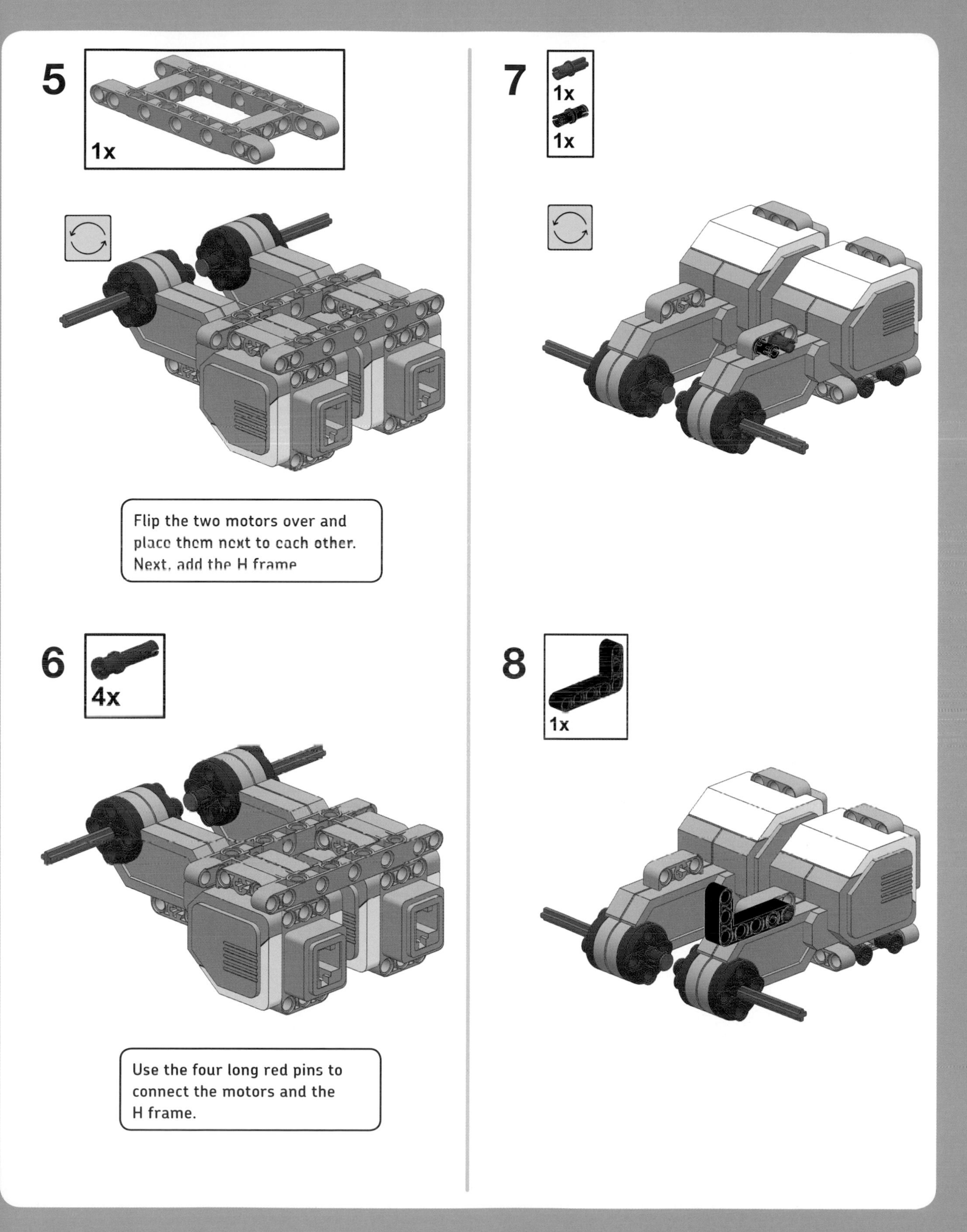
5
1x
Flip the two motors over and place them next to each other. Next, add the H frame
6
4x
Use the four long red pins to connect the motors and the H frame.
7
1x
1x
8
1x

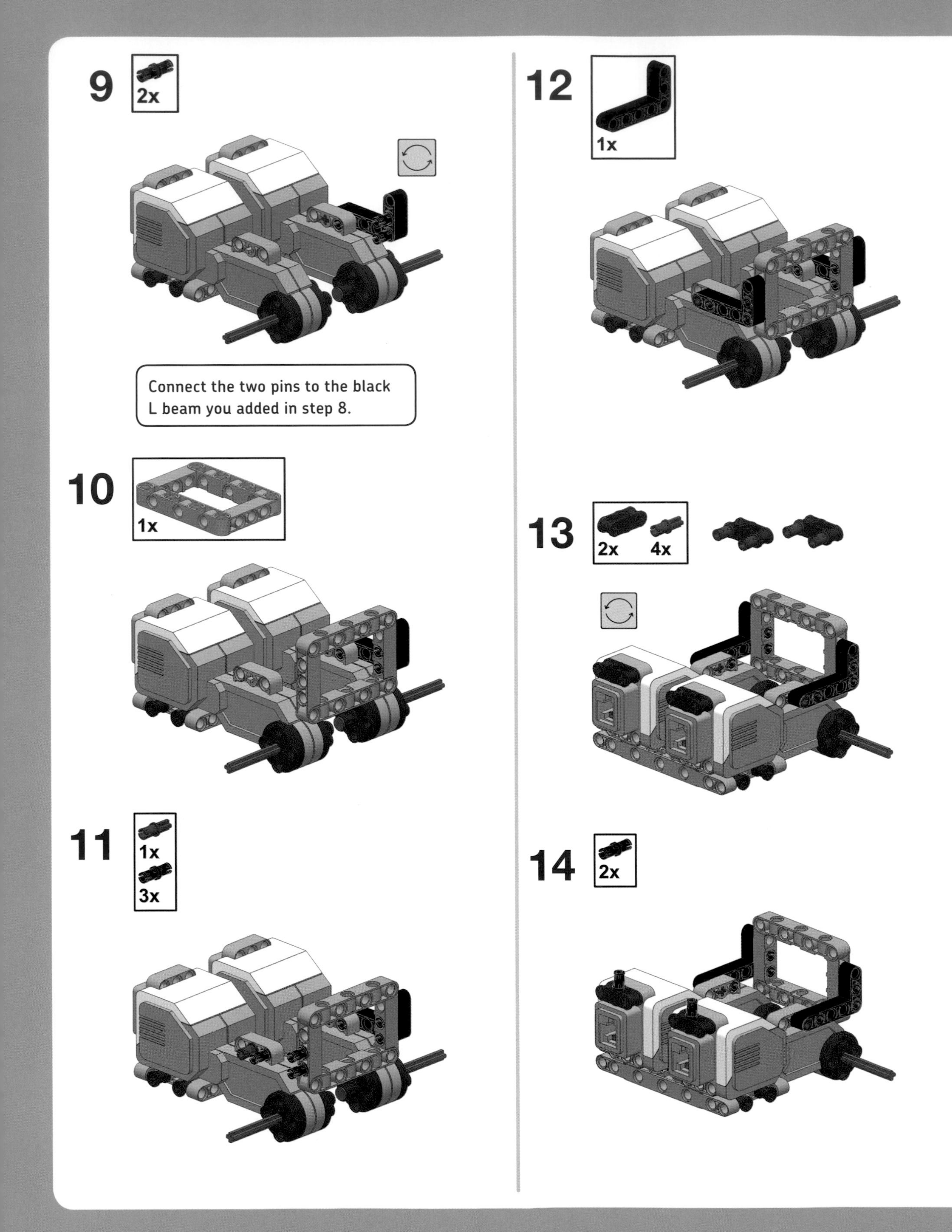
9
2x
Connect the two pins to the black L beam you added in step 8.
10
1x
11
1x
3x
12
1x
13
2x
4x
14
2x

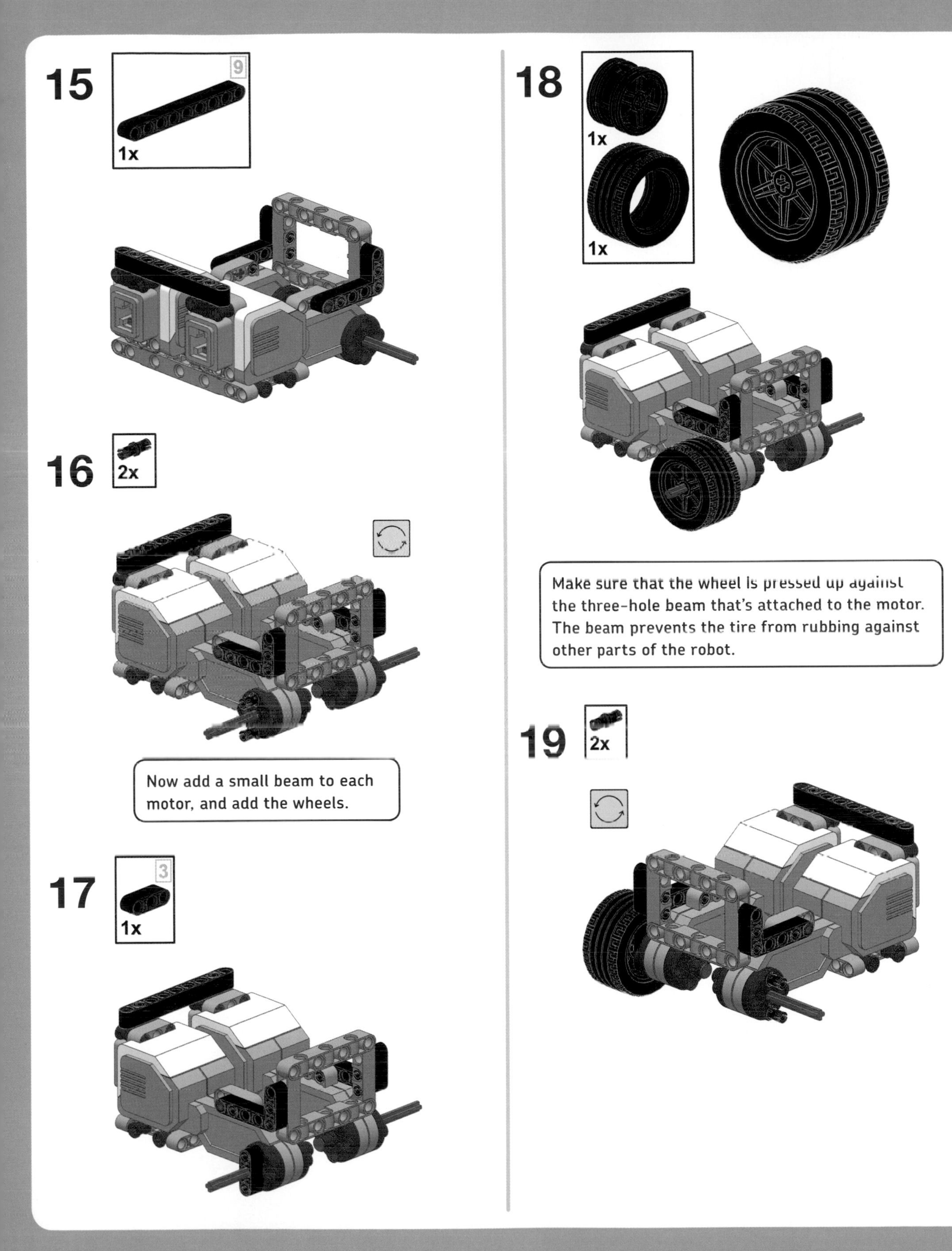
15
1x
16
2x
Now add a small beam to each motor, and add the wheels.
17
1x
18
1x
1x
Make sure that the wheel is pressed up against the three-hole beam that's attached to the motor. The beam prevents the tire from rubbing against other parts of the robot.
19
2x

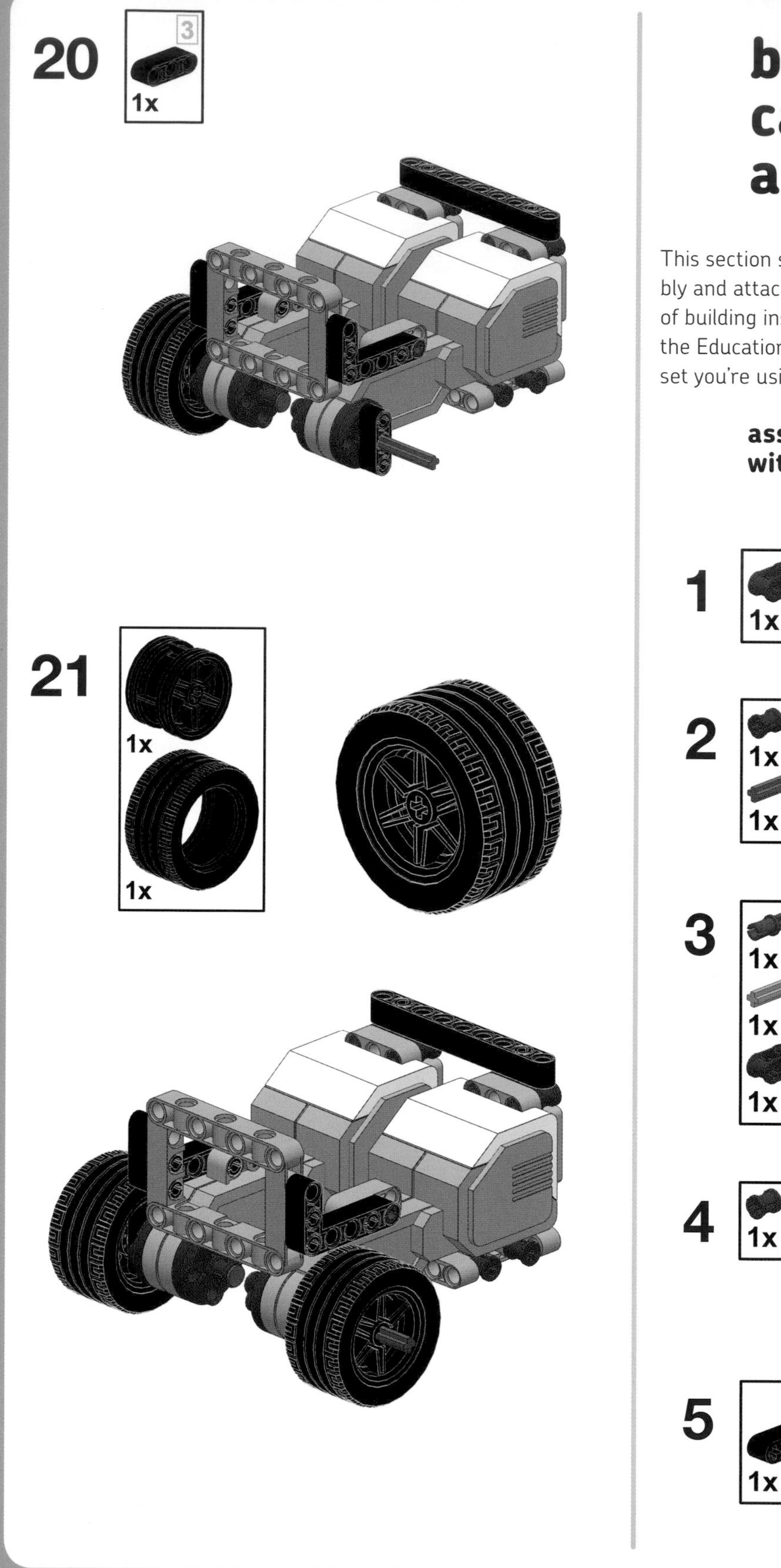

building the caster wheel assembly

This section shows you how to build the caster wheel assembly and attach it to the back of the TriBot. It includes two sets of building instructions: one for the Home Edition and one for the Education Edition. Follow the instructions that apply to the set you're using.

assembling the caster wheel with the home edition

1

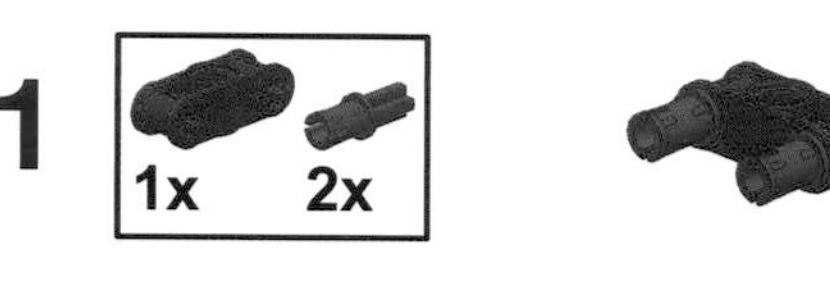

2

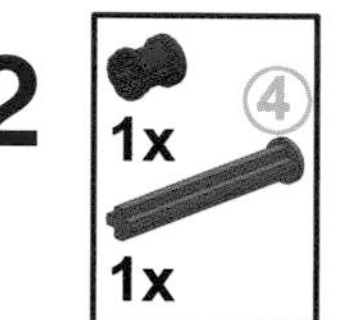

3

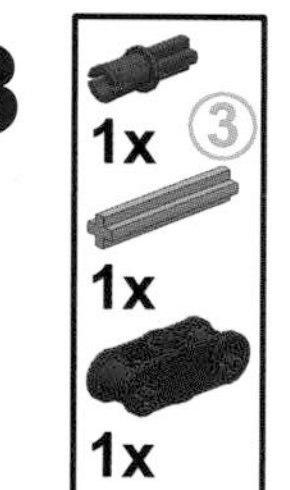

4

5

1x

6

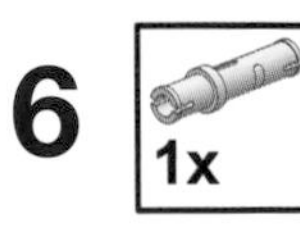

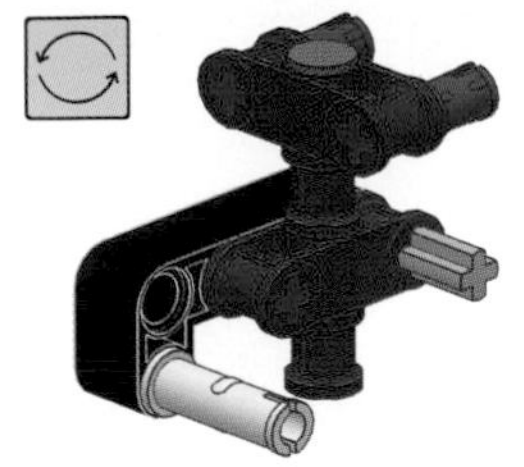

Make sure that you use the tan pin and not a blue one. The tan pin creates less friction, which allows the wheel to spin freely.

7

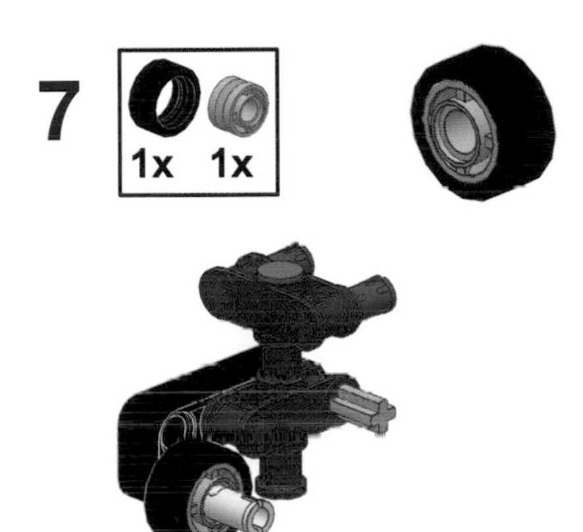

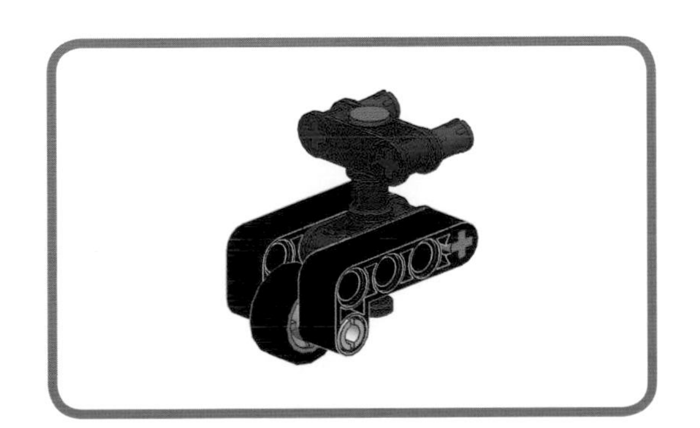

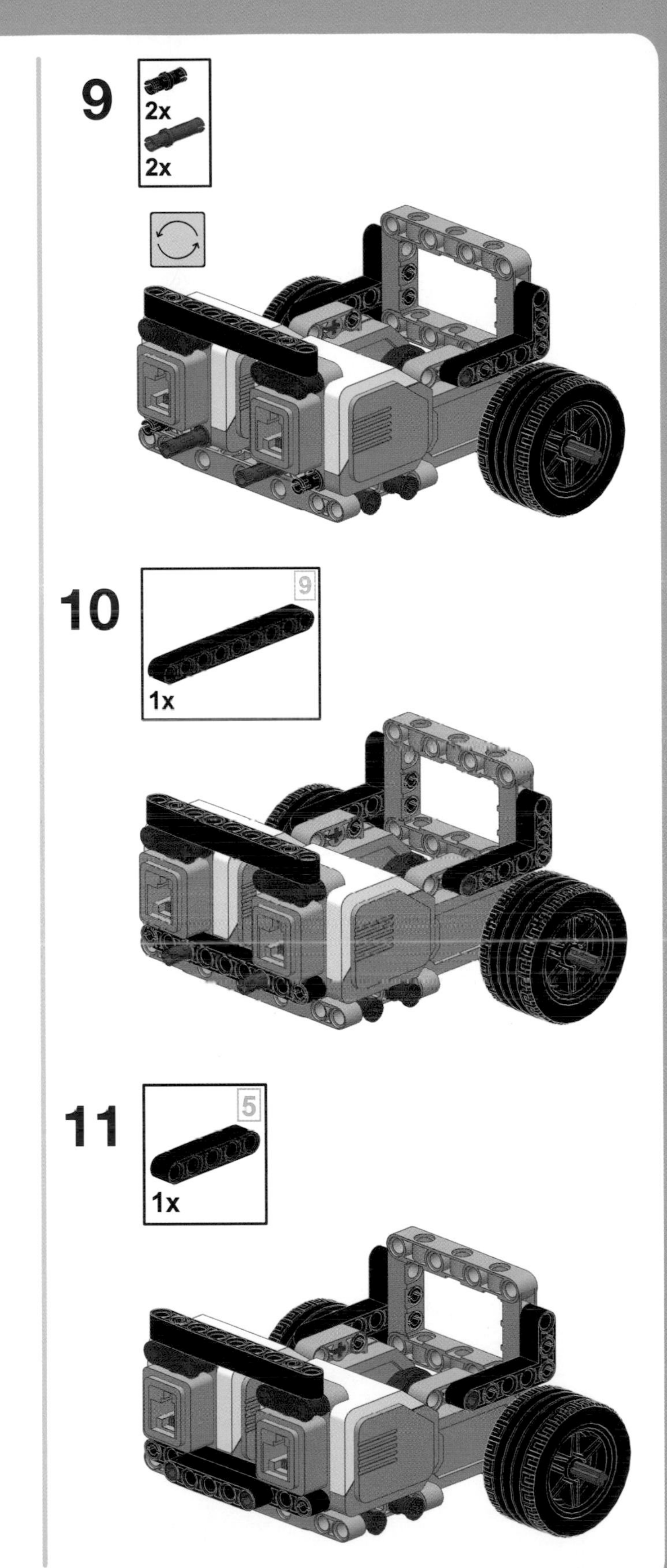

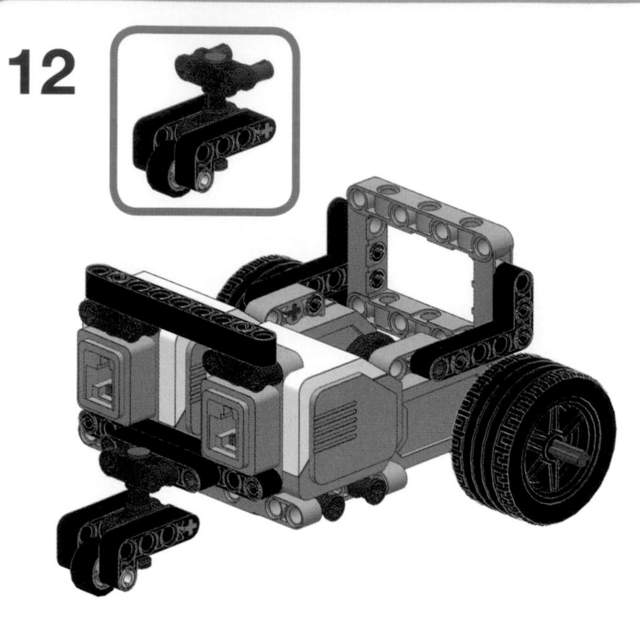

assembling the caster wheel with the education edition

The Education Edition includes a special caster ball part that lets us add a third "wheel" using fewer parts.

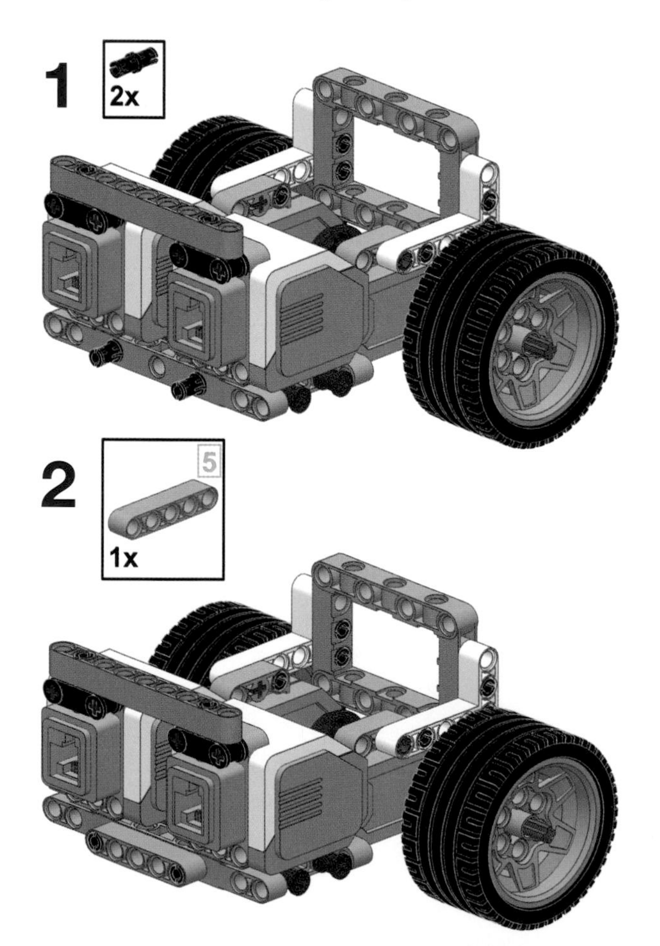

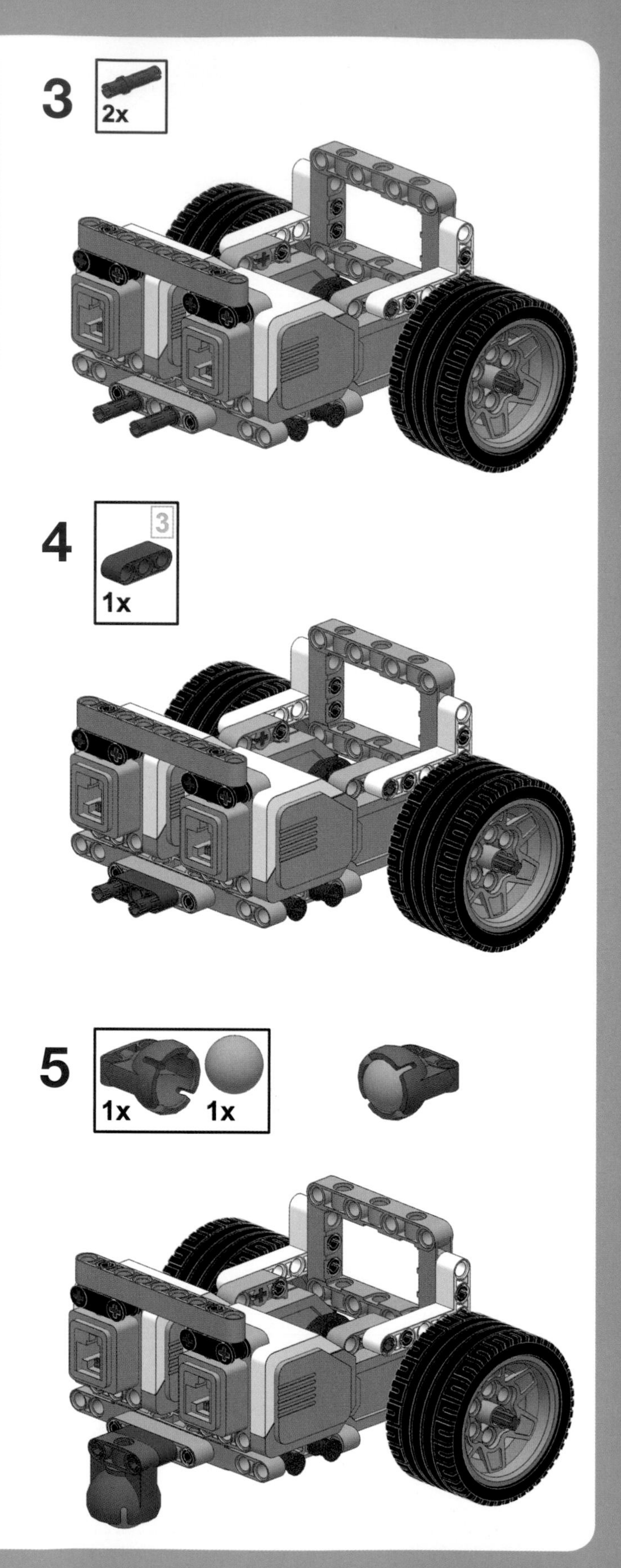

adding the EV3 brick

Now add four pins to the top of the motor assembly and attach the EV3 Brick.

1

4x

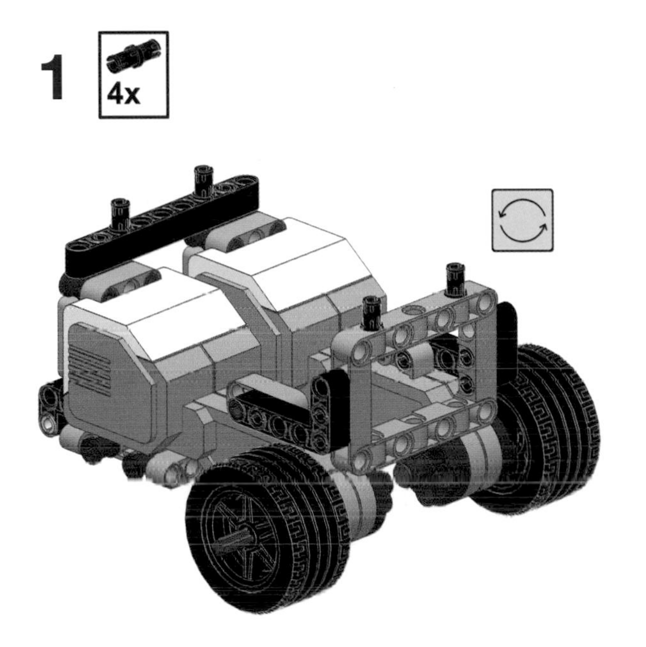

2

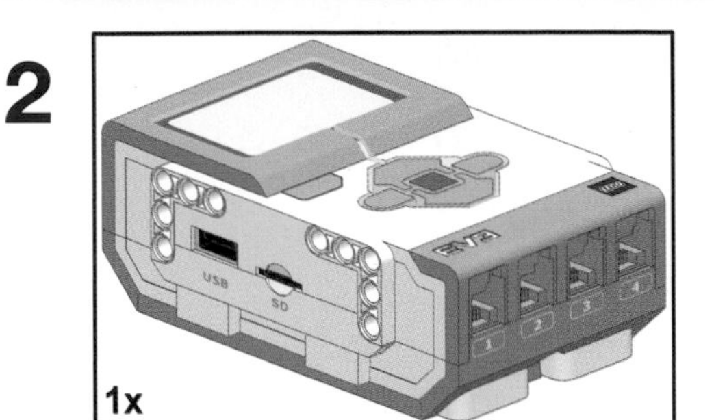

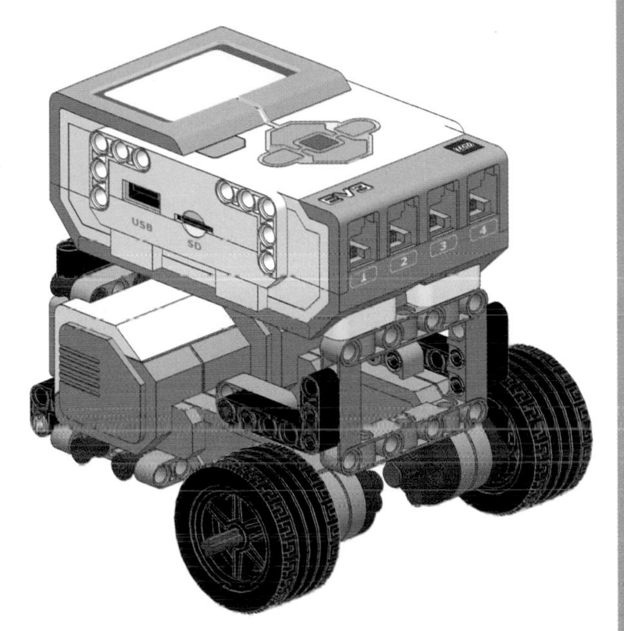

mounting the infrared or ultrasonic sensor

Now mount the Infrared Sensor.

The images in this section show the Infrared Sensor from the Home Edition; substitute the Ultrasonic Sensor if you are using the Education Edition.

1

2x

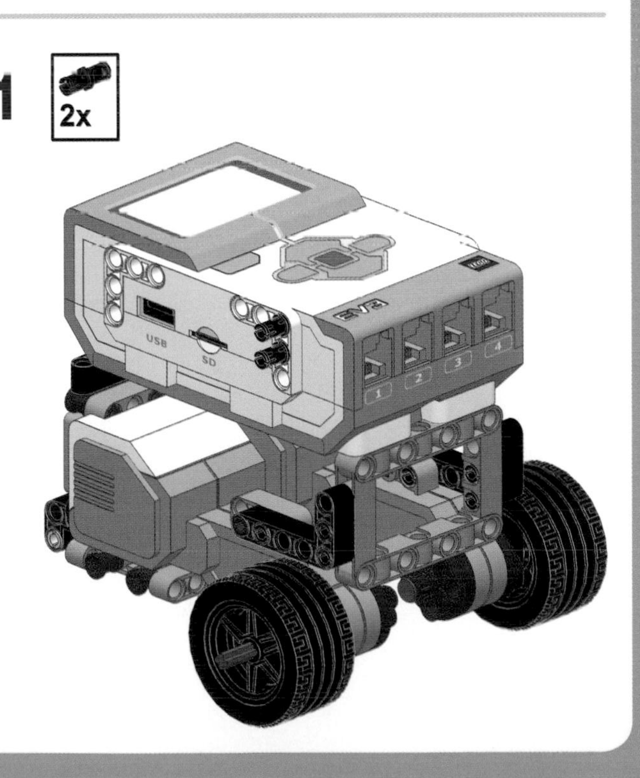

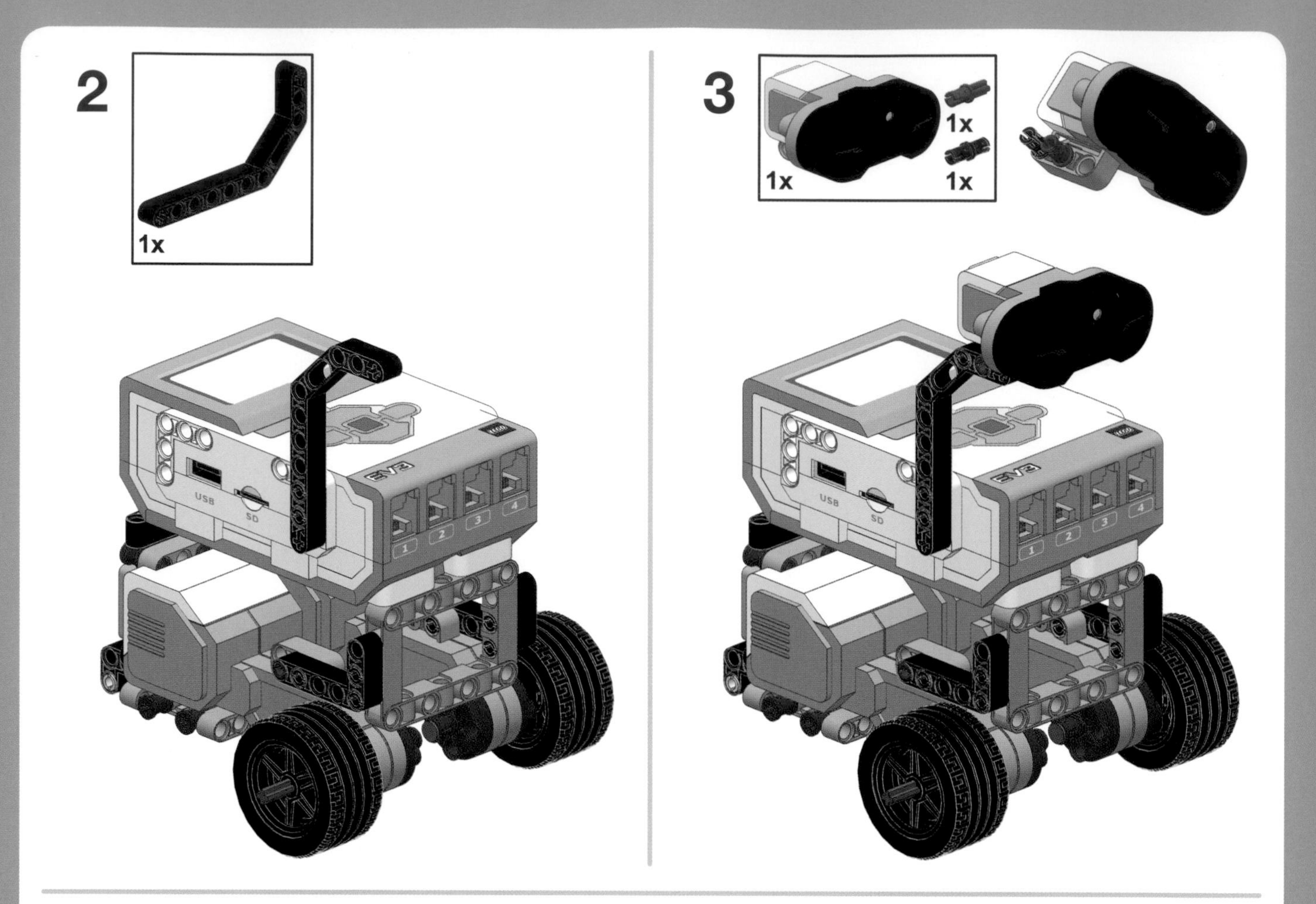

connecting the color sensor

Now connect the Color Sensor. First, create a mounting bracket.

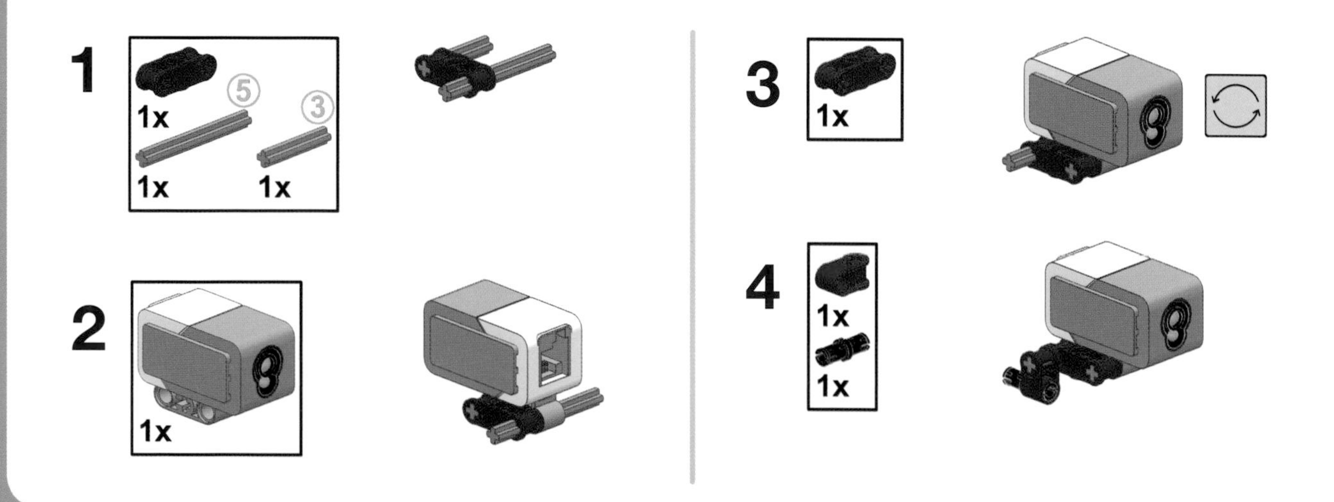

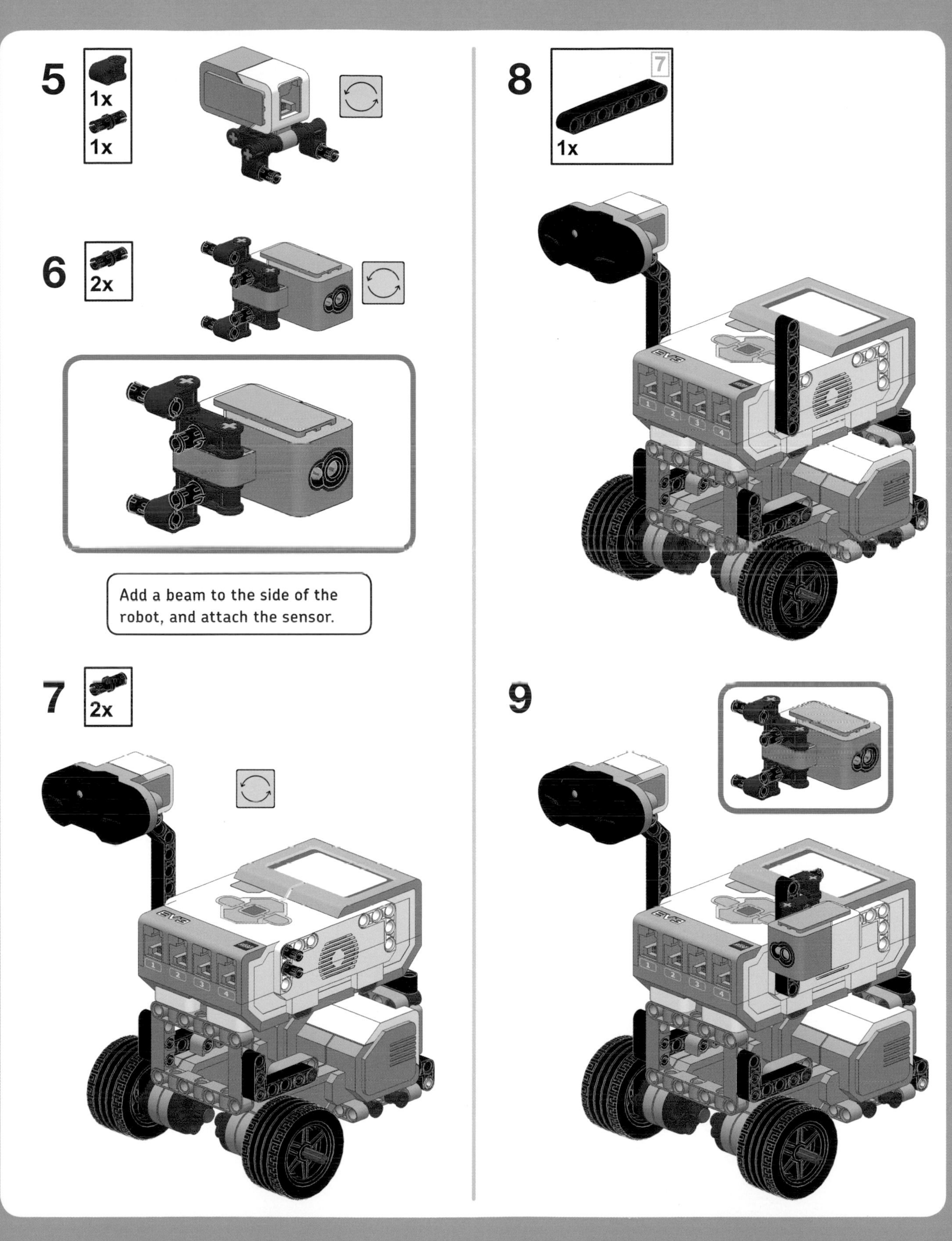
5
1x
1x
6
2x
Add a beam to the side of the robot, and attach the sensor.
7
2x
8
7
1x
9

attaching the gyro sensor (education edition)

If you have the Education Edition, attach the Gyro Sensor to the Brick on the same side as the Ultrasonic Sensor. (Skip these steps if you have the Home Edition.)

1

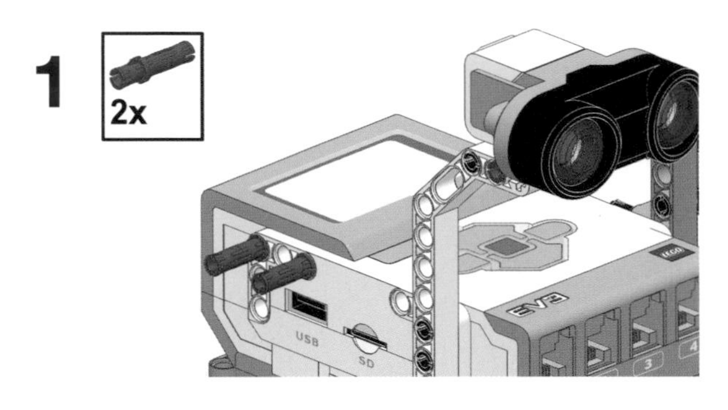

2

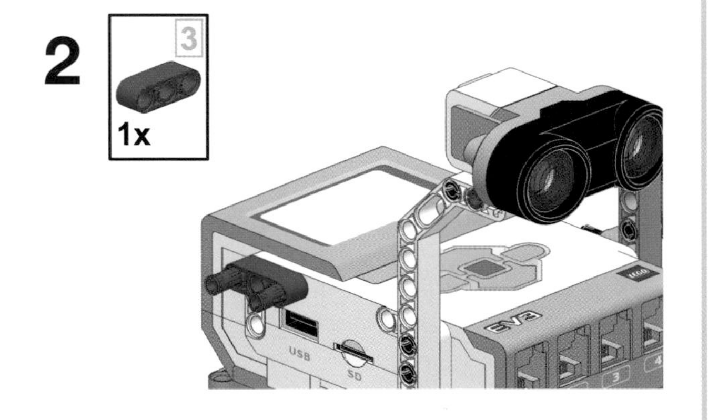

3

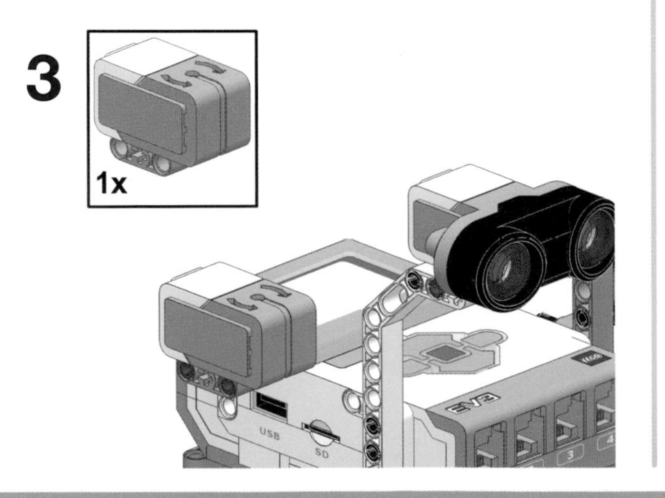

building a touch sensor bumper

Now build a bumper for the Touch Sensor. Start with the arm that hangs in front of the Touch Sensor.

1

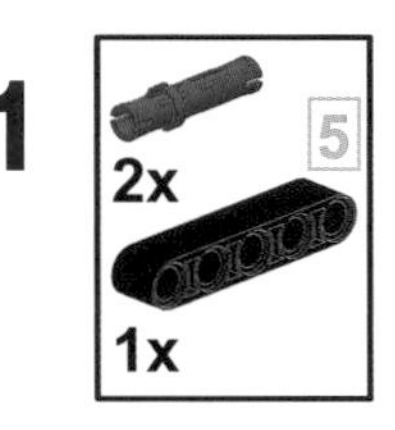

2

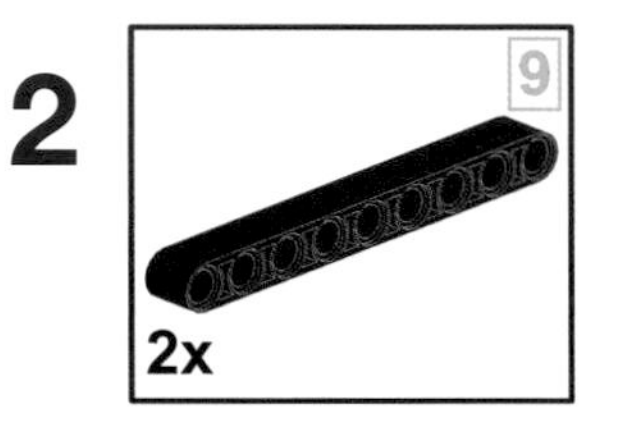

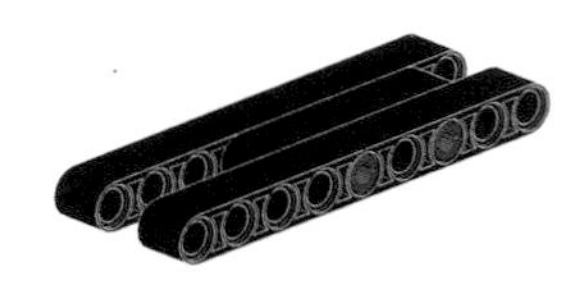

3

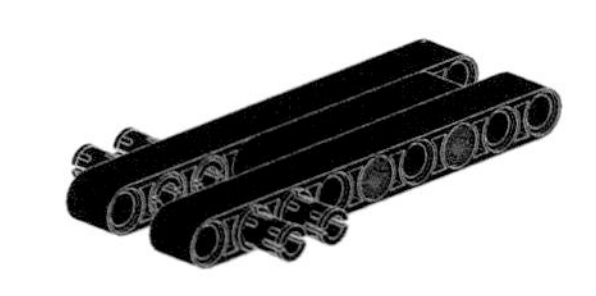

4

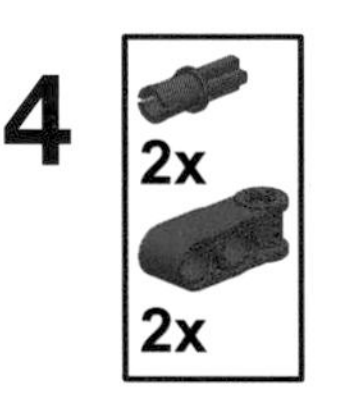

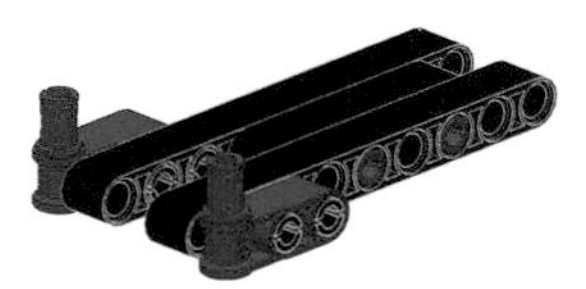

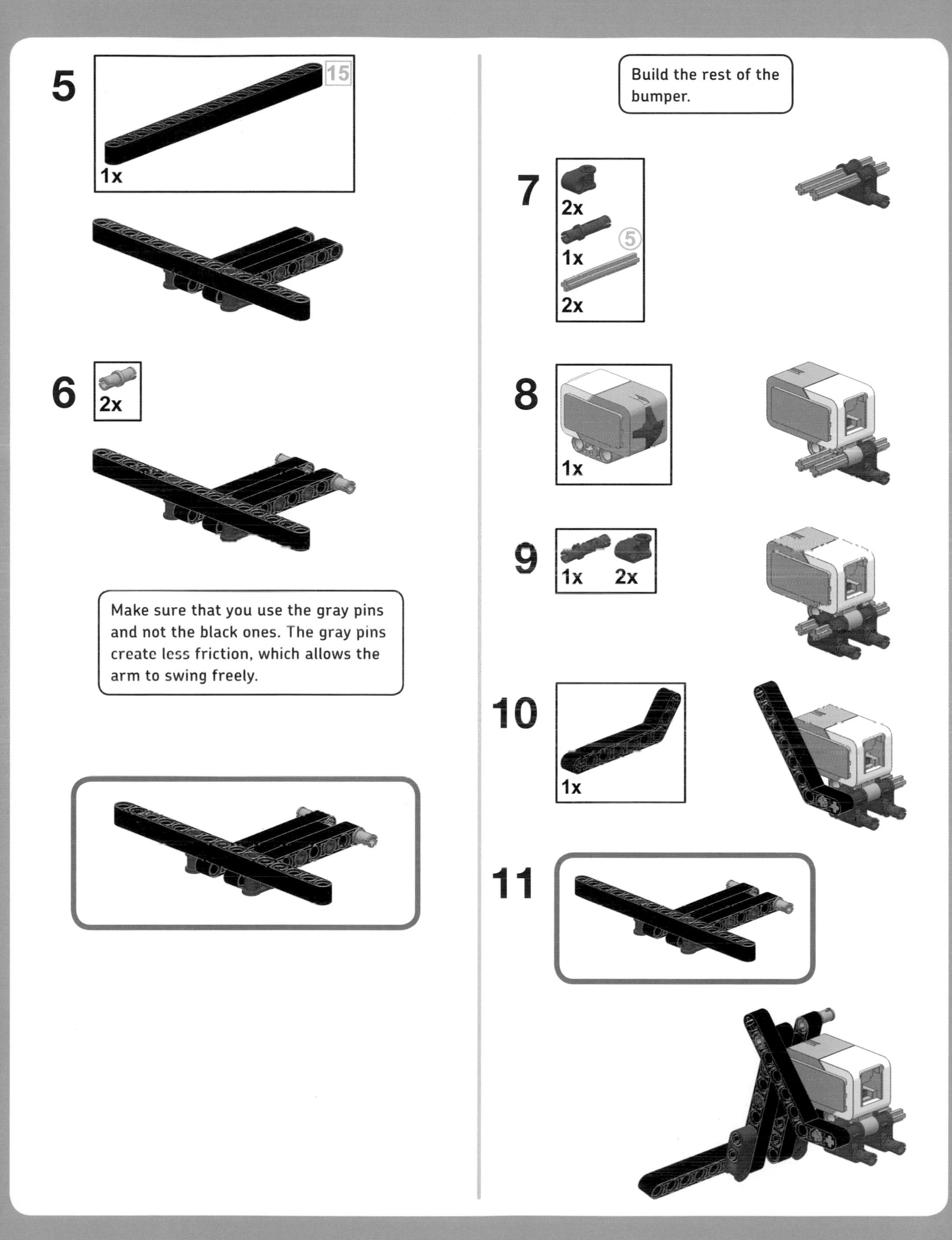
5
15
1x
6
2x
Make sure that you use the gray pins and not the black ones. The gray pins create less friction, which allows the arm to swing freely.
Build the rest of the bumper.
7
2x
1x
5
2x
8
1x
9
1x
2x
10
1x
11

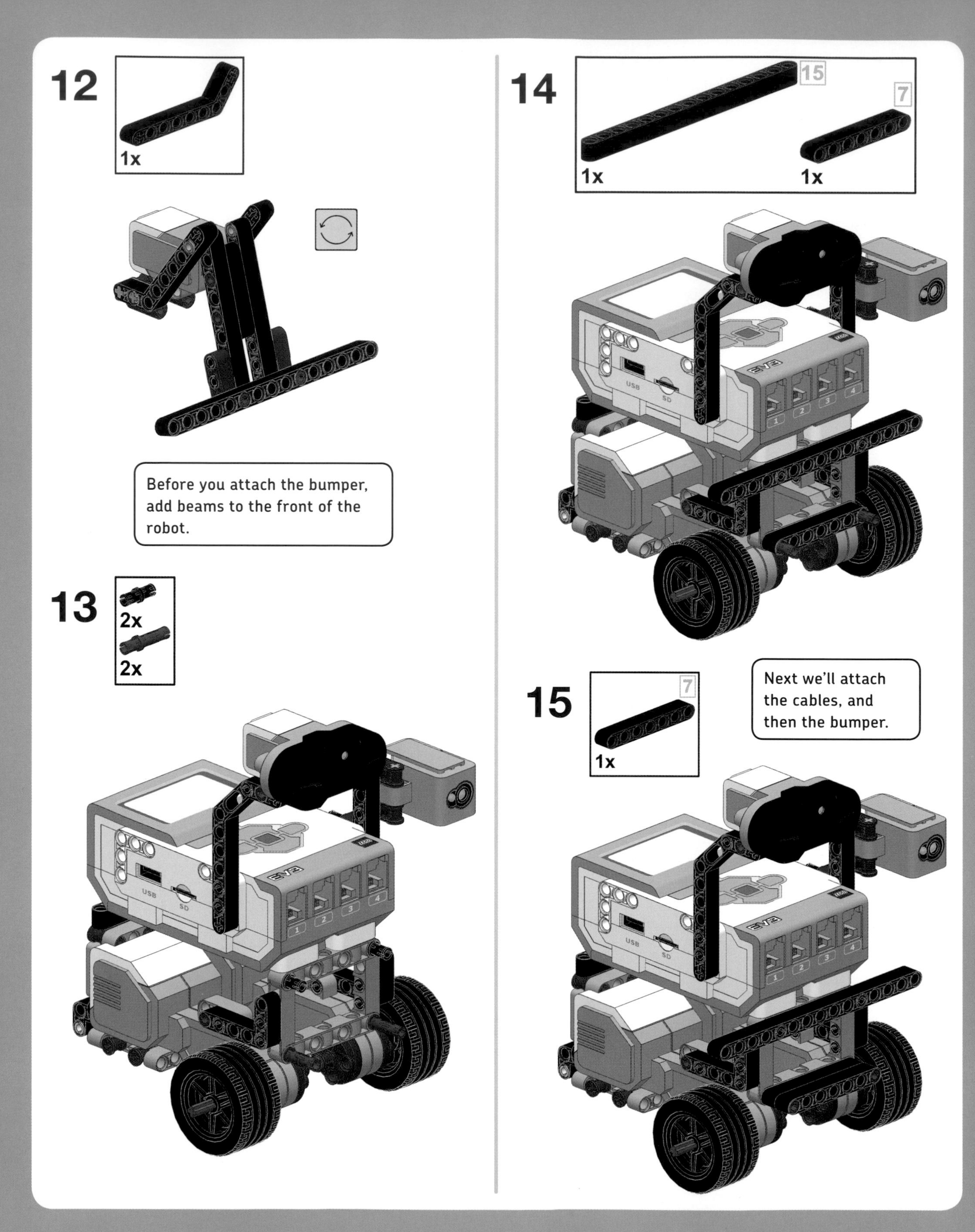
12
1x
Before you attach the bumper, add beams to the front of the robot.
13
2x
2x
14
15
7
1x
1x
15
7
1x
Next we'll attach the cables, and then the bumper.
USB
SD
EV3
1
2
3
4

attaching the cables

It's easier to add the cables before you attach the bumper. Table 3-1 includes the specifications for using cables to attach motors and sensors.

table 3-1: cable connection specifications

Motor or Sensor	Port	Cable Length
Touch Sensor	1	10 inches (25 cm)
Gyro Sensor	2	14 inches (35 cm)
Color Sensor	3	10 inches (25 cm)
Infrared or Ultrasonic Sensor	4	14 inches (35 cm)
Motor on the Infrared (or Ultrasonic) side	C	10 inches (25 cm)
Motor on the Color Sensor side	B	10 inches (25 cm)

Motors use ports B and C to match the default settings for the Move Steering and Move Tank Blocks. Likewise, blocks that use sensors default to the ports listed in the table. Any motor or sensor will work using any port (for example, the Touch Sensor works just as well using port 4 as it does using port 1), but using the default ports makes writing programs easier and less prone to error because you won't need to change port settings. The programs in the rest of this book assume that you connected the motors and sensors according to Table 3-1.

The following steps show how to attach the cables. For clarity, the images show the cable you're working with as green; in reality, all EV3 cables are black.

attaching the touch sensor

Attach a short cable (10 inches or 25 cm) to the Touch Sensor, and feed the cable through the front of the robot and out the side.

Attach the bumper to the beam on the front of the robot, and plug the other end of the cable into port 1.

attaching the infrared or ultrasonic sensor

Use a medium cable (14 inches or 35 cm) to connect the Infrared or Ultrasonic Sensor to port 4.

attaching the color sensor

Use a small cable (10 inches or 25 cm) to connect the Color Sensor to port 3.

attaching the gyro sensor (education edition)

If you have the Education Edition, attach the Gyro Sensor with a medium cable (14 inches or 35 cm) to port 2.

attaching the motors

Use two small cables (10 inches or 25 cm) to attach the motors to ports B and C. In order for the programs in this book to work, the cables must cross over each other. If you look at the robot from the back, the motor on the right is connected to port C, and the motor on the left is connected to port B.

alternate placement for the color sensor

Some programs require that the Color Sensor (instead of the Touch Sensor) be at the front of the TriBot. In these cases, mount the sensor pointing forward as shown.

You can also point the Color Sensor downward for line following. With the Home Edition, the sensor is automatically at the correct height when you attach it to the front of the TriBot. The tires are higher in the Education Edition, so you need to add a few parts; otherwise, the sensor is too far from the floor.

For the Home Edition, attach the Color Sensor as shown here.

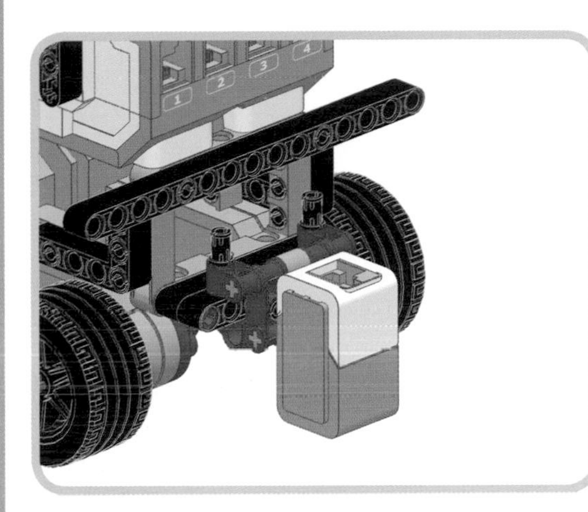

For the Education Edition, follow these steps to attach the Color Sensor.

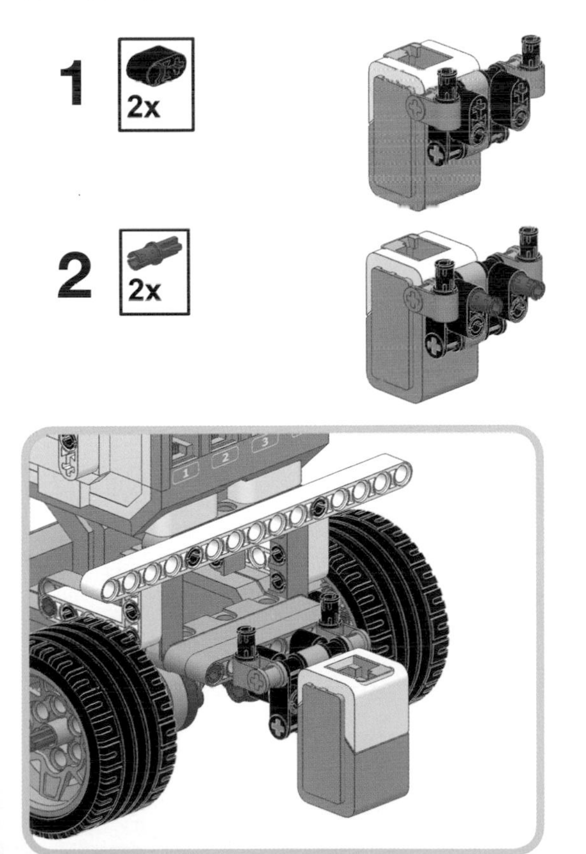

alternate placement for the ultrasonic or infrared sensor

Some programs require that you place the Infrared or Ultrasonic Sensor so that it points to the side of the TriBot. To do so, add two pins to the long beam across the front of the TriBot and then move the sensor.

1

2

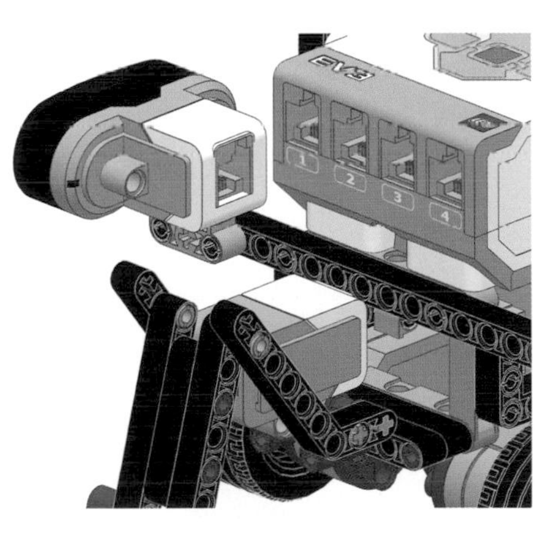

building a lift arm

Now build a Lift Arm using the Medium motor, as shown here:

Here are the parts you need to build the Lift Arm:

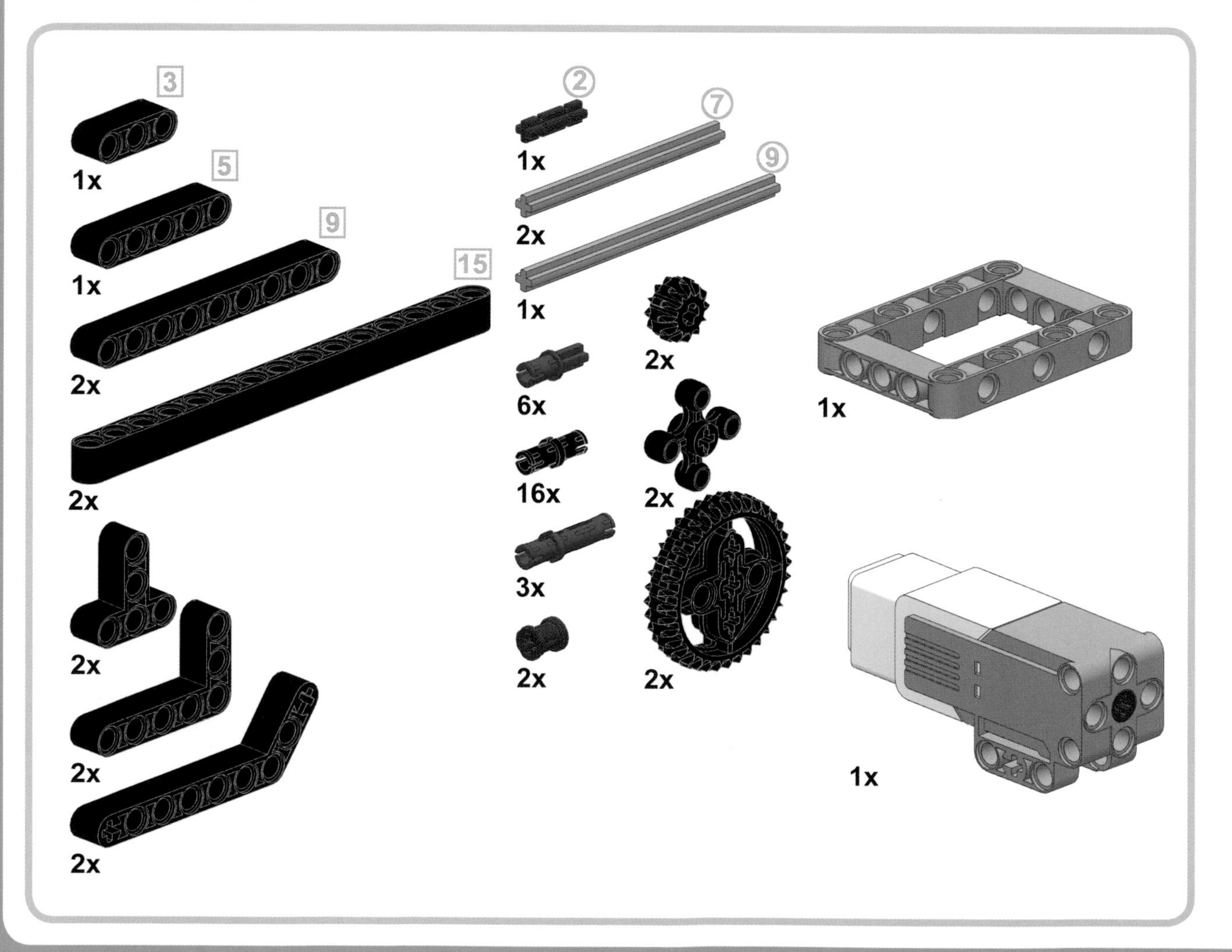

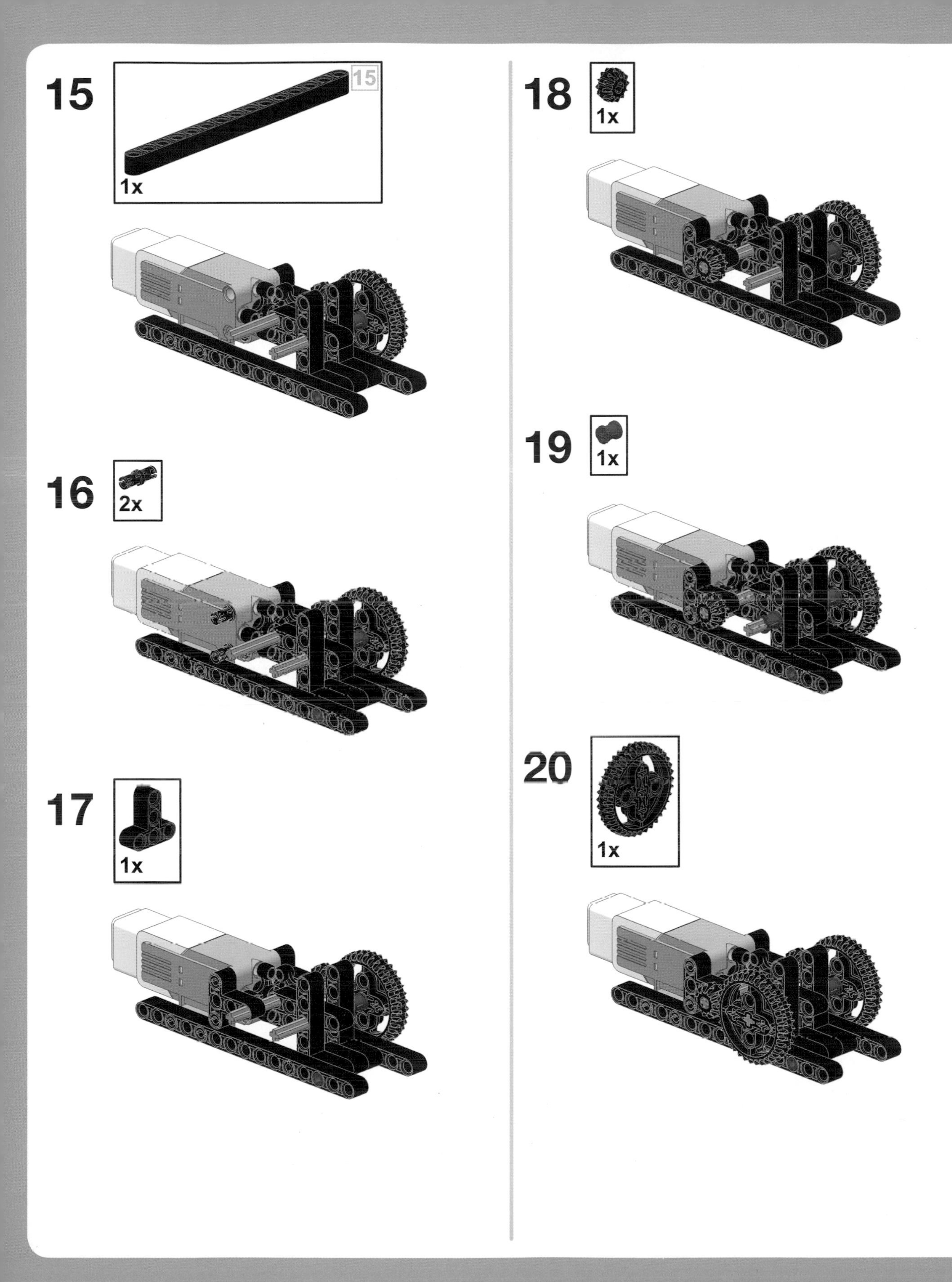
15
15
1x
16
2x
17
1x
18
1x
19
1x
20
1x

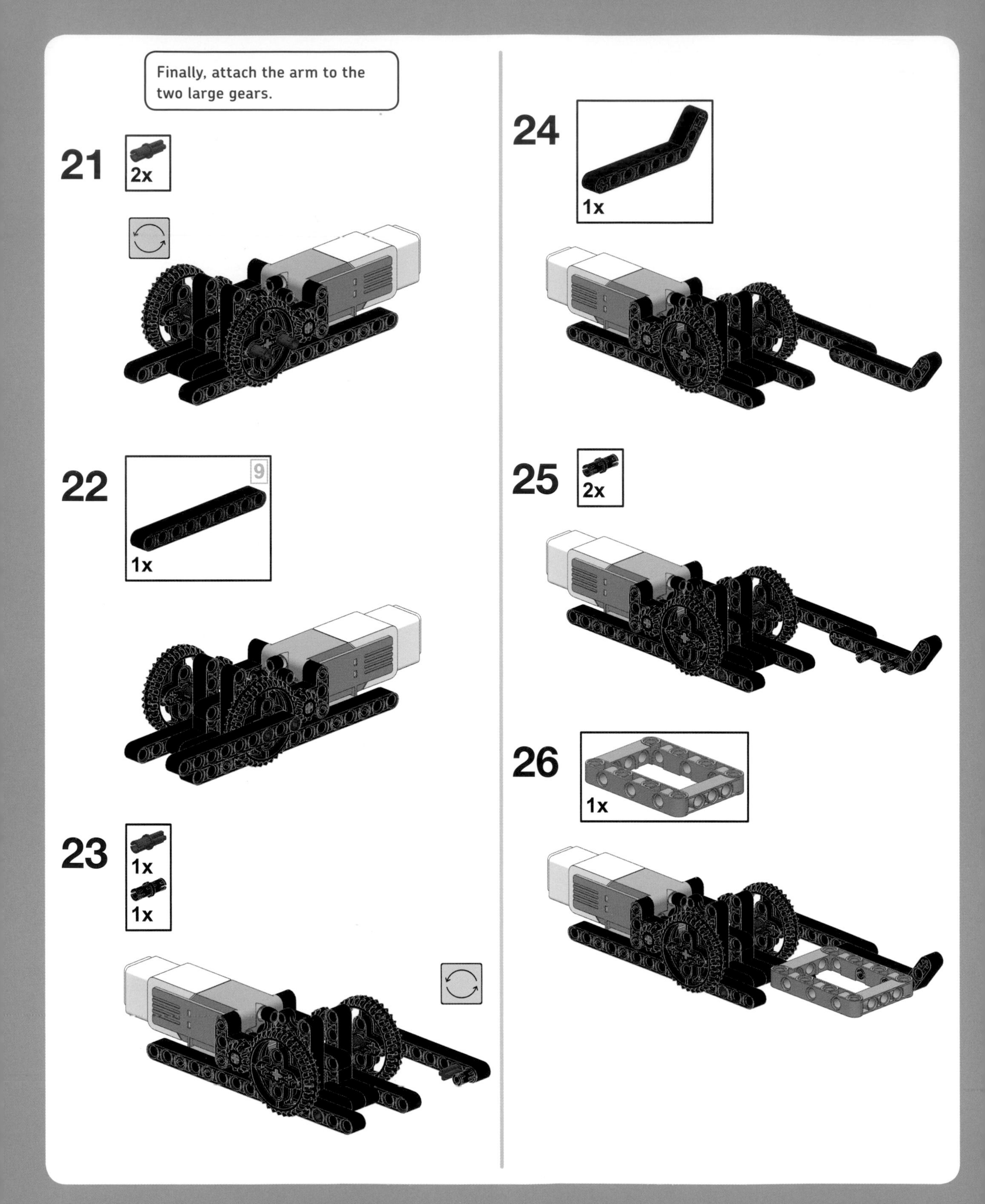

Finally, attach the arm to the two large gears.
21
2x
22
9
1x
23
1x
1x
24
1x
25
2x
26
1x

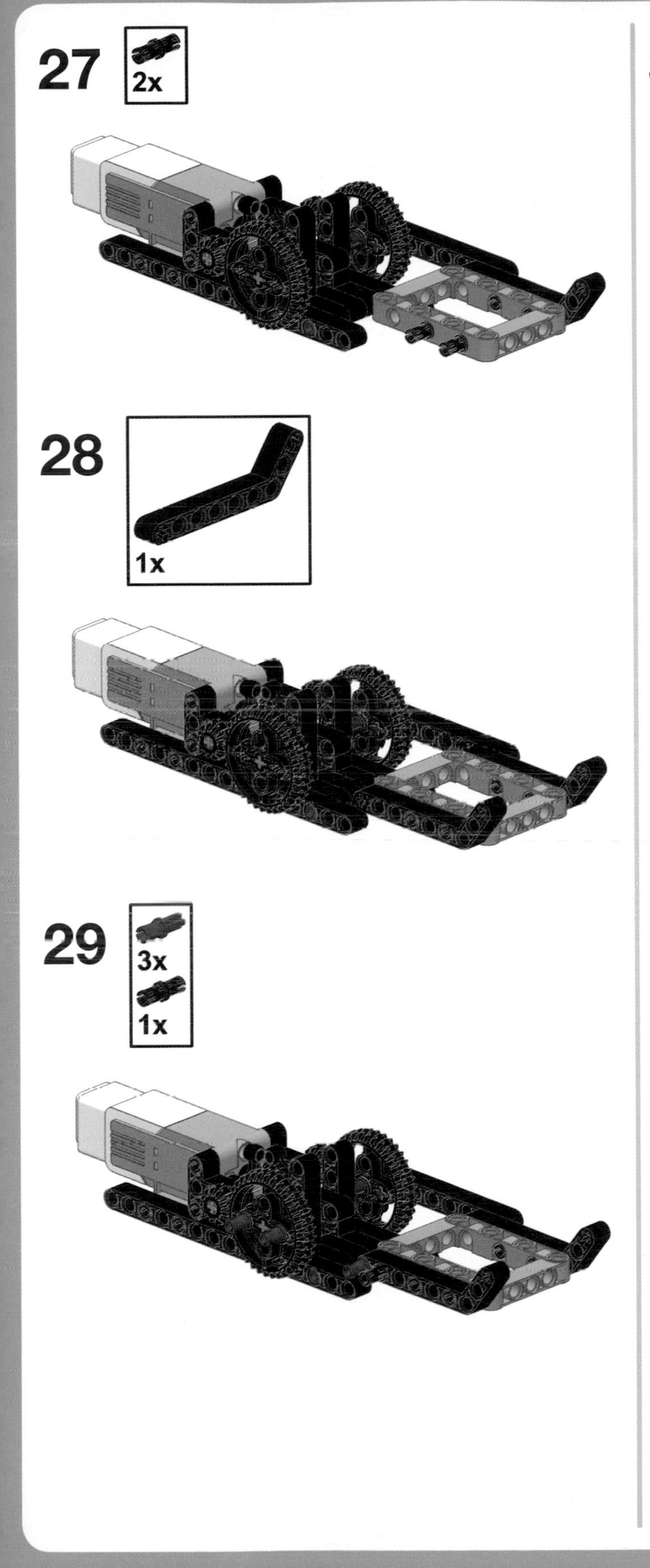

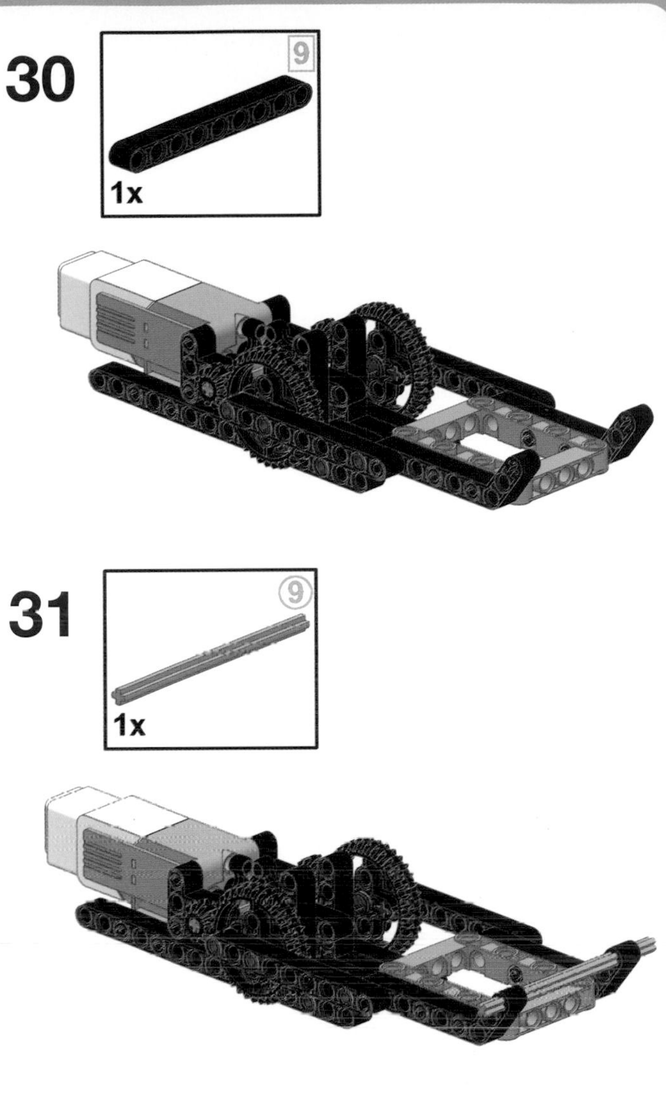

conclusion

You now have a functional TriBot and Lift Arm that you can use for the example programs in the remainder of this book. The first programs use the TriBot in its original configuration, with the Touch Sensor on the front. I'll let you know when to use one of the alternate sensor placements.

motion

Seeing your creation move is one of the most exciting and fascinating things about making a robot. Using the EV3 set, you can create vehicles, robotic arms and grippers, medieval siege engines, and many other moving contraptions. Combine it with your other LEGO sets, and you can create an almost unlimited variety of autonomous or remotely controlled models.

We can thank the EV3 motors for making this movement possible. In this chapter, you'll learn about the motors and the programming blocks that control them. We'll begin with very simple programs and work up to more complex ones.

the EV3 motors

The *Large motor* (Figure 4-1) makes it easy for you to create moving robots. Its case has an unusual shape because it contains a set of gears in addition to the electric motor. The gears adjust the speed and power of the motor's rotation, making it possible to connect a wheel directly to the motor without the need for additional gears.

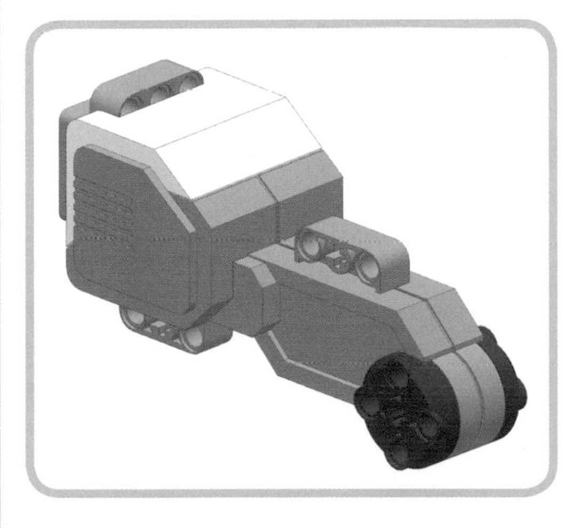

Figure 4-1: The Large motor

The *Medium motor* (Figure 4-2) is smaller and has a more regular shape that makes it easier to incorporate into a robot. This motor turns faster than the Large motor and is ideally suited for moving grippers or arms. On the other hand, it's not as powerful as the Large motor, so it's not as effective for powering a mobile robot.

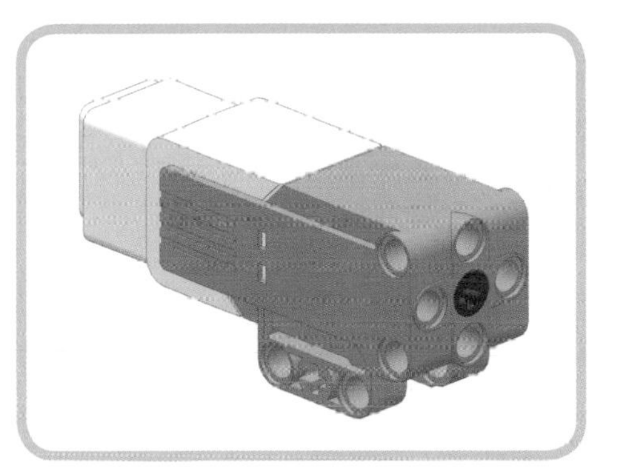

Figure 4-2: The Medium motor

Each EV3 motor contains a built-in Rotation Sensor to measure the rotation of the motor. Because this sensor is part of the motor, you don't have to connect the motor to one of the four sensor ports to use it. The blocks that control the motors (discussed next) automatically use this sensor to make very precise movements. You can also use this sensor to control the robot's movement. (Chapter 5 discusses sensors, including the Rotation Sensor.)

the move steering block

The *Move Steering block* (Figure 4-3) controls the two large EV3 motors. This is the block you'll typically use to make your robot move. It has many options and can perform a wide variety of tasks.

The Move Steering block automatically keeps the two motors synchronized by constantly adjusting how fast each motor moves so that the two wheels work together. It uses the motors' Rotation Sensors to move the robot in a straight line, turn it, or even spin it in place, based on the setting of the Steering control (discussed later in this chapter).

Figure 4-3: The Move Steering block

For example, the following simple program moves the TriBot forward a short distance:

1. Create a new project named *Chapter4*.
2. Change the name of the program to *SimpleMove*.
3. Drag a Move Steering block from the Action palette onto the Programming Canvas. Leave all the settings with the default values. Your finished program should look like Figure 4-4.

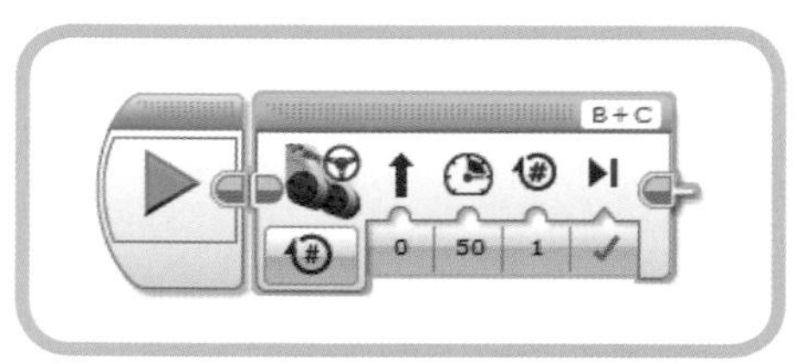

Figure 4-4: The SimpleMove *program*

4. Save the project.
5. Download and run the program. Your robot should move forward a few inches.

Even though the Move Steering block's default values work for the simple program above, you'll probably want to customize it by changing the parameters along the bottom and top of the block. The settings you choose will depend on how you construct the robot and what you want it to do.

NOTE From here on, I won't include a step that tells you to save your project, but you should still save often—especially after making significant changes and before downloading and running your program.

mode

You can tell the block what you want it to do using the *Mode Selector* (Figure 4-5). Use the three On for modes to run the motors for a certain number of rotations, degrees, or seconds before running the next block in your program. The On mode turns the motors on, and then the program immediately runs the next block. The motors stay on until the program ends or they're turned off by another block. The Off mode turns off the motors.

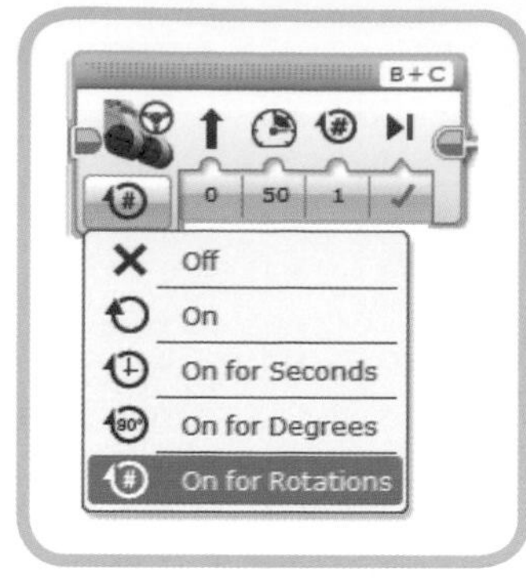

Figure 4-5: The Move Steering block's Mode Selector

After you select a block's mode, the block displays the options that are relevant to that mode. The following sections describe the options available for the Move Steering block and offer some programs to try. Rather than create a new program for each experiment, just create one program and use it for each new test.

1. Create a new program and name it *Tester*.
2. Add a Move Steering block to the program.

At this point, the Tester program should look like the *SimpleMove* program shown in Figure 4-4.

steering

The *Steering parameter* (Figure 4-6) controls the direction in which your robot moves. Possible values range from -100 to 100 and can be set by entering a number in the box or moving the slider.

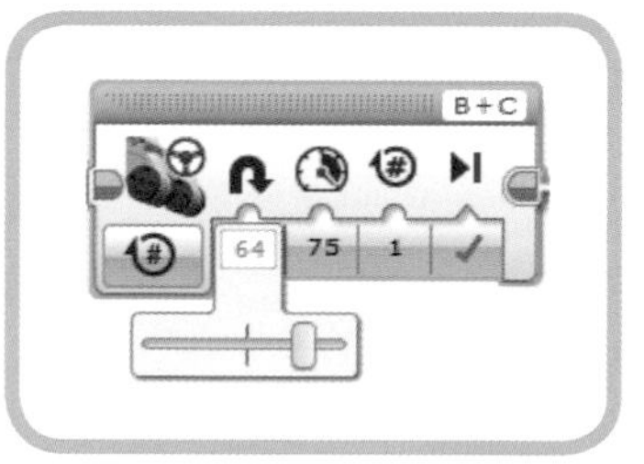

Figure 4-6: The Steering parameter

Move the slider to the middle or enter **0** to make the robot move straight. That is, the EV3 will *try* to make the robot go straight by constantly making small adjustments to the motors to keep them moving at the same speed. Many things can affect how a robot moves, and your program can only control how fast it moves each motor. If your robot is *unbalanced*, meaning there is more weight on one side than the other, it will tend to drift to one side.

NOTE **The type of floor on which your robot is moving will affect its motion, as well as the wheels and the type of caster (third wheel) you use. It's almost impossible to make your robot move perfectly straight, but you can get close enough for most situations.**

If you set the slider all the way to either side or enter -100 or 100, the robot spins in place because the two motors are moving at the same speed but in opposite directions. The distance between the two wheels determines how long the wheels must turn to make your robot spin in a full circle.

If you set the Steering slider somewhere between the middle and one of the ends, the robot makes a gentle turn. The closer the value is to -100 or 100, the tighter the turn will be. To make a turn, the EV3 slows down or stops one of its motors. For very sharp turns, it moves one motor backward and the other forward. If you select the On for Degrees or On for Rotations mode, the duration you set will apply to the faster-moving motor, which is the one on the outside of the curve.

When you are experimenting with different Steering parameters, a single rotation is a little too short to see the full effect of different values. Change the Rotation parameter to five rotations to give yourself a better sense of how the robot moves.

1. Change the Rotations parameter to **5**.
2. Change to Steering parameter to **100**. Here's how the program looks with these changes:

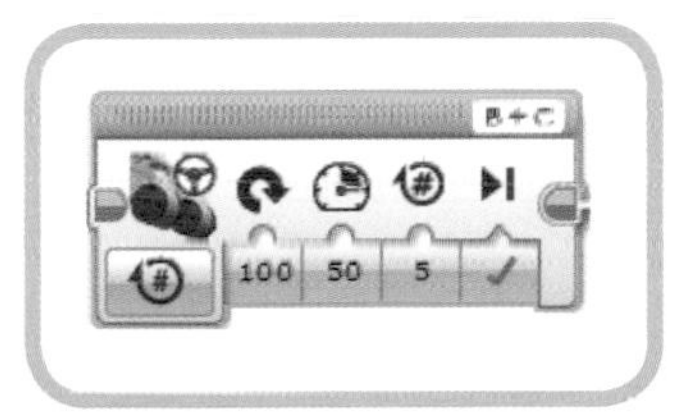

Download and run the program. The TriBot should spin in a circle. Try using several other values for Steering to get a feel for how they affect the TriBot's motion.

power

The *Power parameter* (Figure 4-7) controls how fast the motors move. Set the Power by moving the slider or by entering a value from -100 to 100. Positive values make the motors move forward and negative ones move the motors backward. A setting of 100 makes the motors move as fast as they can, and 0 keeps the motors still.

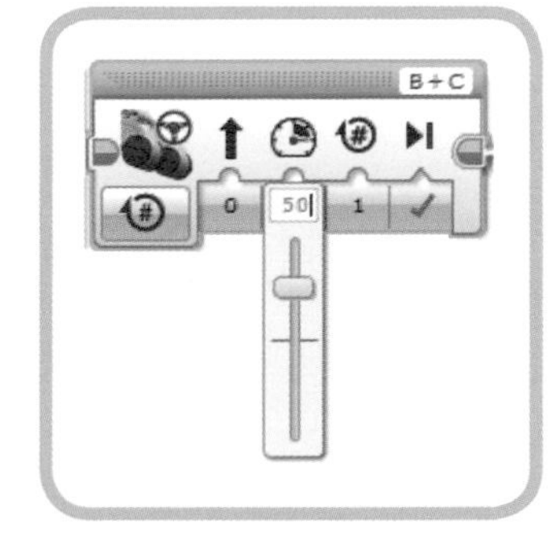

Figure 4-7: The Power parameter

The next set of changes will make the TriBot move in a straight line at full speed.

1. Set the Steering parameter to **0**.
2. Set the Power parameter to **100**.

Download and run the program above to see how fast your robot can go, and then try different Power parameters while keeping the Steering parameter at 0. Next, try adjusting both the Steering and Power parameters. You'll notice that the TriBot can be a little unstable when trying to make a sharp turn while moving very fast.

duration

When you use one of the On for modes, you'll need to set the *Duration parameter* in rotations, degrees, or seconds. Figure 4-8 shows the block with On for Seconds mode selected, and the Duration set to 1 second. Notice that the small icon above the Duration value changes to match the mode. For example, a small clock is shown for the On for Seconds mode.

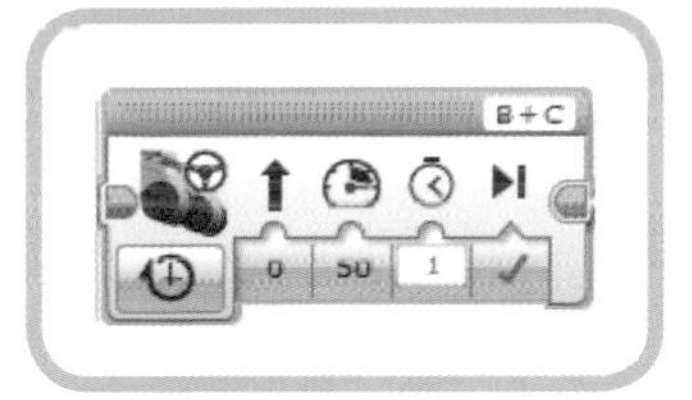

Figure 4-8: The Duration parameter in the On for Seconds mode

The relationship between rotations and degrees is simple: 360 degrees will turn the motor exactly one rotation. It's usually simpler to measure a long move in rotations and a short one in degrees because the numbers are easier to work with.

When using rotations or degrees, you can use a negative number to make the motors move backwards. (You can't use a negative number when setting the Duration in seconds because time travel is not supported by the EV3.)

brake at end

The *Brake at End* parameter (Figure 4-9) tells the block how to stop moving and what to do with the motors after the move is complete. The first option (set with a checkmark) quickly stops the motors and locks them in place. The second option (set with an X) allows the motors to spin freely until they stop on their own. These options are often called Brake and Coast, respectively.

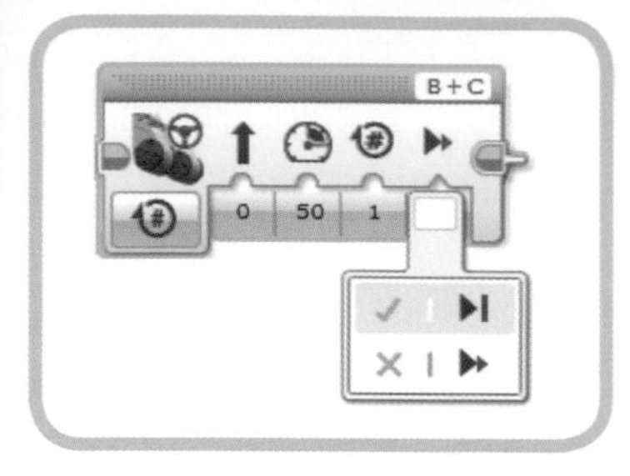

Figure 4-9: The Brake at End parameter

As currently written, the *Tester* program doesn't show the full effect of the Brake setting. After the robot moves forward, the block stops the motor, the program ends, and the EV3 stops trying to hold the motor in place—but the TriBot will still have enough momentum to move forward a bit. To see the real effect of this setting, add a Wait block in Time mode to pause the program for two seconds after the Move Steering block executes. This gives the robot time to come to a complete stop before the program ends.

* To make your robot stop accurately, use the Brake setting when using the **On for Rotations** or **On for Degrees** modes. The motor should stop very close to the duration you set. If you use the Coast option, the motor merely slows down after reaching the duration and consequently moves a little past the target you set.
* Use Brake to keep a motor from moving after the block finishes. For example, if the motor controls a gripper, use Brake to hold the motor still after grabbing an object; this will prevent the object from slipping.
* Holding a motor in place uses extra battery power, so set a motor to Brake only if you need to hold it in place.

port

You can use the *Port parameter* (Figure 4-10) to tell the EV3 which two motor ports to move. Click the left or right motor port (B and C in Figure 4-10) and a menu appears which allows you to select a port. The top choice, marked with the small black block, is for using data wires, which I discuss in Chapter 8.

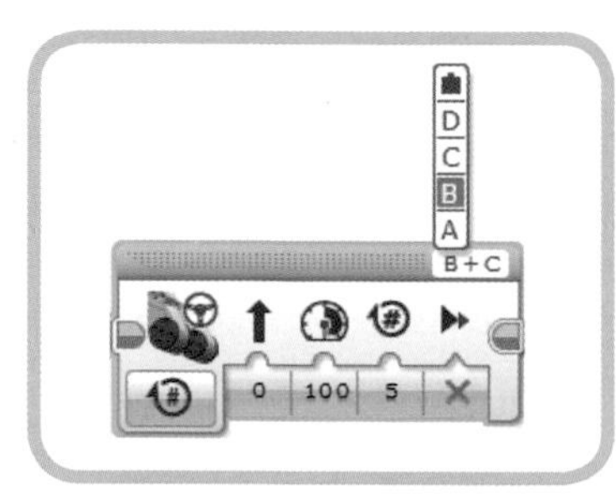

Figure 4-10: The Port parameter

By default, the Move Steering block selects motor port B for the left motor and port C for the right motor. If you don't select the correct ports, and you swap the left and right motors, the Steering parameter will make the robot steer in the opposite direction.

The EV3 software has a feature called Auto-ID, which lets the EV3 software know which motors and sensors are connected to your brick. It can also change the default port that a block uses. For example, when you add a Medium Motor block to your program, it will usually set the port to A. However, if you have the Medium motor plugged into another port when you add a Medium Motor block, the block will use that port by default (as long as your EV3 Brick is connected to the software).

port view

When using the Move Steering block's On for Rotations or On for Degrees mode, setting the Duration and deciding on the correct value can be time-consuming. The *Port View* (Figure 4-11) can be a big help because it can show you how far each motor has moved. Select the middle tab on the Hardware page to show the Port View. (The MINDSTORMS environment must be connected to the EV3 for this window to show you anything useful.)

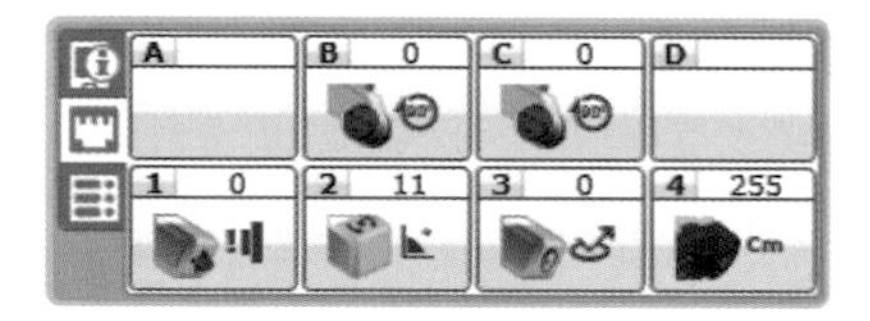

Figure 4-11: The Port View

The Port View shows which ports have motors and sensors attached, as well as the type of motor or sensor. The value above a motor is from the motor's rotation sensor: It represents the total distance the motor moved. As an example, if you move the motor forward 360 degrees and then backward 360 degrees, this value is 0. By default, the value for a motor is displayed in degrees. Click the motor to display a menu that allows you to select rotations. (I'll discuss Port View in more detail in Chapter 5, when I discuss sensor types.)

When the EV3 is not running a program, you can move a motor manually to see how many degrees it moves. This is a particularly useful way to determine how far to move an arm or gripper. You can move a motor by hand to determine the distance you want it to go, and then enter that value into one of the move blocks. To reset the value to 0, click the port letter (A, B, C, or D).

When you run a program, the values are reset to 0 at the start and updated while the program is running. You can add Wait blocks to make the program pause if you want to check the values at various steps.

the EV3 intelligent brick view menu

The Port View has two drawbacks: You need to keep the EV3 connected to the computer, and you need to be sitting in front of your computer to see it. Alternatively, you can measure how far a motor moves using only the menu on the EV3 Brick. To do so, select the third main tab (along the top of the display) and then **Port View**. Select one of the motor ports, and when you move the motor that's attached to the selected port, the EV3 display should show how far it moves.

the ThereAndBack program

Let's use the Move Steering block in a couple of programs. The *ThereAndBack* program will make the TriBot move forward three feet, turn around, then return to where it started. Some measuring is required to solve this problem, so if your rulers are metric, feel free to change the problem to use one meter instead of three feet.

This program uses three Move Steering blocks: one to move forward, one to spin the robot around, and one to move the robot back to where it started.

moving forward

The first block needs to move the robot forward three feet, but feet isn't one of the duration options. How do you figure out how far three feet is in degrees or rotations? Let's consider rotations.

One way is to write a program that moves the robot a long distance, say, 10 rotations. Before running the program, mark your robot's starting position on the floor (a piece of tape should do), run the program, and measure how far the robot moves in inches. Divide the distance moved by the number of rotations to determine how far the robot moves in one rotation. For example, my robot moves 51 inches in 10 rotations, or 5.1 inches in a single rotation. The robot needs to go 36 inches, so I divide 36 by 5.1 to find the Duration setting needed to move 36 inches. Start with that calculated value and adjust the setting until you get what you need.

Be aware that the number you need will vary slightly if you change the type of floor the robot is moving on, the wheels you use, or the block's Power parameter. A duration of 6.8 rotations at a Power of 50 worked well for me.

NOTE Because the tires from the Education Edition are bigger than those in the Home Edition, they roll farther in one rotation. A Duration of 5.4 will move the robot 36 inches when using these tires.

When you know the duration, you can start writing the program. Here's the first set of steps:

1. Create a new program called *ThereAndBack*.
2. Add a Move Steering block next to the Start block. The mode will be set to **On for Rotations** by default.
3. Set the Rotations parameter to the number you calculated.

   Figure 4-12 shows the program at this point.

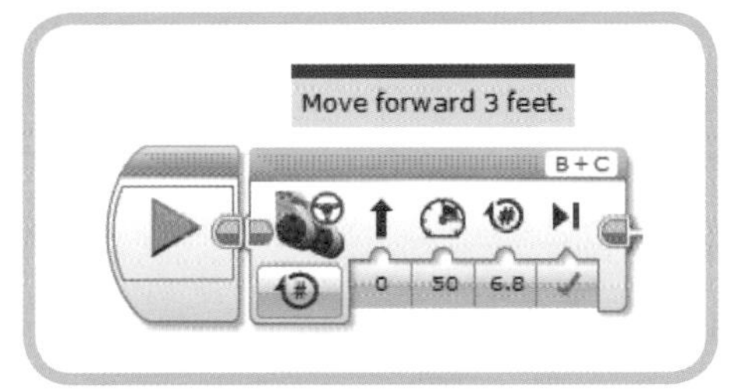

Figure 4-12: Step 1 of ThereAndBack

Test these settings by running the program several times across a measured distance. Your robot should stop very close to the same spot each time. If it doesn't, try lowering the Power parameter. Adjust the Duration if the robot stops at the same spot but doesn't travel the correct distance.

turning around

A second Move Steering block will turn the robot around for the return trip. Move the Steering slider all the way to one side to make the TriBot spin. (Choose either side; the direction isn't important.)

Once again, the challenge in configuring this block lies in figuring out the duration value. After some testing, I found that a duration setting of 425 degrees worked well (325 degrees using the Education Edition tires). I turned the Power down to 40 to get a good, consistent turn because with the Power at 50, it turned a little too far about half the time. This is a typical trade-off between speed and accuracy.

Here are the next steps in building this part of the program:

4. Drag a Move Steering block after the existing Move Steering block.
5. Set the mode to **On for Degrees**.
6. Drag the Steering slider all the way to one side.
7. Set the Degrees to **425** and the Power to **40**. Use these for initial values; you can adjust them as needed during testing.

Figure 4-13 shows how the program should look with this new block added.

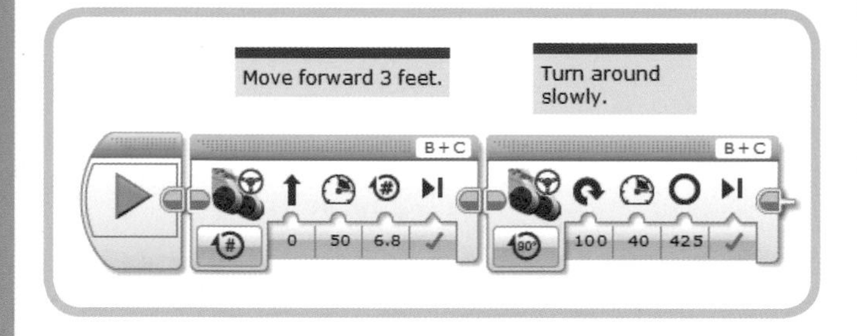

Figure 4-13: ThereAndBack *with a second Move Steering block*

testing a single block

To fine-tune the Degrees and Power parameters, it's easier to test the block by itself over and over instead of waiting for the robot to travel three feet and then seeing whether it turns around correctly. The Run Selected button (Figure 4-14) will run a single block or a group of blocks. If you select the second Move Steering block and then click Run Selected, your robot should only turn around (instead of moving three feet first). Adjust the Degrees and Power parameters as needed until your robot consistently turns all the way around.

moving back to the start

To move the TriBot back to where it started, add a third Move Steering block that travels the same distance as the first. The final steps in the program are as follows:

8. Place a Move Steering block at the end of the program.
9. Set the Rotations and Power to the same values you used for the first Move block.

Figure 4-15 shows the final program.

Test your completed program. A slight error in the duration of the second block may show up after the TriBot travels back the three feet, so you may need to adjust the Power or Degrees parameter of the second Move block. When things are working well, try increasing the speed to find the point where the consistency of the program starts to suffer. You should notice that the robot starts to turn too far or fails to get back to the starting point if it's going too fast. When moving very fast, the wheels tend to slip more, and other issues, such as imbalance in the robot, can cause more apparent errors.

the AroundTheBlock program

The *AroundTheBlock* program will make the TriBot travel in a square pattern and end up where it started. For this example, I'll use a square that is three rotations long on each side. At the corner, the robot can move in a gentle curve to produce a smoother motion, rather than just spinning in place.

To travel in a square, the robot needs to move along a side and turn a corner, move along the next side and turn a corner, and continue like this for all four sides.

Figure 4-14: The Run Selected button

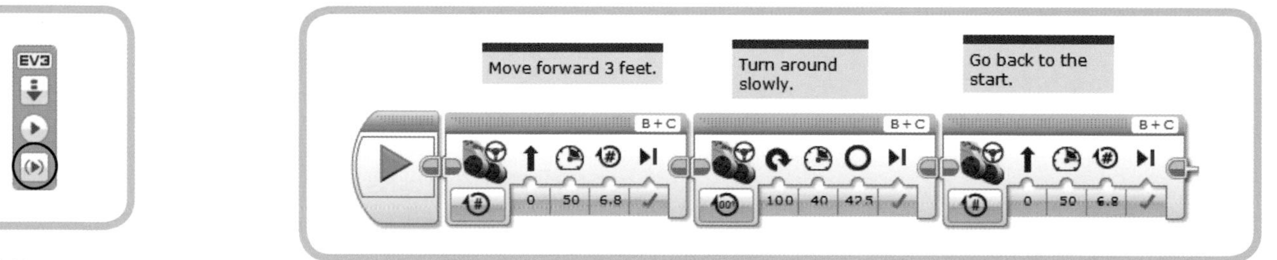

Figure 4-15: Final ThereAndBack *program*

the first side and corner

The first part of the program uses two Move Steering blocks to move the TriBot straight and make the first turn. To move along the edge, just set the duration to three rotations. To turn the corner with a gentle curve, set the Steering to 25.

The next step is to find the Duration value that gives you an accurate turn around the corner. I found that 2.4 rotations worked well (or 1.8 using the Education Edition tires). Your setting might differ due to factors like the Steering parameter and surface you use. Here's how to build this part of the program:

1. Create a new program called *AroundTheBlock*.
2. Add a Move Steering block after the Start block.
3. Set the Rotations parameter to **3**.
4. Add a second Move Steering block to the program.
5. Set the Rotations parameter to **2.4**.
6. Set the Steering parameter to **25**.

   Figure 4-16 shows the program at this point.

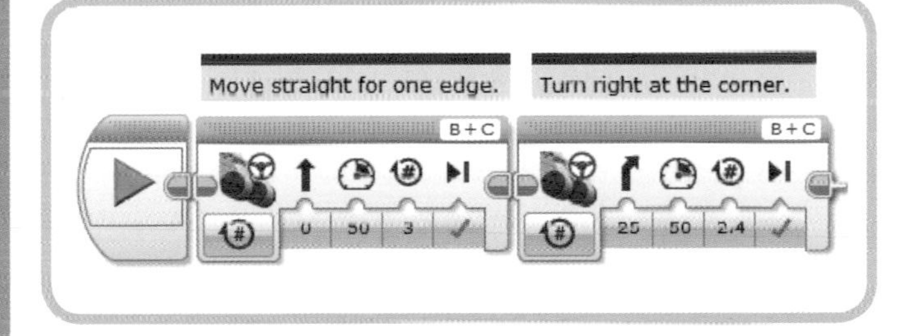

Figure 4-16: The AroundTheBlock *program moving straight for three rotations and making a turn at the corner*

the other three sides and corners

Now extend the program to go all the way around the square. You could add six more Move Steering blocks and use the same settings for the other three edges and corners, but that would be tedious. And imagine what would happen if you wanted to go around the square 10 times—you would need to add 78 more blocks! An easier way is to use the Loop block.

The *Loop block* (Figure 4-17) lets you run a group of blocks multiple times. You'll find the Loop block on the Flow Control palette, next to the Wait block. You can run the two Move blocks four times (once for each side of the square) by placing them inside a Loop block. (We'll learn more about the Loop block in Chapter 6.)

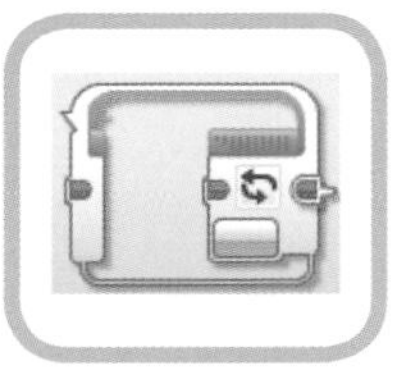

Figure 4-17: The Loop block

7. Drag a Loop block onto the end of the program. Your program should look like this:

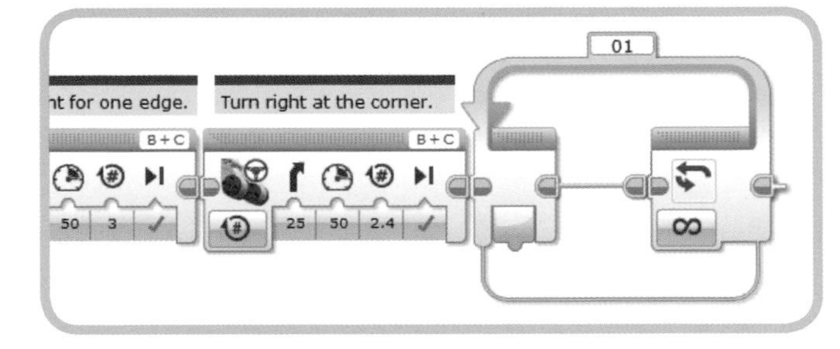

8. Drag the two Move blocks to the middle of the Loop block. The Loop block should expand as you drop in the Move blocks. Now the program should look like this:

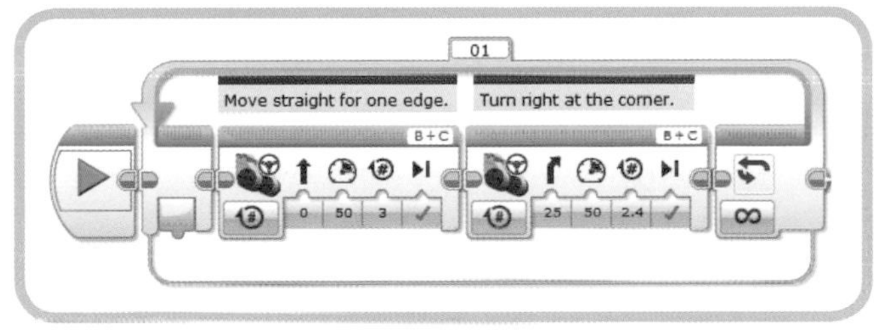

NOTE If you added comments above the two Move Steering blocks, be sure to move them so that they stay above the blocks they describe. Comments don't automatically move with the blocks, so another option is to make a habit of adding them after you edit your program.

The Loop block has lots of different modes, including one or more for each type of sensor. When you first add the Loop block to your program, it should have the **Unlimited** mode selected, as indicated by the infinity symbol on the bottom of the right side of the block (Figure 4-18). For this program, you need to change the mode to **Count**, which will let you configure the block to repeat four times.

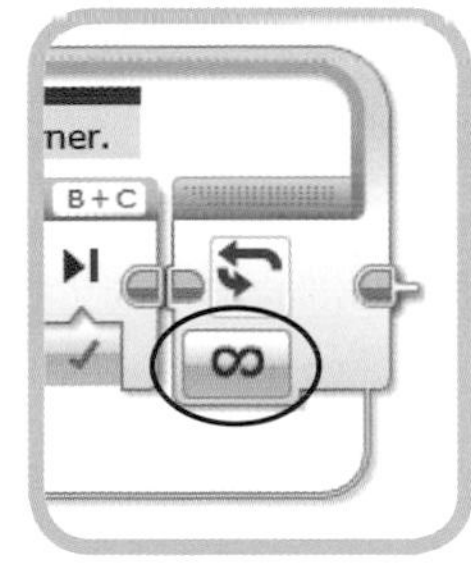

Figure 4-18: The Loop block's Mode Selector

Follow these steps to make the loop execute four times:

9. Click the Mode Selector and choose **Count**. A box appears to the right of the Mode Selector for entering the number of times the loop should repeat.
10. Enter **4** for the Count parameter.

    Figure 4-19 shows the complete program.

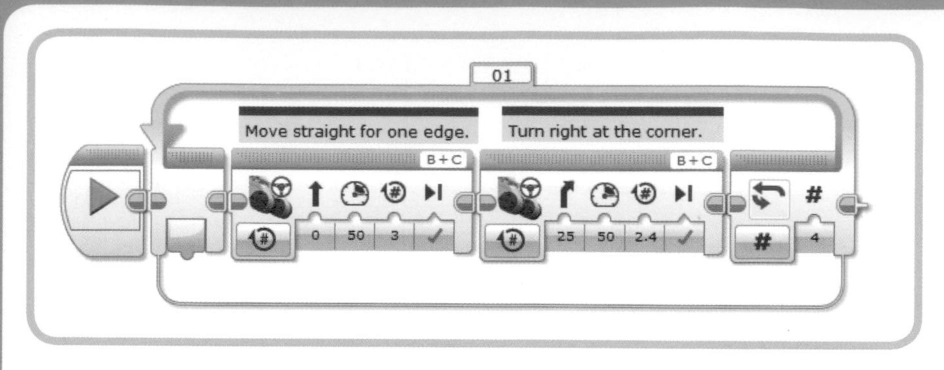

Figure 4-19: The complete AroundTheBlock *program*

testing the program

When you run the program, your TriBot should make a complete square. If it doesn't end up where it started, adjust the duration for the second Move Steering block. This "turning" block runs four times, and errors will accumulate as the robot moves around the square. The goal is to make the margin of error small enough so that it has little impact on your program, which for this program means getting reasonably close to the starting position.

the move tank block

The *Move Tank block* (Figure 4-20) is like the Move Steering block except that it has one Power parameter for each motor.

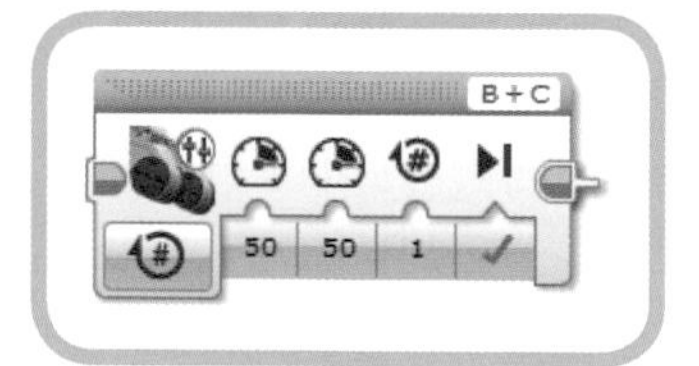

Figure 4-20: The Move Tank block

The robot's direction of movement depends on how the two Power parameters are set. If both have the same value, the robot will move straight. If the values differ, the robot will turn, and the sharpness of the turn will depend on how different the values are. If they have opposite values (one positive and one negative), the robot will spin.

As you've seen, you can get similar behavior with the Move Steering block, but by explicitly setting the Power level of each motor, the Move Tank block gives you a little more control. For example, when you use a Move Steering block with the Power set at 50 to turn the corner, the outside motor keeps moving at Power of 50 and the other motor slows down. On the other hand, you could use the Move Tank block to speed up the motor on the outside of the curve, or slow one motor down and speed the other up.

NOTE The Move Steering block is modeled after the way a car is driven: The Power parameter is the accelerator, and the Steering parameter is the steering wheel. The Move Tank block is modeled after the way tracked vehicles like tanks and bulldozers work. It can be more natural to use the Move Tank block in programs for robots that use treads instead of wheels.

the large motor and medium motor blocks

Use the *Large Motor block* or the *Medium Motor block* (Figures 4-21 and 4-22) when you want to move a single motor. The difference between these two blocks is the type of motor they control and the default Port setting. The default port is D for the Large Motor block and A for the Medium one. The modes and other parameters for both blocks are identical to those of the Move Steering block, except that there is no Steering control because these blocks only control one motor.

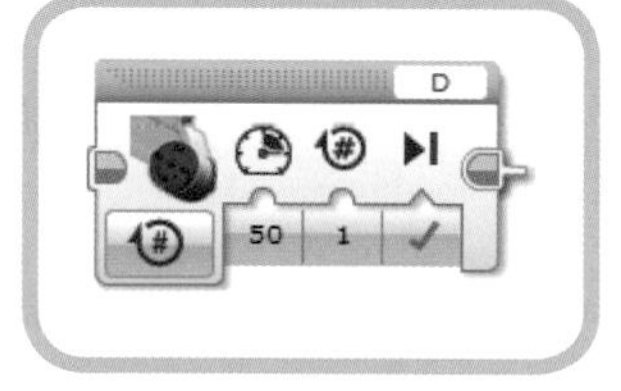

Figure 4-21: The Large Motor block

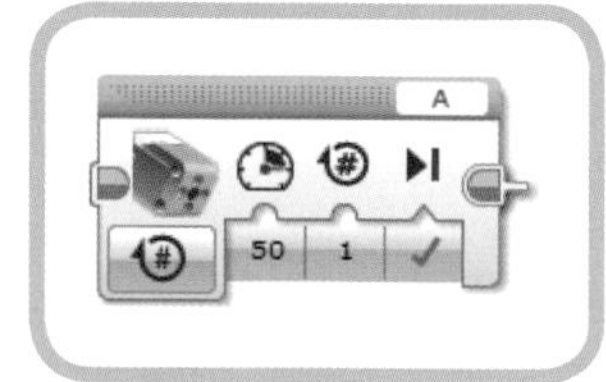

Figure 4-22: The Medium Motor block

the lift arm

Let's use the Lift Arm (Figure 4-23) to experiment with the Medium Motor block. This Lift Arm could be used to make a robot lift beams or blocks—or, turn it over and it can capture a ball. With a little adjustment in the gears, you could also turn it into a small catapult.

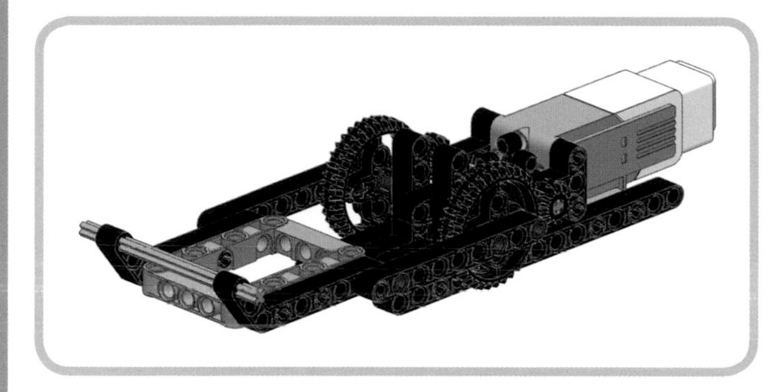

Figure 4-23: The Lift Arm

Start with the arm pointing out horizontally as in Figure 4-23. To move the arm by hand, gently turn the gear attached to the motor. If you try to move the arm directly, the resistance from the motor tends to make the parts separate or the gears slip.

Let's start by seeing how a Medium Motor block behaves with the default settings.

1. Create a new program named *LiftArm*.
2. Connect the Medium motor to port A on the EV3 using the long (20 inches or 50 cm) cable.
3. Add a Medium Motor block to the program. The program should look like this:

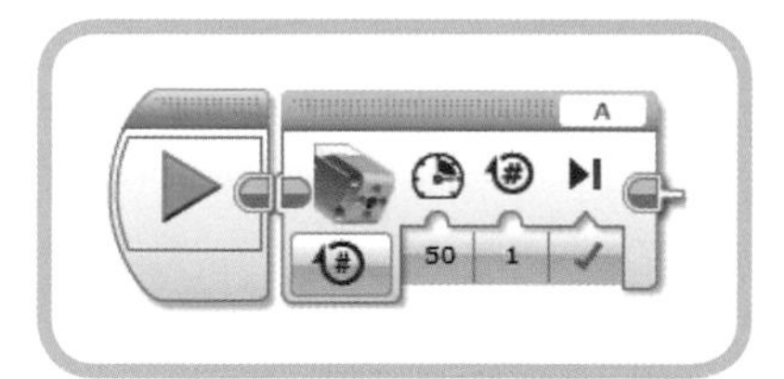

Run the program and you should see the hook connected to the large gears move down and press against the surface below it, lifting up the motor and the rest of the mechanism. From this test, you can tell one important thing: A positive Power value moves the arm downward. Move the arm back to horizontal manually and then try the program with a negative Power value. What happens?

4. Change the Power parameter to **-50**. The program should now look like this:

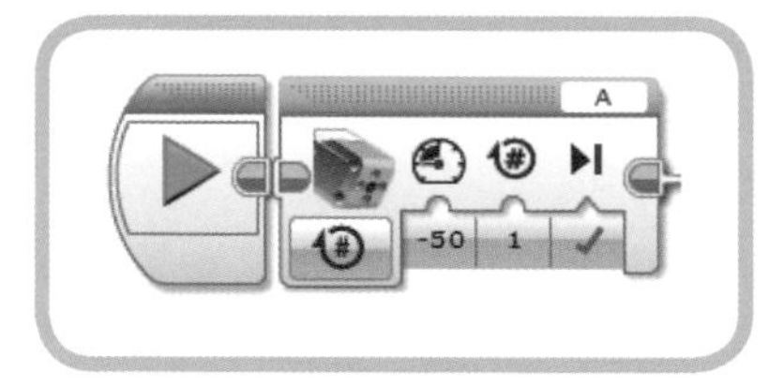

Run the program now, and the Lift Arm should move up and then stop a little past the point where it's pointing straight up. The arm should have moved about a third of the way around a complete circle, or 120 degrees.

The block is set to move the motor one rotation (360 degrees), so why does the arm move only one-third of this distance? The reason lies in the fact that the smaller gear has 12 teeth and the larger one has 36 teeth. Because the smaller gear has to go around three times to make the larger one go around once, you can use a small Power level in the Medium Motor block to make the arm move very smoothly.

Move the arm back to the horizontal position and try a move with a low Power value.

5. Change the Power to **-10** and the Rotations to **0.75**. The program should now look like this:

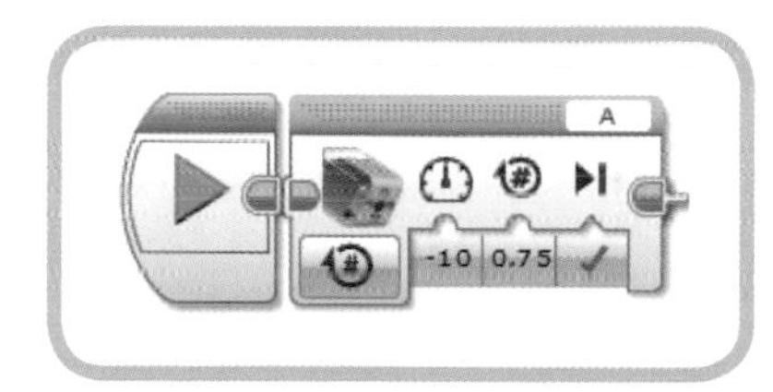

Run the program and you should see the Lift Arm move smoothly from a horizontal to a vertical position.

the invert motor block

The Lift Arm moves down when a positive Power level is used because of the way the four-tooth gear on the motor connects to the four-tooth gear on the axle. Of course, it's more intuitive for a positive Power level to move the arm upward. You could rebuild the Lift Arm (just move the gear on the axle to the other side of the motor), but why not fix it with a block? The *Invert Motor block* (Figure 4-24) on the Advanced palette will do just that.

Figure 4-24: The Invert Motor block

Figure 4-25: CoastTest *step 1*

The Invert Motor block reverses the meaning of the Power parameter for the motor connected to the selected Port.

6. Add an Invert Motor block between the Start block and the Medium Motor block. The Port parameter should default to A, so you shouldn't need to make any changes.

7. Change the Power parameter on the Medium Motor block to **10**. The final program should look like this:

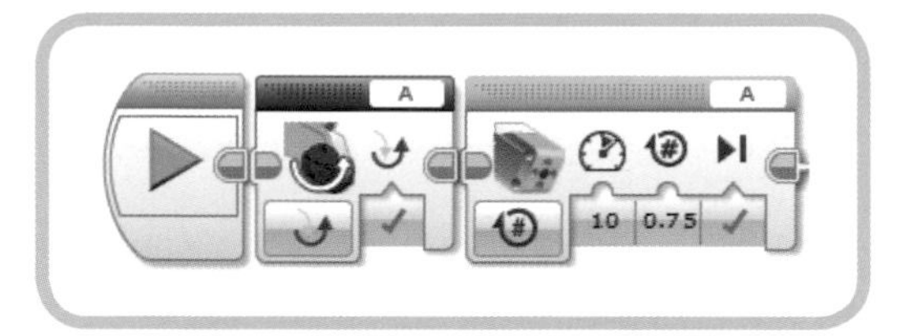

Run the program and it should behave exactly like the previous version, except that now the number for the Power parameter is positive.

a problem with coasting

All the motor control blocks have a Brake at End parameter with the choices Brake and Coast, which we looked at earlier. When the Coast setting is combined with either the On for Rotations or On for Degrees mode, these blocks behave in a way that might be unexpected: After coasting to a stop, the next movement seems to be a little shorter or a little longer than what you specified.

When you run a motion block using these settings, the *firmware* keeps track of how far the motor actually moved (the firmware is the program that runs on the EV3 and executes the program you write). Because the motor coasts to a stop, it will move a little more than the duration you specified, and the next motion block that runs will adjust its duration to account for the extra distance.

The next program, *CoastTest*, will demonstrate this. You'll use a couple of Move Steering blocks to move the motors and use Port View to see how far the motors go. Finally, I'll show you how to avoid this behavior if it's causing a problem for your program.

1. Create a new program named *CoastTest*.

2. Add a Move Steering block and keep all the default settings.

3. Add a Wait block after the Move Steering block. Set the time to **5** seconds.

4. Add another Move Steering block to the end of the program.

5. Add another Wait block to the end of the program. Set the time to **5** seconds.

   The program should look like Figure 4-25.

   At this point, both Move Steering blocks are set to brake at the end of the move. Run the program and watch the motor positions using the Port View. Figure 4-26 shows the Port View after the first Move Steering block, and Figure 4-27 shows it after the second one. Notice that both blocks worked as expected: Each one moves the motors almost exactly 360 degrees.

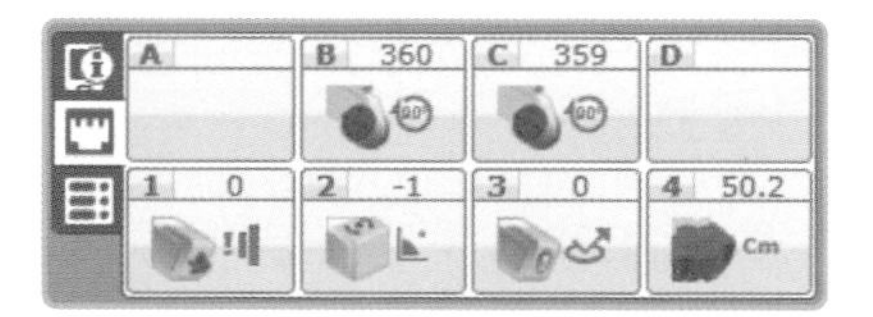

Figure 4-26: Positions after the first Move Steering block

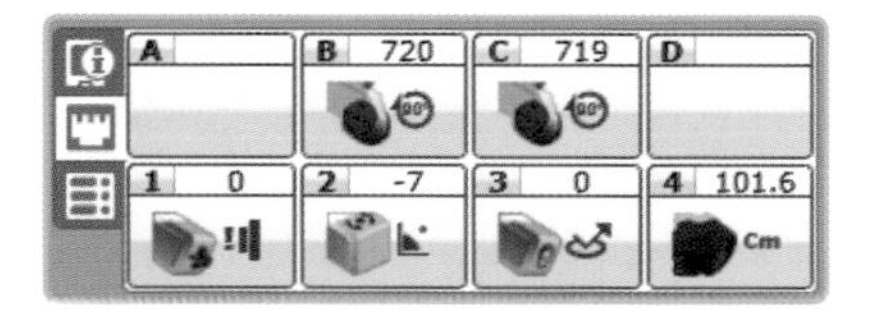

Figure 4-27: Positions after the second Move Steering block

NOTE While a program is running, a moving pattern of white diagonal strips appears on the colored bar at the top of the currently executing block.

Now modify the program so that the first Move Steering block coasts to a stop.

6. Change the Brake at End parameter for the first Move Steering block to **False**. The block should now look like this:

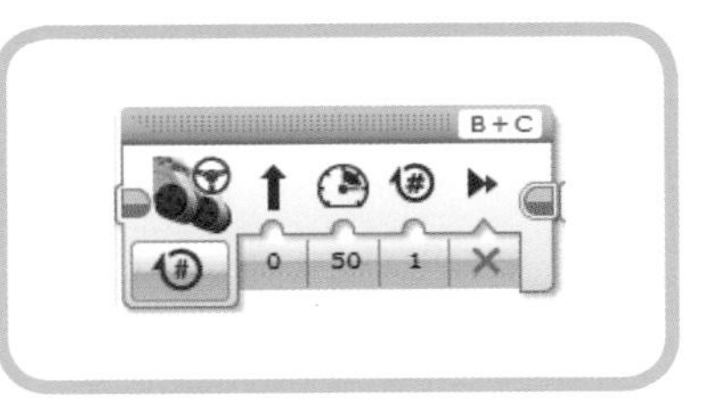

Run the program again and watch the Port View. Figures 4-28 and 4-29 show the Port View after the first and second Move Steering blocks, respectively.

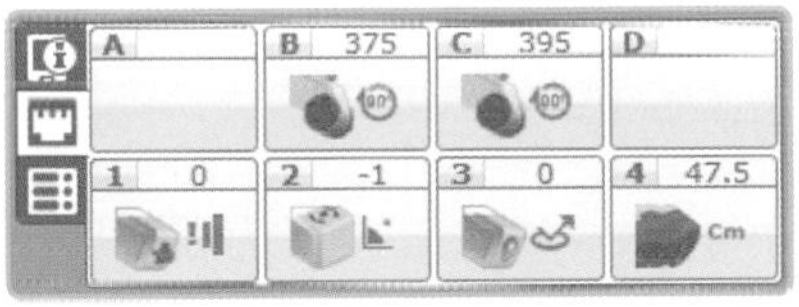

Figure 4-28: After the first Move Steering block coasts to a stop

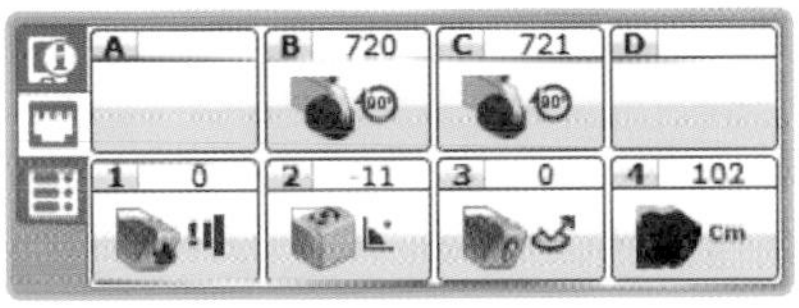

Figure 4-29: After the second Move Steering block coasts to a stop

The first Move Steering block moved the motors one rotation, and then they coasted a bit farther (an extra 15 to 35 degrees in this case). Next, instead of making the motors rotate an additional 360 degrees, the second Move Steering block moved the motors just enough so that the total distance covered was two rotations, or 720 degrees (to within one degree). Whether this behavior is "correct" or not really depends on what you wanted the program to do. If you wanted the robot to move a total of 720 degrees, this program works fine, but if you want it to move an additional 360 degrees with the second block no matter how much the robot has moved so far, what do you do?

The problem we need to solve is that the firmware tracks how far the motors coast after a move completes and then uses that information for the next move. To cancel this offset, use a Move Steering block in Stop mode with the Brake at End parameter set to Brake.

Place this block just before the second Move Steering block in the *CoastTest* program. Because the TriBot isn't moving at this point in the program, the new block should only cancel the offset caused by the previous coasting.

7. Add a Move Steering block between the first Wait block and the second Move Steering block.

8. Set the new block's mode to **Off**.

The program should now look like Figure 4-30. Run the program and watch the Port View.

Figures 4-31 and 4-32 show the Port View after the first and second Move Steering blocks, respectively. This time, the first Move Steering block moved the motors 360 degrees, and then it coasted a little farther. The second Move Steering block moved the motors an additional 360 degrees. Adding the new block changed the behavior so that the final Move Steering block moved the full distance.

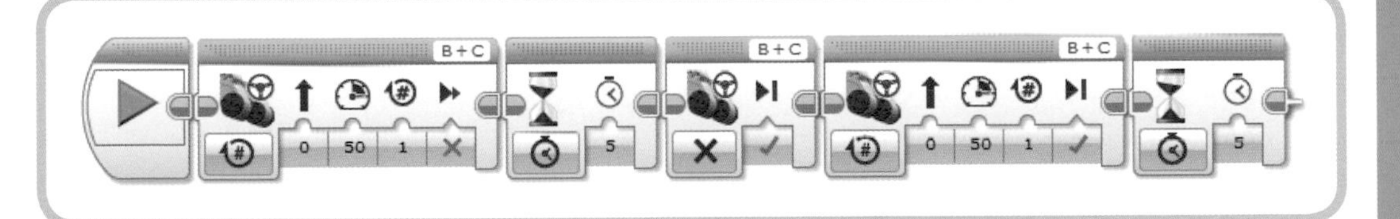

Figure 4-30: The final CoastTest *program*

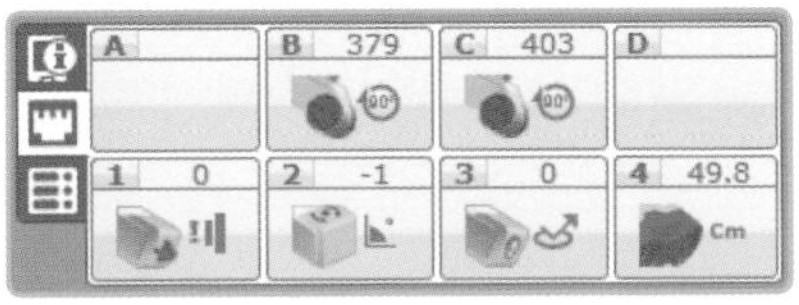

Figure 4-31: Positions after the first Move Steering block

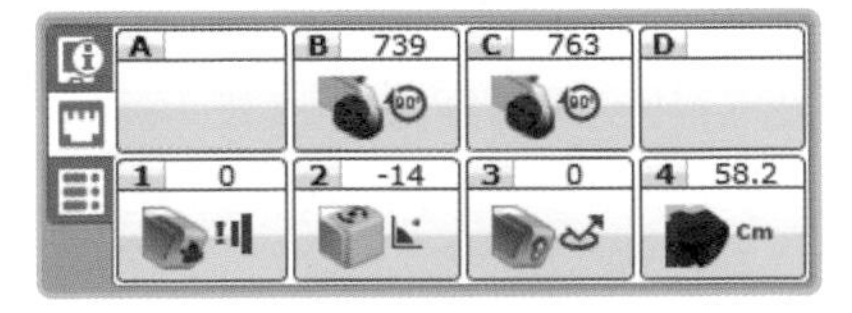

Figure 4-32: Positions after the second Move Steering block

further exploration

Try these activities for more practice using the blocks introduced in this chapter:

1. Use the coast option with the *ThereAndBack* program. Do you notice any change in the transitions between the moves? How does this affect how closely the TriBot gets back to the starting point?
2. Modify the *AroundTheBlock* program to use Move Tank blocks instead of Move Steering blocks. First try to replicate the existing turning motion, and then experiment with different combinations of Power settings for the two motors.
3. Write a new program that uses the same method as the *AroundTheBlock* program to travel in a triangle instead of a square.
4. Set up a simple obstacle course and program the TriBot to go through it using a sequence of Move Steering or Move Tank blocks. Notice that getting the TriBot to repeat the same pattern gets more difficult as you add more steps.

conclusion

The EV3 set comes with three motors designed to make it easy to build a wide variety of robots. The EV3 software provides several blocks for controlling the motors, giving you lots of flexibility when deciding how your robot should move. The Move Steering block is the most common choice because it's simple to use and its ability to synchronize two motors makes it easy to program a two-wheeled robot. The Move Tank block gives you a little more control over each motor. The Large Motor and Medium Motor blocks allow you to control a single motor.

The example programs showed you a few different ways to use these blocks, including one that moves the TriBot around a square. At this point, you should be able to program your robot to follow any course by combining motion blocks. You'll get plenty more practice using the motors in subsequent chapters.

5

sensors

In this chapter, you'll learn how to use the EV3 sensors to make your robot react to what's happening around it. We humans learn about the world around us using our five senses (touch, sight, sound, smell, and taste). A robot uses sensors in a similar way to gather information about its environment. By using the EV3 sensors, you can program your robot to avoid obstacles, follow a line on the floor, react to light, identify objects based on their color, and much, much more!

In most cases, your program will use the data from the sensors to make decisions about what to do next, but it can also use sensors to collect data as part of an experiment. This chapter describes how to operate sensors and how to use them to make decisions about your robot's next move.

NOTE The EV3 Education software supports two types of projects; programs and experiments. An experiment project is specifically designed to collect and present data. While this is an excellent tool for classroom use, it is outside the scope of this book. When I use the word *experiment*, I mean it in the generic sense and not in reference to an EV3 Education software experiment project.

using the sensors

Three programming blocks have built-in support for sensors: Wait, Loop, and Switch. You can use these blocks to make a program wait until something happens, run a group of blocks repeatedly until something happens, or choose which blocks to run based on the data from a sensor. You'll use all three blocks in this chapter.

Figure 5-1 shows the Mode Selector for the Wait block, which includes modes for all of the EV3 sensors, some of which can be used in more than one way. For example, the Color Sensor can either measure the amount of light it detects or the color of an object. The first program in this chapter (*BumperBot*) uses the Wait block with the Touch Sensor so you can see how this works.

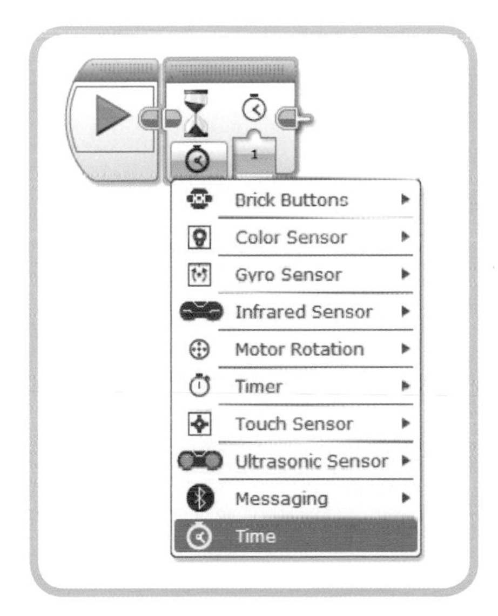

Figure 5-1: The Wait block's Mode Selector

NOTE The Home Edition of the EV3 software won't include options for the Gyro Sensor or Ultrasonic Sensor. If you purchase one of these sensors separately, you can download the programming blocks from *http://www.lego.com/*.

the touch sensor

The *Touch Sensor* (shown in Figure 5-2) has a small button on the front. The Wait, Loop, and Switch blocks use input from this sensor to tell whether the button is pressed, released, or bumped (pressed and then released). The Touch Sensor is often used to control a program or to detect when a robot runs into something. For example, you can have a robot wait until you press the Touch Sensor before starting to move.

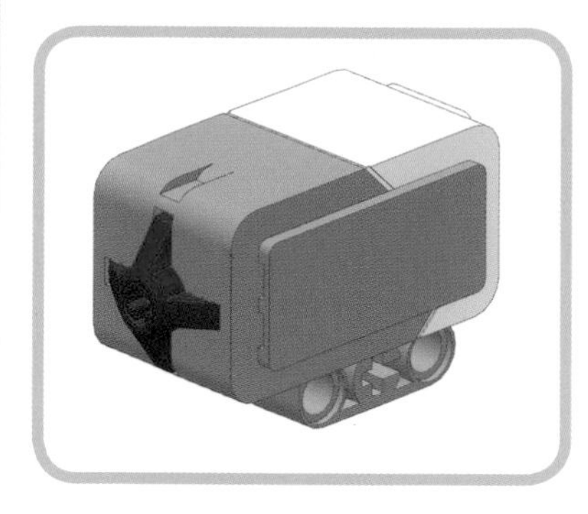

Figure 5-2: The Touch Sensor

You can use the Touch Sensor with the Wait block in two ways. In Compare mode, you name a state (Pressed, Released, or Bumped) and the Wait block pauses the program until the Touch Sensor matches that state. In Change mode, the Wait block checks the state of the sensor when the block starts executing and then waits for the state to change either from Pressed to Released or from Released to Pressed. Figure 5-3 shows these two choices on the Mode Selector menu for the Wait block.

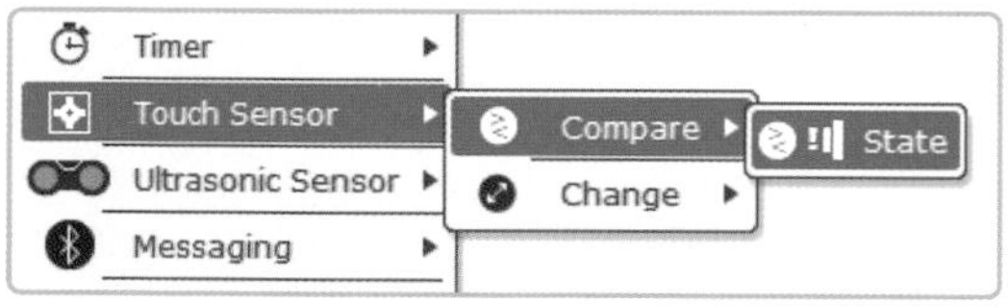

Figure 5-3: Touch Sensor modes for the Wait block

Figure 5-4 shows the Wait block with the Touch Sensor Compare mode selected and options to set the port.

Figure 5-4: Selecting the port

Each sensor has a default port. If your EV3 is connected to the software, the Auto-ID feature should warn you if you select a port that's not the one the sensor is connected to by showing a caution sign as shown in Figure 5-5. It's generally safer to use the default ports whenever possible.

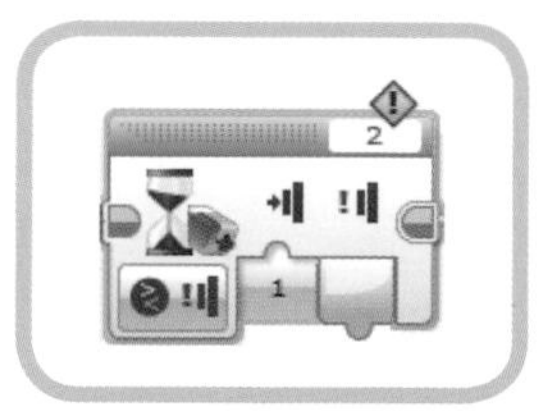

Figure 5-5: Auto-ID can let you know if you select the wrong port.

The other configuration item sets the state that you want to wait for: Released, Pressed, or Bumped, as shown in Figure 5-6.

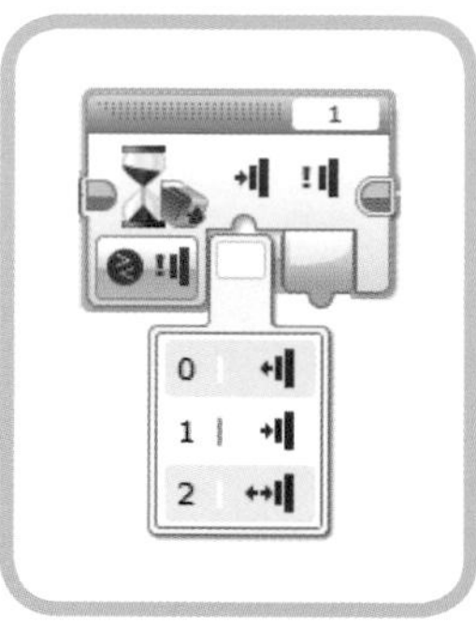

Figure 5-6: Selecting the state

The numbers at the left of each state (0 for Released, 1 for Pressed, and 2 for Bumped) can be useful when using data wires.

the BumperBot program

In this section, you'll build the *BumperBot* program using the Touch Sensor with the TriBot's front bumper to make the robot wander around a room. When the robot runs into something, the Touch Sensor is pressed, and the program makes the TriBot back up, turn around, and then move forward again. The robot will keep going until you stop it by pressing the Back button on the EV3 Brick.

moving forward

The first part of the program makes the robot move in a straight line until it runs into something. A Move Steering block with the mode set to On keeps the robot moving forward, and a Wait block uses the Touch Sensor to tell the robot when it runs into something. When the Touch Sensor is pressed, the program should stop the motors and make the TriBot back up a bit, turn in a different direction, and then start moving again until it runs into another obstacle. We'll place the entire program in a Loop block so that the TriBot keeps going until you stop it. Follow these steps to complete the first part of the program:

1. Create a new project named *Chapter5*.
2. Create a new program named *BumperBot*.
3. Drag a Loop block from the Flow Control palette onto the program. This block makes the program repeat until you stop it. (By default, the Loop block is set to **Unlimited** mode, so you don't need to make any changes.)
4. Drag a Move Steering block onto the program and place it inside the Loop block. The Loop block will expand to make room for the Move block.
5. Set the mode to **On** and the Power to **25.**

We set the Power to a moderate speed of 25 to get everything working before you speed things up. (You can increase the speed later, of course.)

Figure 5-7 shows the program with the Move Steering block added to the Loop.

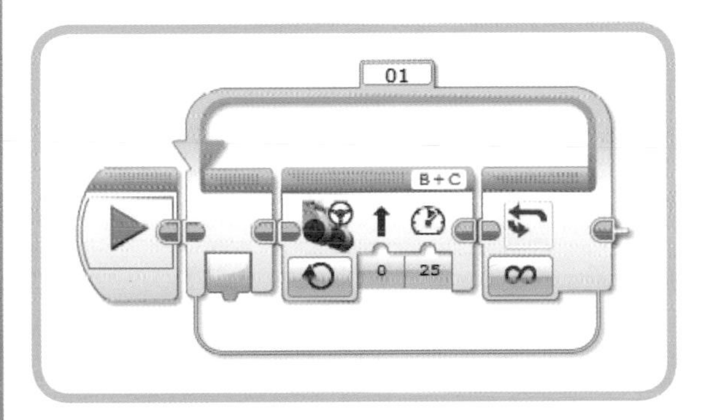

Figure 5-7: BumperBot *program with the Move Steering block added to the Loop block*

detecting an obstacle

The next part of the program uses input from the Touch Sensor to stop the motors when the TriBot runs into something (see Figure 5-8).

6. Drag a Wait block into the Loop block to the right of the Move Steering block.
7. Click the Mode Selector and choose the **Touch Sensor Compare** mode. By default, the State is set to **Pressed**, so keep that setting.
8. Drag another Move Steering block into the Loop block. Set the mode to **Off**.

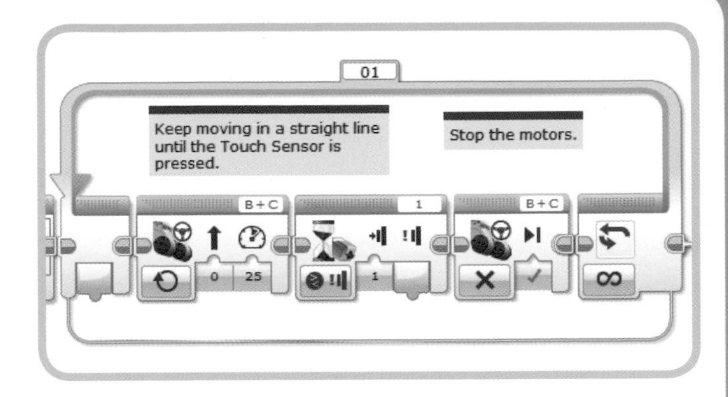

Figure 5-8: Waiting until the Touch Sensor is pressed and then stopping

backing up and turning around

Now we'll make the TriBot back up and turn in a different direction (see Figure 5-9).

9. Add another Move Steering block to the Loop block. Set the mode to **On for Degrees**.
10. Set the Power parameter to **25**.
11. Set the Duration parameter to **-300** to make the TriBot back up.
12. Add another Move Steering block to the Loop block. Drag the Steering slider all the way to the right or set the value to **100** to make the TriBot spin.
13. Set the mode to **On for Degrees** and the Power parameter to **25**.
14. Set the Duration parameter to **250**. You can experiment with different values to see how turning affects the program. (I find it useful to make at least a quarter-turn so that the TriBot doesn't take several tries to move away from a wall.)

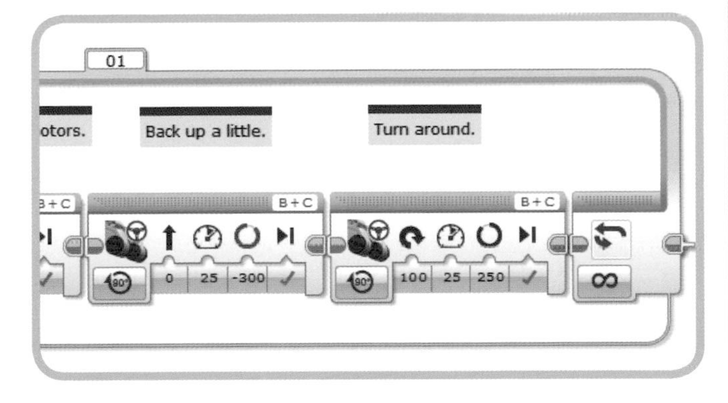

Figure 5-9: Backing up and turning around

CHALLENGE 5-1

When the program is working, try increasing the Power parameter of the Move Steering blocks. How fast can you make the TriBot move around a room before it becomes unstable? Does the turn start to spin a little out of control at some point? It may or may not, depending on the surface.

CHALLENGE 5-2

Modify the *BumperBot* program to make the TriBot play a game of ball toss.

Take a small, soft ball, and roll it at the bumper. The program should wait for the bumper to be pressed and then quickly move forward a small distance, which will push the ball back to you. The TriBot should then reverse back to its starting point and wait for your next toss. You'll need to use a ball that's big enough to engage the Touch Sensor and small or soft enough not to damage your EV3. (A ping-pong ball is too small, and a baseball is too hard. A tennis ball works well.)

testing

Next, download the completed program to your EV3 and test it. Remember, when your program runs, the TriBot should move forward in a straight line until it runs into something, at which point it should back away, turn around, and start again. The TriBot should keep running the program until you stop it by pressing the Back button on the EV3. To find a combination that works, experiment with the durations for the last two Move Steering blocks to change how far the robot backs up and turns.

the color sensor

The *Color Sensor* (Figure 5-10) measures the color or brightness of the light entering the small window on the front of the sensor. This sensor can be used in three different modes: Color, Reflected Light Intensity, and Ambient Light Intensity. After learning how each mode works, we'll make a color-detecting program and a line-detecting one to see the Color Sensor in action.

Figure 5-10: The Color Sensor

color mode

In *Color mode*, the sensor can detect the color of an object placed in front of it. The sensor can detect black, blue, green, yellow, red, white, and brown. If it can't determine the color in front of it, it uses a No Color value or chooses a color close to the one it sees. For example, if you place an orange object in front of the sensor, it might read red, yellow, or No Color, depending on the shade of orange. To get an accurate color reading, place the sensor very close to the object (without touching it) in order to reduce the influence of other sources of light.

As shown in Figure 5-11, the Wait block can use the Color Sensor in a Compare or Change mode. In Change mode, the block determines the color in front of the sensor when the block starts and then waits for the color to change.

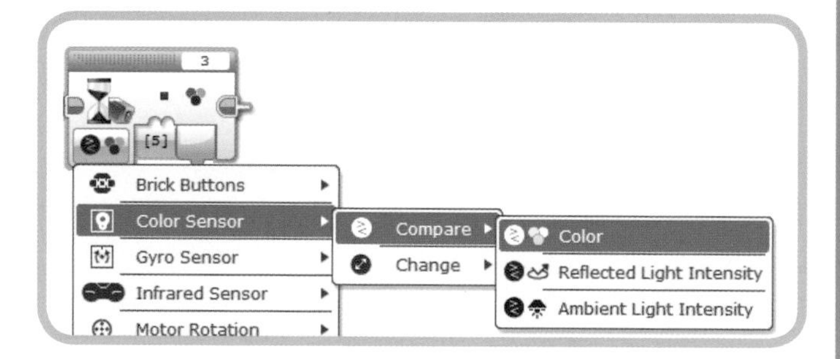

Figure 5-11: Selecting the Color Sensor

In Compare mode, the block waits until the sensor detects a color selected from a list, as shown in Figure 5-12. The white box with the red line through it represents No Color. The squares above the color menu show which colors the block is waiting for. For example, the block shown in Figure 5-13 is waiting for the sensor to detect green, blue, or red. (You'll use the sensor in this mode when you build the *IsItBlue* program later in this chapter.)

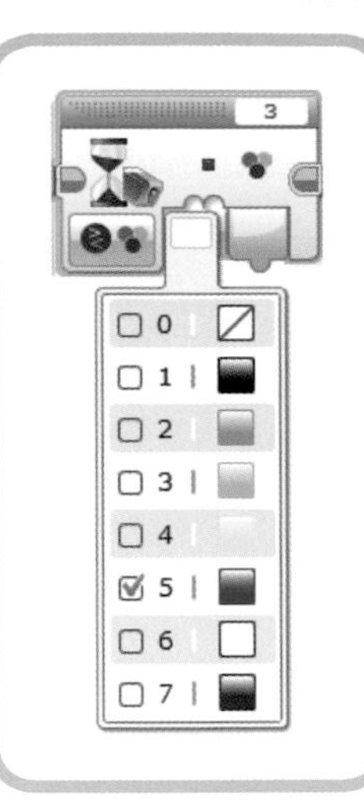

Figure 5-12: Selecting the color or colors to wait for

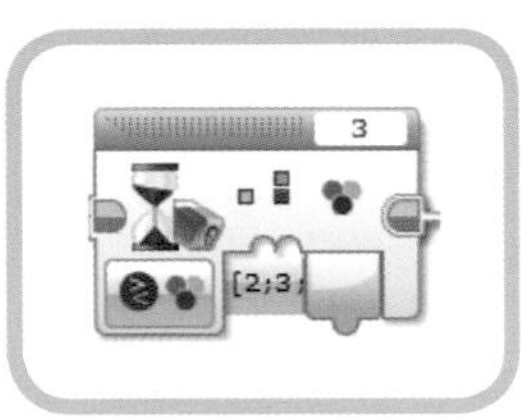

Figure 5-13: Waiting for green, blue, or red

reflected light intensity mode

In *Reflected Light Intensity mode* (shown in Figure 5-14), the sensor turns on its red LED and measures the amount of light reflected back to it from an object. The values range from 0 to 100, with 0 meaning very dark and 100 meaning very bright. This mode is useful for line following. As with Color mode, position the sensor close to the object in order to block other light sources that could interfere with the reading. The items to configure in this mode are the Compare Type and Threshold value. The Compare Type tells the block how to compare the sensor reading with the Threshold value (the value that will trigger a certain behavior). The choices from top to bottom are Equal To, Not Equal To, Greater Than, Greater Than or Equal To, Less Than, and Less Than or Equal To. The block shown in Figure 5-14 is set to wait until the sensor reading is less than 50.

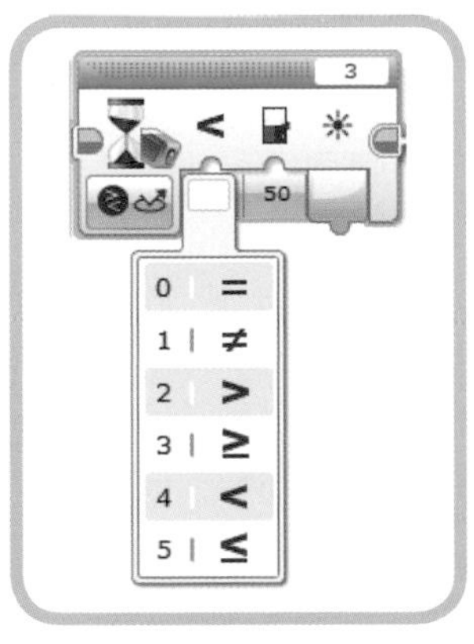

Figure 5-14: Selecting the Compare Type

The Change - Reflected Light Intensity mode lets you configure the amount and direction of change you want to wait for. You can wait for the value to increase, decrease, or change in direction. For example, the block shown in Figure 5-15 will wait for the light intensity to either increase or decrease by 10. If the sensor had a reading of 55 when this Wait block started, the program would pause until the sensor read 65 or greater, or 45 or less.

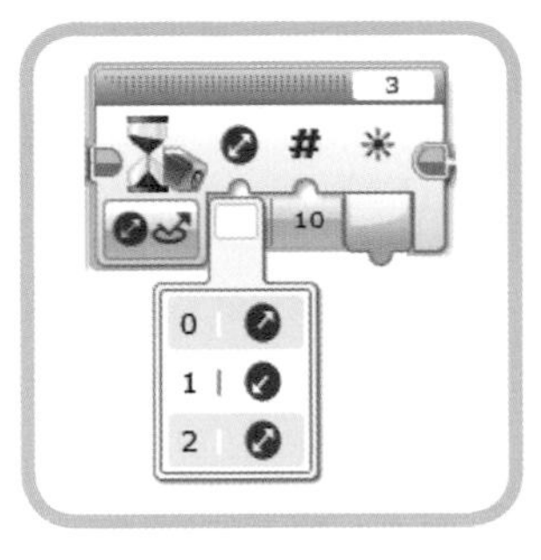

Figure 5-15: Setting the direction of change

NOTE It might seem like there are an overwhelming number of settings for each sensor, but many sensor modes (especially ones that measure numeric values) are configured in exactly the same way.

ambient light intensity mode

The third way to use the Color Sensor is to measure *ambient light*, which is light in the robot's environment. Figure 5-16 shows the Wait block in Color Sensor - Compare - Ambient Light Intensity mode. When measuring ambient light, the sensor offers Compare and Change modes, with the same configuration options that we saw earlier.

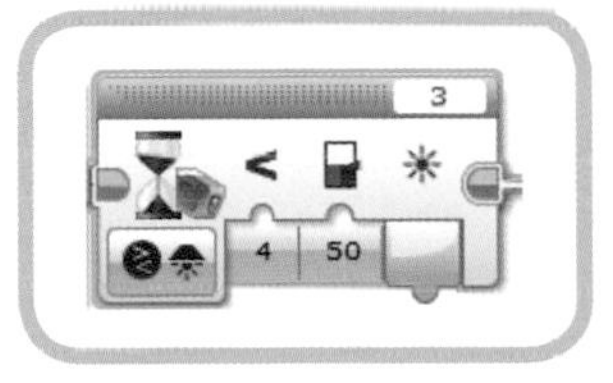

Figure 5-16: Color Sensor - Compare - Ambient Light Intensity mode

NOTE You can tell which mode the Color Sensor is using from the LED on the front of the sensor. In Color mode, you'll see a bright, multicolored light; in Reflected Light Intensity mode, you'll see a bright red light; and in Ambient Light Intensity mode, you'll see a dim blue light. Noticing how the sensor is lit can help you quickly identify when you've chosen the wrong mode, which is a very common bug.

port view

Port View can show you a sensor value using any of the modes the sensor supports. Click the sensor in the Port View window to select a mode. For example, Figure 5-17 shows the menu that appears when you click the Color Sensor.

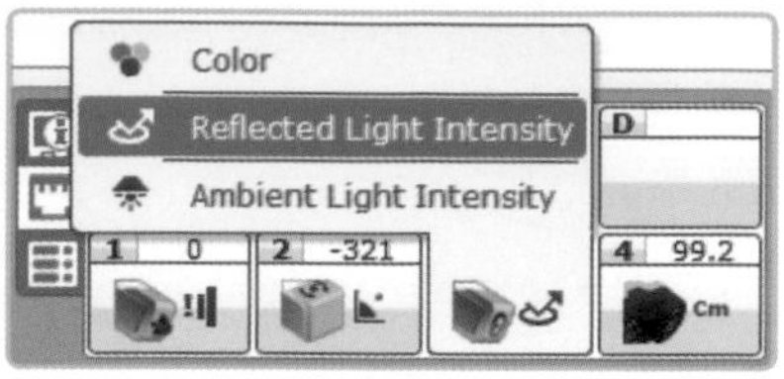

Figure 5-17: Selecting the Color Sensor mode

The value from the sensor is always displayed as a number. This is perfect when using the Color Sensor in Reflected Light Intensity mode, but is less user-friendly when using Color mode. Figure 5-18 shows the Port View with Color mode selected. Notice that the value displayed for the Color Sensor is 6 instead of White. You can see which number corresponds to each color in the Wait block's color selection list (shown in Figure 5-12).

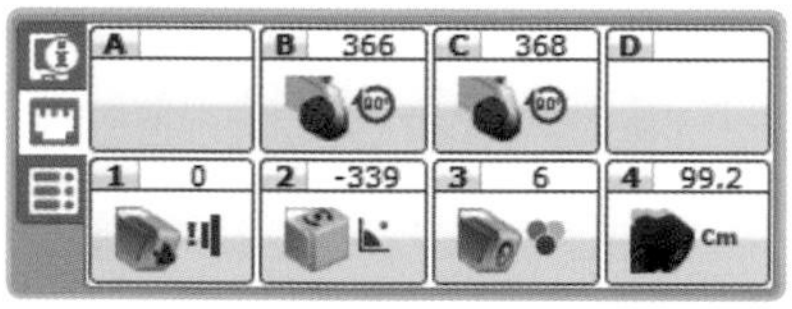

Figure 5-18: Port View with Color mode selected

the IsItBlue program

In this section, you'll build the *IsItBlue* program, which uses the Color Sensor to identify blue objects. When you run the program, the robot will say "Yes" or "No" depending on the color of the object placed in front of the sensor.

the switch block

In order for our program to decide what to say based on input from the Color Sensor, we'll use the Switch block. This block tells the program to run one set of blocks if the sensor sees a blue object and another set of blocks if it doesn't.

Here is how to create this program:

1. Create a new program named *IsItBlue*.
2. Drag a Switch block onto the canvas. The Switch block is on the Program Flow palette.
3. Select **Color Sensor - Measure - Color** mode.

CHALLENGE 5-3

Use Port View to experiment with the different Color Sensor modes. Try to get a feel for how the darkness and finish of a surface affect the reflected light reading. Use Color mode to see the colors that the sensor is good at identifying and those it has trouble with.

At this point, the Switch block should look like Figure 5-19. The black box at the top of the block indicates that the program will run the blocks in the top section if the sensor reads Black. The white box with the red line through it indicates that the blocks in the bottom section will run if the sensor reads No Color. If the sensor reads a value other than Black or No Color, the black box at the top will run. (The Switch block will choose the group of blocks marked with the dot by default.)

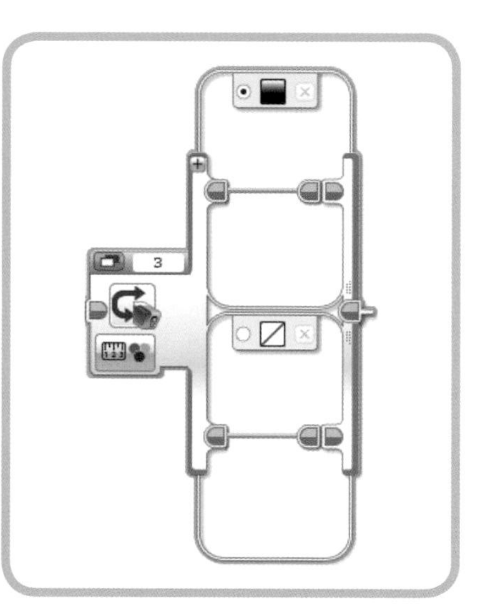

Figure 5-19: Color Sensor - Measure - Color mode

4. Click the box at the top of the lower section and choose **Blue**, as shown in Figure 5-20.
5. Drag a Sound block onto the upper area inside the Switch block.
6. Select **No** from the Sound File list.
7. Drag another Sound block onto the lower area inside the Switch block.
8. Select **Yes** from the Sound File list for this block. The program should look like Figure 5-21.

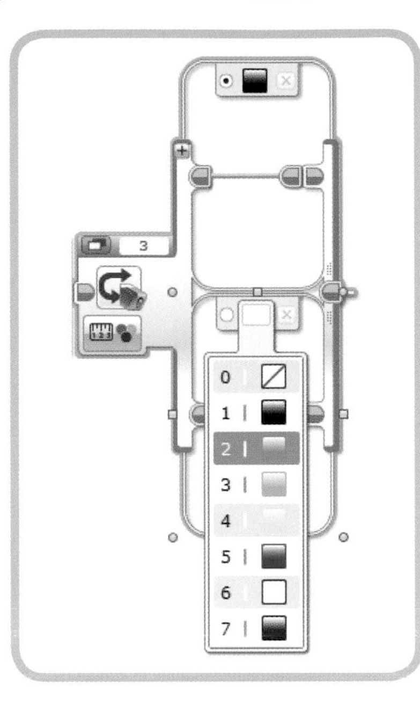

Figure 5-20: Selecting the color

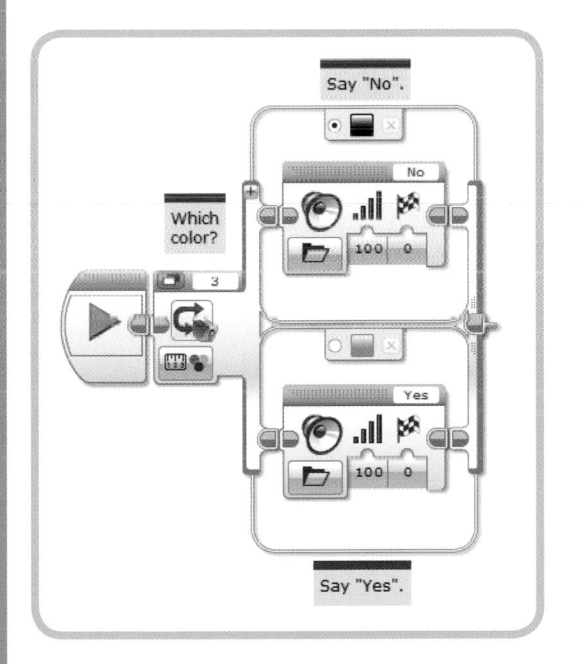

Figure 5-21: The complete IsItBlue *program*

Before running the program, place the test object in front of the Color Sensor. When you run the program, it should tell you if the object is blue. The program should end after saying "Yes" or "No," so run it again to test it on another object.

improving the program

The *IsItBlue* program works in its current form, but it's a little inconvenient to use. You can improve it by adding a way to tell the robot when the object to identify is in place. You can also make the program keep running until you stop it.

using the touch sensor

First, we'll add a block that uses the Touch Sensor to let the program know when you are ready to use the Color Sensor.

1. Drag a Wait block onto the program to the left of the Switch block.
2. Set the mode to **Touch Sensor - Compare - State**.
3. Set the state to **Bumped**, as shown in Figure 5-22.

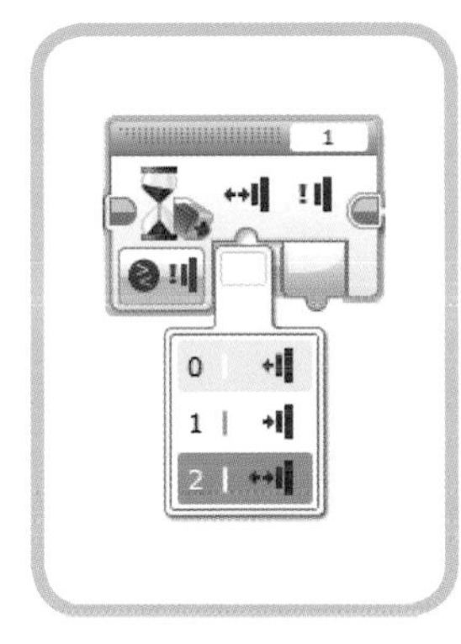

Figure 5-22. Selecting the State

The program should now look like Figure 5-23. When you run it, it should wait for you to press and release the Touch Sensor before saying "Yes" or "No" and will still exit after checking the color of one object.

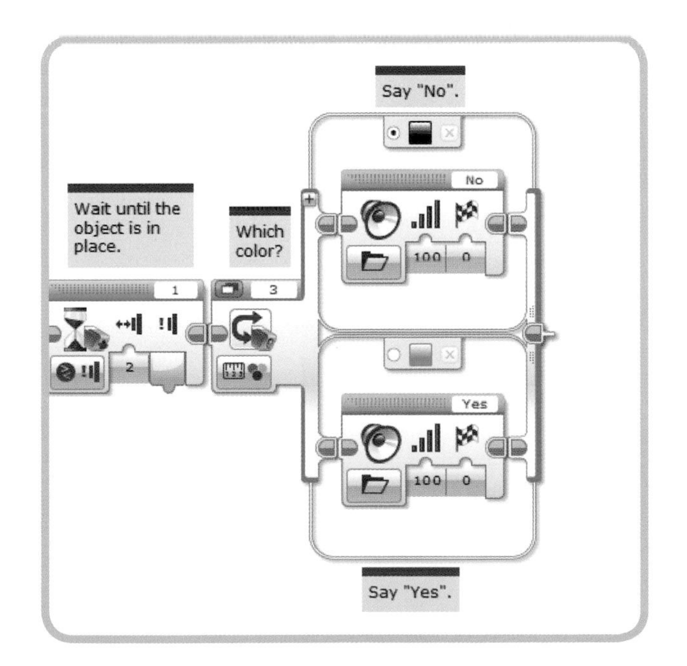

Figure 5-23: Waiting until the object is in place

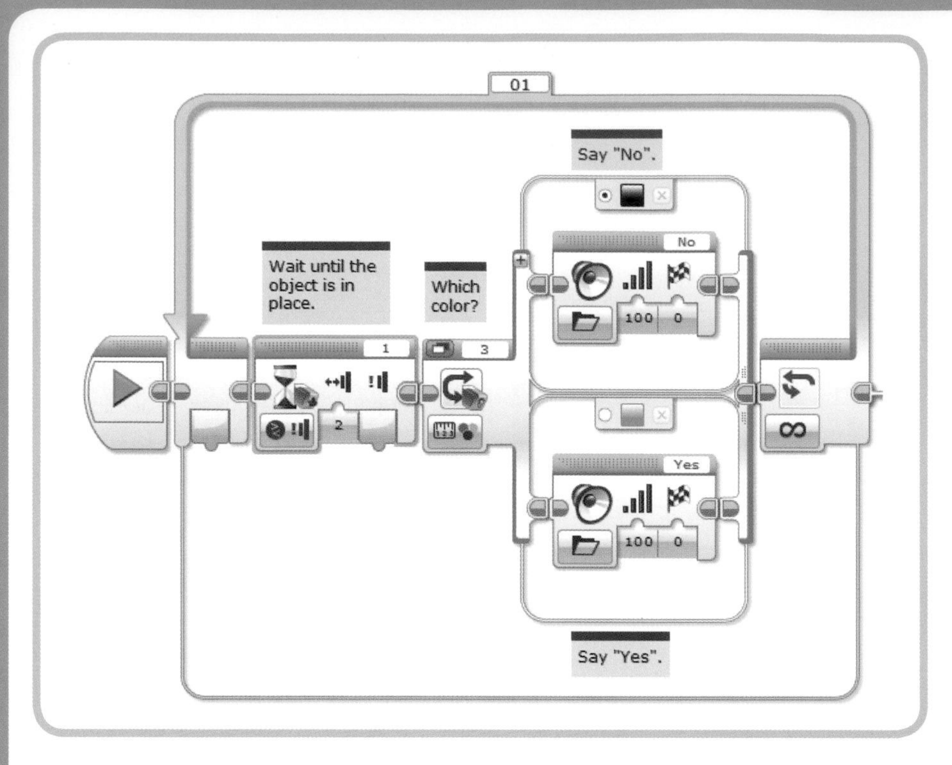

Figure 5-24: The improved version of IsItBlue

adding a loop

To make the program keep running, add a Loop block and then move the existing blocks inside it.

4. Add a Loop block to the end of the program.
5. Select the Wait, Touch, and Switch blocks, and drag them into the Loop block.

Figure 5-24 shows the new version of the program. When you run it, it should wait for you to hit the bumper, say "Yes" or "No," and then go back to waiting again. Press the Exit button on the EV3 to end the program.

the LineFinder program

In this section, you'll write a program that makes the TriBot move forward until it finds a line on the floor using the Color Sensor in Reflected Light Intensity mode. The *LineFinder* program starts the TriBot moving forward and stops it when a dark line is detected. But first, remove the Touch Sensor bumper from the front of the TriBot and replace it with the Color Sensor mounted so that it points downward, as shown in Figure 5-25.

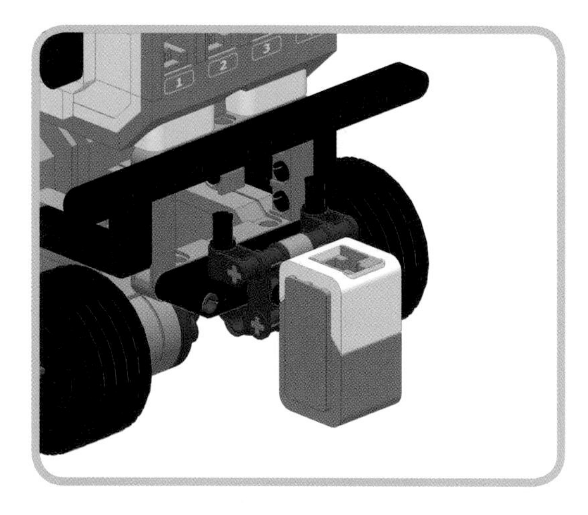

Figure 5-25: Color Sensor position for the LineFinder *program*

NOTE To test this program, you need a light surface with a dark line. You could use a white poster board and either a magic marker or a piece of black electrical tape to make the line.

The basic idea of this program is similar to the first part of the *BumperBot* program. A Move Steering block in On mode makes the TriBot start moving, and another one in Stop mode makes it stop. Between those two blocks, we'll place a Wait block in Color Sensor Reflected Light Intensity mode. The only new challenge is finding the Threshold value for the Wait block.

using port view to find a threshold value

Use Port View (Figure 5-26) to find a reasonable Threshold value to make the robot stop. Make sure the EV3 is connected to the software and Port View is set to show the Color Sensor in Reflected Light Intensity mode. In Figure 5-26, this value is shown in port 3. Place the TriBot over the white area and note the sensor value. Now move the robot so the sensor is on the line and note the new value. In my case, the sensor values were 74 and 6, respectively.

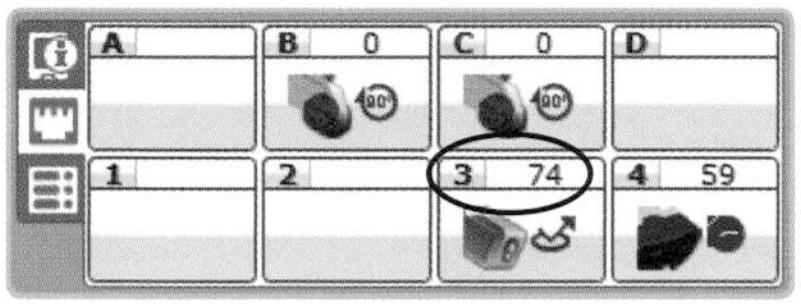

Figure 5-26: The Color Sensor value on a light surface

If the sensor is straddling the line and the white area, its value will be between the two extremes (6 and 74 here). But then what Threshold value should we have the robot wait for before it stops moving? If we set the Threshold to 6, the robot won't stop until the sensor is completely over the black. Because we want the TriBot to stop when it gets to the line, we should set a Threshold value higher than 6.

A simple but effective way to get a reasonable Threshold value is to take the middle value between the light and dark reading. This way, the robot will stop soon after it detects the darkness of the line instead of waiting until the sensor is completely over the line and seeing only black. Adding 74 plus 6 and then dividing by 2 gives a value of 40, so we'll use 40 as the Threshold value in the code shown here. (Be sure to calculate your own value based on your Color Sensor.)

Now we'll build the program shown in Figure 5-27.

1. Create a new program named *LineFinder*.
2. Add a Move Steering block to the program and set the mode to **On**.
3. Set the Power parameter to **25**.
4. Add a Wait block and set the mode to **Color Sensor - Compare - Reflected Light Intensity**.
5. The Compare Type will already be set to **Less Than**. Set the Threshold value to the target you determined.
6. Add a Move Steering block after the Wait block. Set the mode to **Off**.
7. Add a Wait block. Set the time to **5** seconds.

The second Move Steering block and the Wait block make sure that the robot comes to a full stop before the program ends so that the robot's momentum won't carry it past the line. Run the program and adjust the Threshold value in the first Wait block if needed. Try increasing the Power on the first Move Steering block to see how fast you can go and still accurately stop at the line.

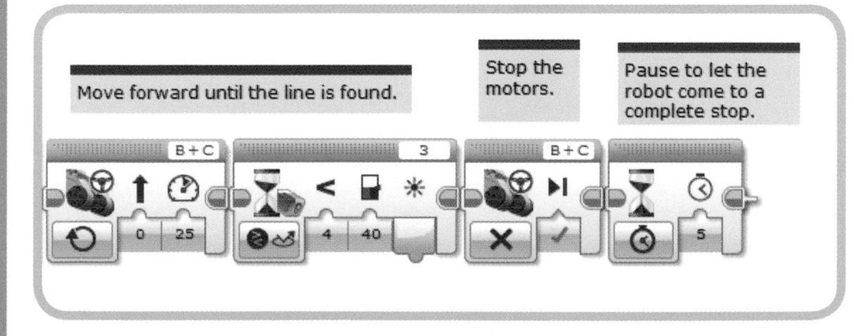

Figure 5-27: The LineFinder *program*

CHALLENGE 5-4

The *AroundTheBlock* program from Chapter 4 moves the robot forward for three rotations to make it travel around a square. Change the number of times the Loop block repeats from 4 to 40 (so the TriBot makes 10 trips around the square) and you should see the TriBot veer more and more off course because any errors from each move—such as wheels slipping, or turning slightly too far or not far enough—accumulate and eventually become large enough to notice. One way to address this is to use the blocks from the *LineFinder* program in the *AroundTheBlock* program so that the TriBot will travel forward until it finds the line and then turn. Set up a test area with four dark lines, one at the end of each side of the square. The program should make the TriBot move forward until it sees the first line, turn 90 degrees, move until it sees the second line, and so on. This won't make each individual move any more accurate, but it will help keep the errors from accumulating.

CHALLENGE 5-5

You can combine pieces of the *AroundTheBlock*, *LineFinder*, and *RedAndBlue* programs to create a simple path-following program that uses a series of colored lines to tell the robot how to move from one line to the next. Use blocks from the *LineFinder* program to make the robot move forward until a line is found. Then modify the Switch block from the *RedAndBlue* program to make the robot turn left or right depending on the color of the line. You can use the Move Steering block from *AroundTheBlock* to make the turn, with the Steering parameter set appropriately.

the infrared sensor and remote

Figure 5-28 shows the Infrared Sensor and the Infrared Remote. You can use the *Infrared Sensor* to measure the distance to an object, or the distance and direction to the remote. (The Infrared Sensor is not included in the Education Edition.) You can often substitute the Ultrasonic Sensor (using one of the Distance modes) for the Infrared Sensor to measure distance.

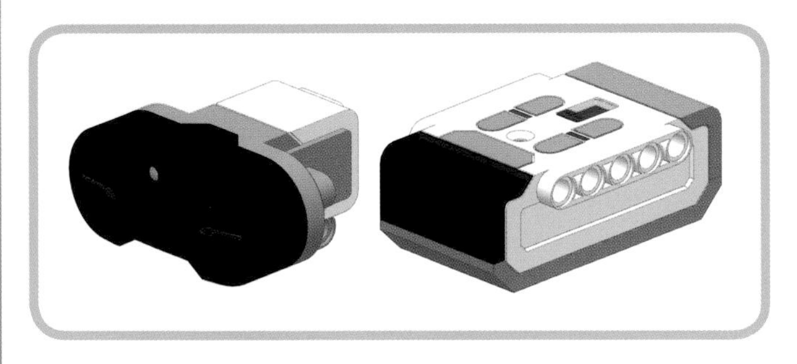

Figure 5-28: The Infrared Sensor and Remote

proximity mode

Use *Proximity mode* to measure the distance to an object in front of the sensor. In this mode, the sensor emits an infrared signal and then measures the strength of the returned signal, which gives your program a rough indication of the distance to the nearest object. The sensor value will be between 0 (near) and 100 (far or when no object is detected). Many factors influence how an object reflects an infrared signal, such as its color and hardness. For example, a soft, round object such as a tennis ball can produce a larger sensor reading than a hard, flat object like a book, even if they are at the same distance.

Figure 5-29 shows the Wait block in Infrared Sensor - Compare - Proximity mode, which lets your program wait until an object is detected at a certain distance. The Change Proximity mode allows your program to wait for the distance of the nearest object to change by a certain amount.

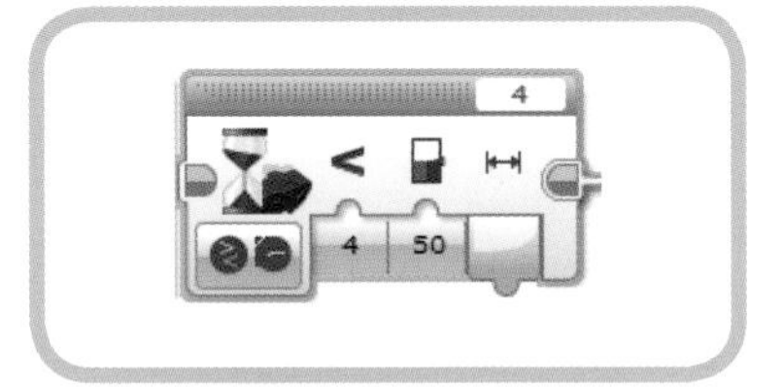

Figure 5-29: Wait block in Infrared Sensor - Compare - Proximity mode

beacon heading and beacon proximity mode

In the Beacon Heading and Beacon Proximity modes, the Infrared Sensor indicates the direction or distance to the Infrared Remote. To use these modes, put the remote in Beacon mode by pressing the Beacon button (see Figure 5-30).

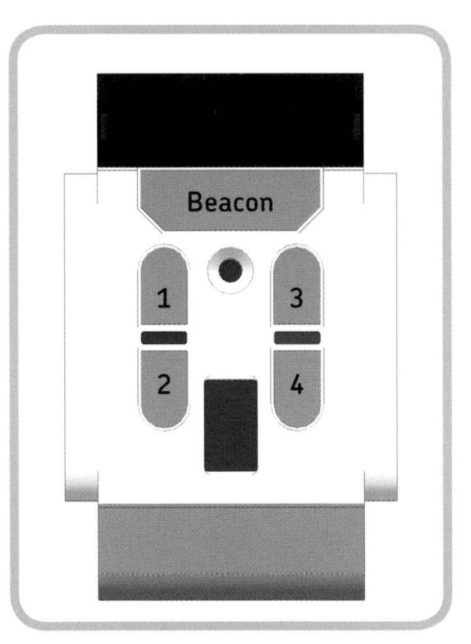

Figure 5-30: The Infrared Remote buttons, including the Beacon button

Figure 5-31 shows the Wait block in Compare - Beacon Heading mode. The Infrared Remote can operate on four channels, which use slightly different infrared signals to communicate with the sensor. By default, the channel is set to 1, but you may need to select one of the others to avoid interference from your TV or stereo remote, or another EV3 Infrared Remote. If your robot doesn't react to the EV3 remote or reacts to your TV remote, try using a different channel by adjusting the red slider on the EV3 remote, and then changing the channel setting in the programming block to match.

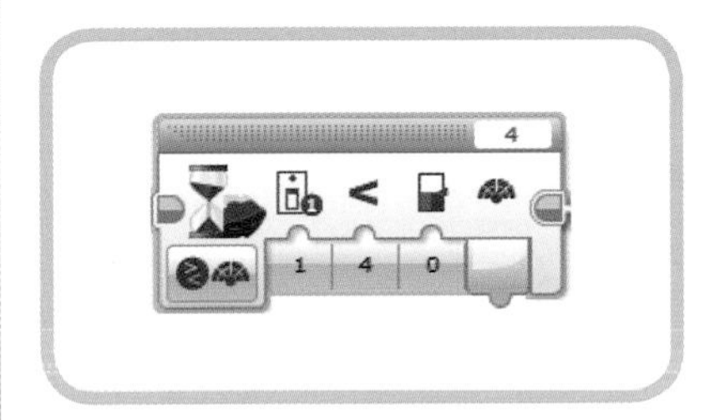

Figure 5-31: Wait block in Infrared Sensor - Compare - Beacon Heading mode

In *Beacon Heading mode*, the Infrared Sensor reads from -25 to 25. A value of 0 means the beacon is directly in front of the sensor. Positive values indicate that the beacon is to the right of the robot, and negative ones indicate that it's to the left. (You won't get an exact measurement in degrees.)

In *Beacon Proximity mode*, the sensor measures the relative distance to the beacon from 0 (near) to 100 (far).

NOTE The infrared signal used between the sensor and the remote is a "line of sight" signal, meaning that there has to be a direct line without obstructions between the remote and the sensor in order to get a measurement.

remote mode

Remote mode allows you to use the buttons on the remote to control your program. You select the channel of the remote and the button or buttons you want to wait for, as shown in Figure 5-32.

CHALLENGE 5-6

To see how the remote works in Beacon mode, use Port View to display the Beacon Heading and Beacon Proximity values one at a time. Move the beacon around, toward, and away from the sensor. Notice that readings are only stable when the beacon is within a few feet and more or less in front of the sensor.

You can choose a single button, a pair of buttons, or no buttons at all. You can select more than one option to make the block wait for any combination of the five buttons. If you select more than one option in the list, the block will stop waiting when any one of the selected options is met. (We'll use this block in the next section to control the *BumperBot* program.)

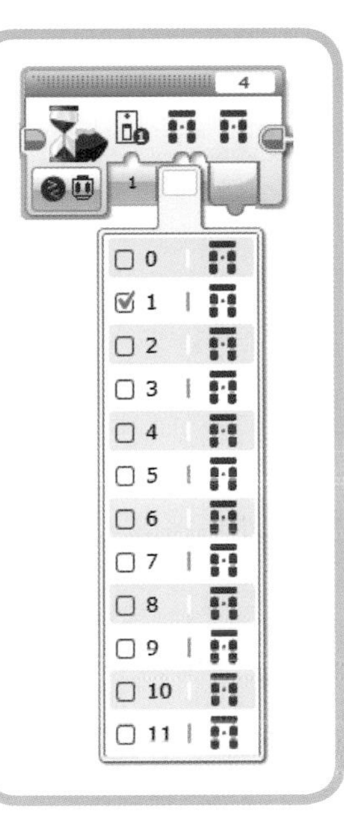

Figure 5-32: Wait block in Infrared Sensor - Compare - Beacon Heading mode

the BumperBot-WithButtons program

In this section we'll add the Infrared Sensor and Remote to the *BumperBot* program, so that it will wait for a button on the Infrared Remote before it starts moving. (If you have the Education Edition with no Infrared Sensor, you can skip this section.) Follow these steps:

1. Open the *BumperBot* program.
2. Add a Wait block at the start of the program, between the Start block and the Loop block.
3. Set the mode of the Wait block to **Infrared Sensor - Compare - Remote**.

When you run the program (shown in Figure 5-33), the TriBot should wait for you to press button 1 on the remote before starting. (The remote must be in front of the robot, pointing at the sensor.)

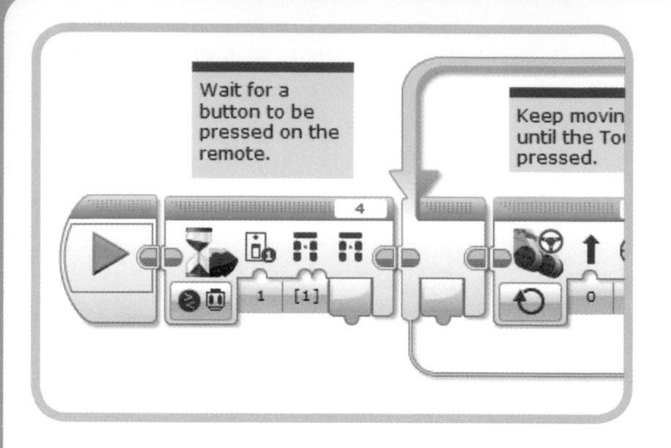

Figure 5-33: Waiting for a button to be pressed in Remote mode

the ultrasonic sensor

The *Ultrasonic Sensor* (shown in Figure 5-34) measures the distance to an object by measuring the time it takes for high-frequency sound waves to be reflected from the target object—a kind of sonar. (This sensor is only included with the Education Edition. Most programs that use it to measure distance can also be written using the Infrared Sensor in Proximity mode.)

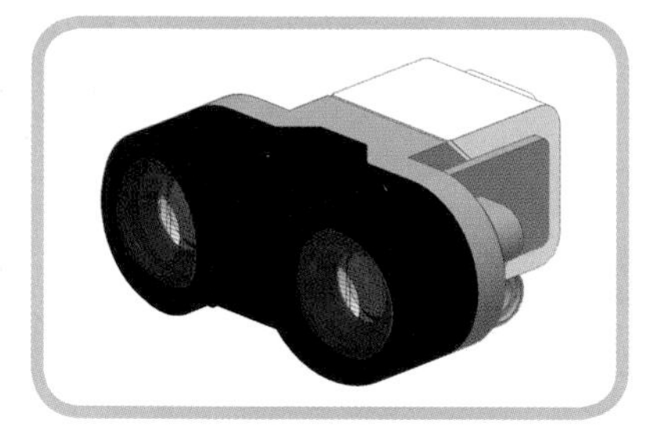

Figure 5-34: The Ultrasonic Sensor

An object's shape and texture greatly affect how well it will be detected by the Ultrasonic Sensor; some surfaces reflect sound waves better than others and are therefore easier to detect. Flat, hard surfaces are the easiest to detect because they reflect most of the sound waves, while curved surfaces reflect some sound waves but scatter others, and soft objects tend to absorb sound waves instead of reflecting them. Consequently, the sensor will be able to detect hard, flat objects at a greater distance than with soft, round ones.

distance inches and distance centimeters modes

The Ultrasonic Sensor is typically used in either the *Distance Inches mode* or the *Distance Centimeters mode*. These modes are identical in every way except their units of measurement. Figure 5-35 shows the Wait block with the mode set to Ultrasonic Sensor - Compare - Distance Inches.

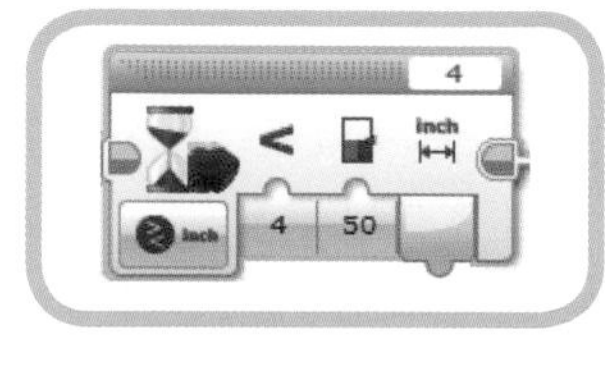

Figure 5-35: Wait block in Ultrasonic Sensor - Compare - Distance Inches mode

The configuration items are the same as those of the Color Sensor, except that the Threshold value is a distance instead of a light intensity. Next, we'll use the Ultrasonic Sensor to build the *DoorChime* program.

presence/listen mode

The Ultrasonic Sensor can detect the presence of another Ultrasonic Sensor using *Presence/Listen mode*, which can be useful for games or challenges involving multiple robots. The only configuration to set in the Wait block in Ultrasonic Sensor - Compare - Presence/Listen mode is the port that the sensor is attached to (Figure 5-36). The block waits until another Ultrasonic Sensor is detected.

Figure 5-36: Wait block in Ultrasonic Sensor - Compare - Presence/Listen mode

the DoorChime program

Now we'll create is a simple door chime. Place the TriBot in a doorway with the Ultrasonic or Infrared Sensor facing across the opening of the doorway (see Figure 5-37), and it will chime when someone walks through the door. You'll put the main part of the *DoorChime* program in a loop so it automatically starts over in preparation for the next person.

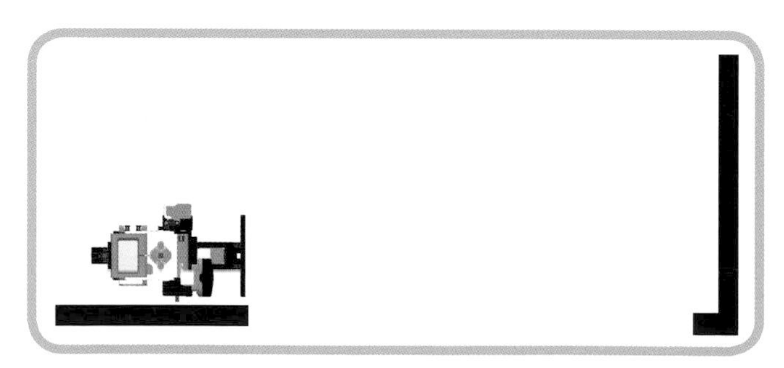

Figure 5-37: Positioning the TriBot in a doorway for the DoorChime program

detecting a person

You can use either the Ultrasonic or Infrared Sensor in this program to detect a person walking past the robot, because both can measure the distance of an object. When you first start the program, the sensor should measure the width of the doorway. When a person passes the sensor, it reads the distance between the robot and the person. To determine the trigger Threshold value, use Port View to see what the sensor reads when you place it in the doorway. (In my case, the doorway is 32 inches wide, and the Infrared Sensor in Proximity mode reads 61). When people walk through the door, they'll be closer to the robot, so I'll use a value of 55 for the Infrared Sensor, and 30 inches for the Ultrasonic Sensor. Follow these steps to build the program:

1. Create a program with the name *DoorChime*.
2. Add a Loop block to the program. The loop should continue until the program is stopped, so you don't need to change any of the settings.
3. Add a Wait block inside the Loop block.
4. Set the mode to either **Ultrasonic Sensor - Compare - Distance Inches** or **Infrared Sensor - Compare - Proximity**, depending on the sensor you are using.
5. Set the Threshold value to the value that works for your doorway.

Figure 5-38 shows the program using the Infrared Sensor.

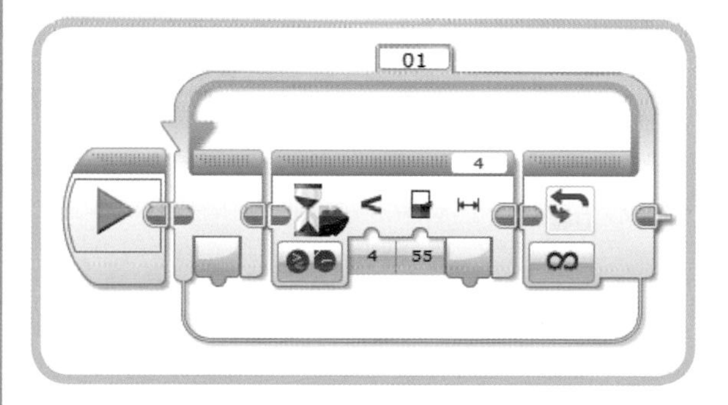

Figure 5-38: Waiting for a person, using the Infrared Sensor

playing a chime

To play the chime, you can use two Sound blocks with the mode set to **Play Note** to play any two notes. Experiment with different combinations or even add more notes if you want.

6. Add a Sound block inside the Loop block.
7. Set the mode to **Play Note**.
8. Click the Note configuration item and select a note from the keyboard that appears.
9. Add another Sound block inside the Loop block.
10. Set the mode item to **Play Note** and select a different note.

Figure 5-39 shows the two Sound blocks added to the program. Download and test the program, and experiment with different distances to find the trigger value that works best. Try different duration and volume settings for the Sound blocks and see how creative you can be with the chime.

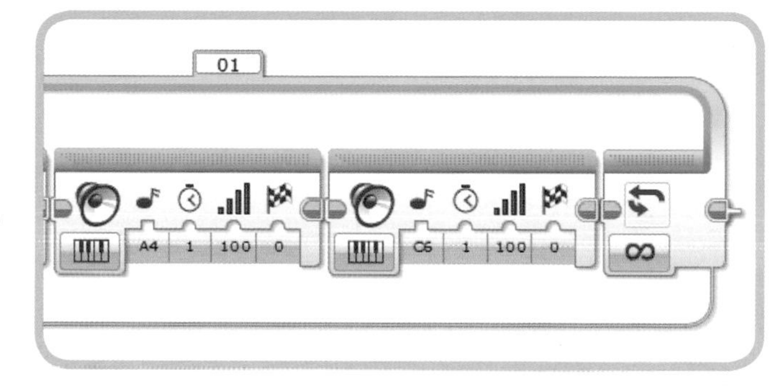

Figure 5-39: Playing a chime

stopping the chime

As currently written, the TriBot program chimes when a person walks through the doorway and keeps chiming until the person moves away. This has the potential to be very annoying! How can we make it chime only once each time a person walks through the doorway?

To solve this problem, we'll add another Wait block after the two Sound blocks to pause the program until the person moves beyond the doorway. The Wait block tells the program to wait for the Ultrasonic or Infrared Sensor to give a reading that is greater than the original trigger value.

11. Drag a Wait block inside the Loop block to the right of the Sound blocks.
12. Set the mode to either **Ultrasonic Sensor - Compare - Distance Inches** or **Infrared Sensor - Compare - Proximity**, depending on the sensor you are using.
13. Change the Threshold value to be greater than the trigger value you used for the first Wait block.
14. Set the Compare Type item to **Greater Than**.

Figure 5-40 shows the program with the Infrared Sensor. Test it to see whether it behaves as expected when you move through the doorway, and experiment with different distances to see how the program behaves if more than one person moves through the doorway.

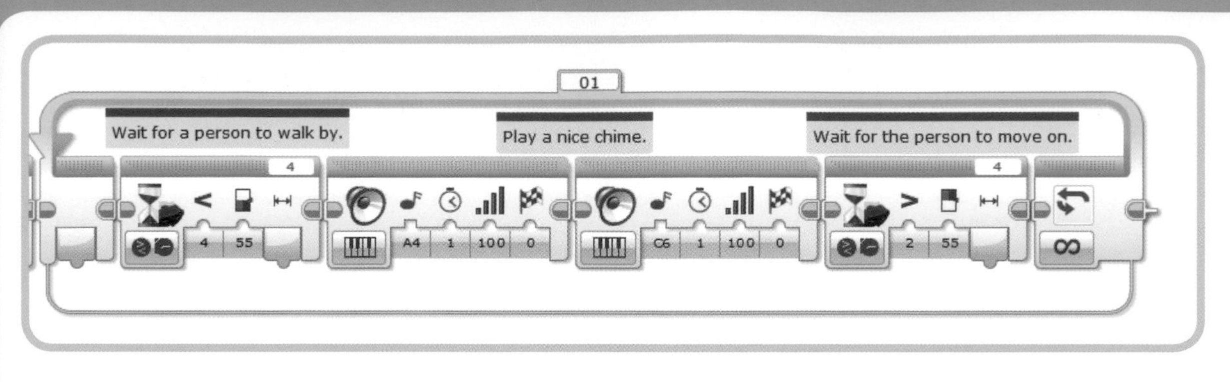

Figure 5-40: The completed DoorChime *program using the Infrared Sensor*

the gyro sensor

The *Gyro Sensor* (included only in the Education Edition), shown in Figure 5-41, measures rotational motion. It can tell your program how fast the sensor is rotating in the direction of either of the two arrows shown on its cover. Because the sensor knows how fast it's moving in degrees per second, it can also determine how far it has moved in degrees.

Figure 5-41: The Gyro Sensor

The Gyro Sensor only measures motion in one plane, so be sure to mount it accordingly. For example, as mounted in Figure 5-42, it can be used to measure how fast or far the TriBot rotates to the left and right but not to detect when the TriBot tips sideways or tumbles forward or backward.

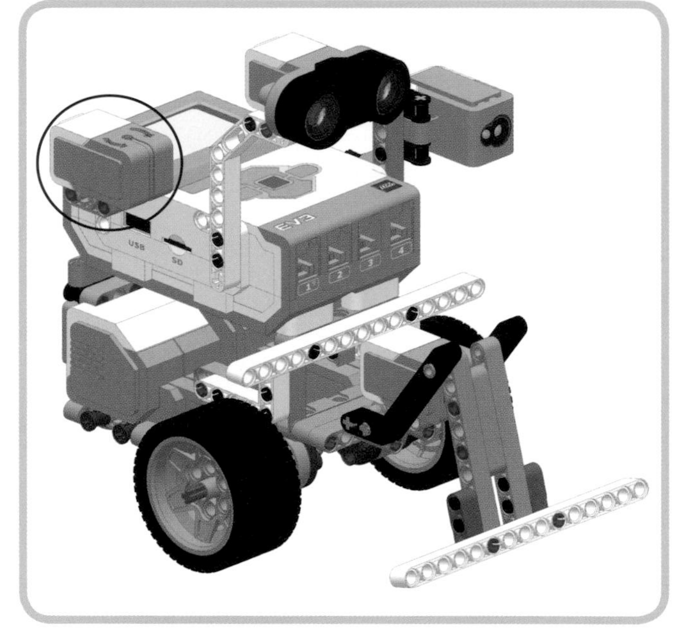

Figure 5-42: The Gyro Sensor on the TriBot

rate mode

In *Rate mode*, the Gyro Sensor measures the speed of rotation in degrees per second. This can be useful when you want your robot to turn at a set speed. The Wait block in Gyro Sensor - Compare - Rate mode is shown in Figure 5-43. The Gyro Sensor reads a positive number for clockwise rotation and a negative one for counterclockwise rotation.

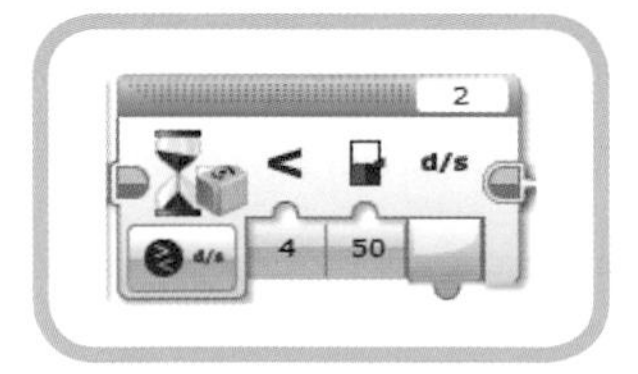

Figure 5-43: Wait block in Gyro Sensor - Compare - Rate mode

You could use the sensor to detect changes in rotation to tell you that a robot is tipping over, or to perform experiments in motion. Imagine recording data from the sensor while the robot travels through an obstacle course or maze. You could then examine the data to learn how the TriBot's motion changes with time.

angle mode

In *Angle mode* (Figure 5-44), the sensor reads the distance the robot has turned in degrees from the last time the sensor was reset. Positive values mean clockwise rotation, and negative values are counterclockwise.

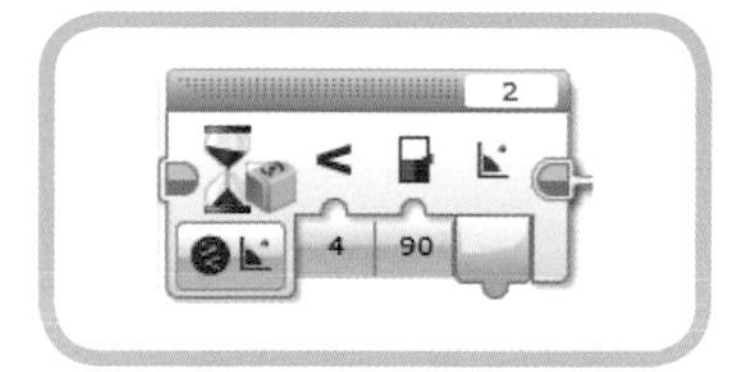

Figure 5-44: Wait block in Gyro Sensor - Compare - Angle mode

The Change mode lets you do things like tell the robot to turn 90 degrees from wherever it is, regardless of the starting point. This is particularly useful for turning corners.

NOTE The Gyro Sensor measures the rate of motion and derives the Angle value from the rate of movement and the time elapsed. Move the sensor too fast and you may disrupt this process. Also note that the Angle measurement tends to drift a little over time due to small inaccuracies in the measurements. When you connect the Gyro Sensor to the EV3, hold the sensor still for a few seconds to allow for calibration. (I find it works well if I unplug the sensor, wait a few seconds, and then plug it back in with the robot sitting still.)

resetting the angle

When your program starts, the Gyro Sensor resets the Angle value to zero. To reset it while your program is running, use the Gyro Sensor block in Reset mode, as shown in Figure 5-45. (This block is on the Sensor palette.)

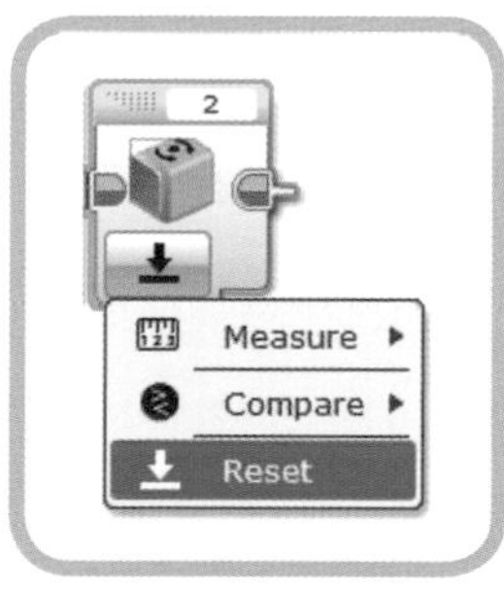

Figure 5-45: Resetting the Gyro Sensor

the GyroTurn program

The Gyro Sensor is particularly useful for controlling how the TriBot moves so that it can make accurate turns.

The goal of the *GyroTurn* program is to have the robot make a 90-degree turn, using the Gyro Sensor to determine when to stop the motors. Here's how to get started:

1. Create a new program named *GyroTurn*.
2. Add a Move Steering block to the program.
3. Set the mode to **On** and the Power parameter to **30**.
4. Set the Steering parameter to **25**.
5. Add a Wait block to the program. Set the mode to **Gyro Sensor - Change - Angle**.
6. Set the Threshold value to **90**.
7. Add a Move Steering block to the end of the program. Set the mode to **Off**.
8. Add a Wait block to the program. Set the time to **5** seconds.

Figure 5-46 shows the full program. The first Wait block will wait for the Angle measurement from the Gyro Sensor to change by 90 degrees. The second Wait block makes sure the robot comes to a complete stop before the program ends. Five seconds is more than enough time for the robot to stop moving. (I typically use five seconds anytime I think one second might not be enough.)

Run the program and watch Port View to see how close the final angle is to 90 degrees. With the Move Steering block's Power item set at 30, the turn is pretty accurate—usually to within 1 degree. Now try increasing the Power parameter to 50, 70, and 90. As the robot moves faster, the accuracy worsens and the robot often goes too far. This happens because there's a small delay between when the program recognizes that the sensor has met the threshold and when the motors are stopped. When the motors are moving fast, the TriBot can go several degrees beyond the target before it stops moving. I'll show you how to deal with this issue in Chapter 13, after you learn more EV3 programming techniques.

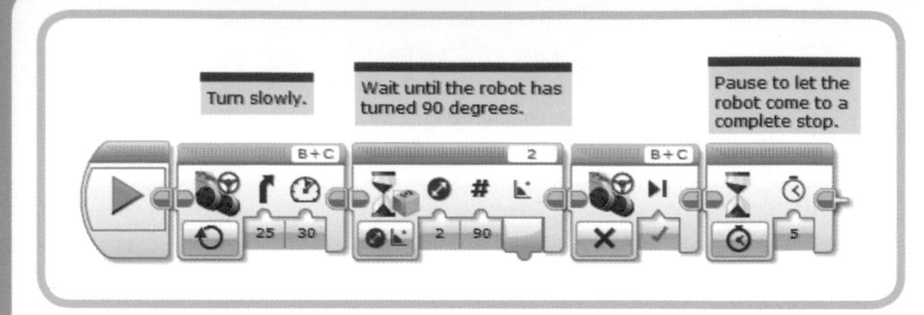

Figure 5-46: The GyroTurn *program*

the motor rotation sensor

Each EV3 motor has a built-in *Motor Rotation Sensor* that measures how far the motor has turned in degrees or rotations. You can use this sensor to control how far the robot moves, and to read a motor's current Power parameter, which can be useful for experimenting with the Move Steering and Move Tank blocks or for measuring the speed of a motor being turned by an outside force (like a windmill or treadmill). Figure 5-47 shows the Wait block with the mode set to Motor Rotation - Compare - Degrees.

You can read the value at any time, or reset it with a Motor Rotation block in Reset mode, as shown in Figure 5-48. (You'll often use two blocks with the Rotation Sensor: one to read the value and one to reset it.)

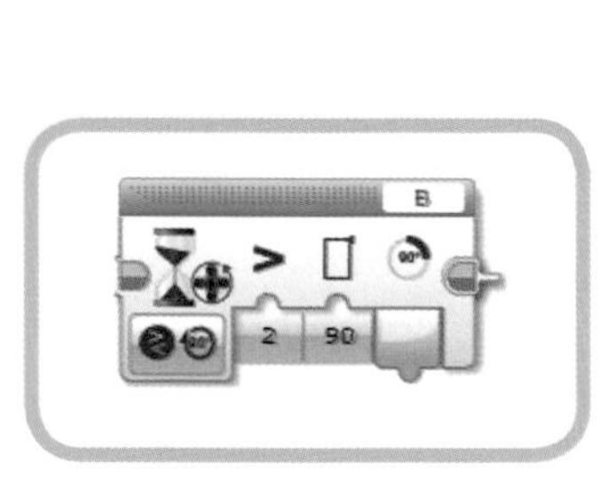

Figure 5-47: Wait block in Motor Rotation - Compare - Degrees mode

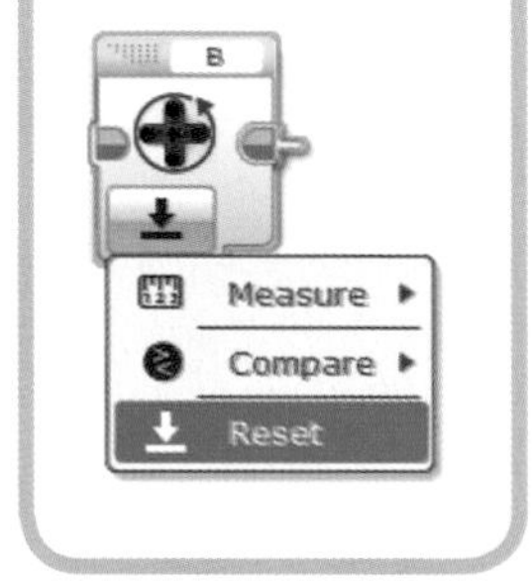

Figure 5-48: Resetting the rotation to zero

CHALLENGE 5-7

Modify the *AroundTheBlock* program to use the blocks from the *GyroTurn* program to make the 90-degree turn at each corner. Try both versions of the program using different speeds and different floors. Using the sensor should improve the repeatability of the turns when the conditions change.

the BumperBot2 program

In the original *BumperBot* program, a Move Steering block with the duration item set to -300 degrees made the TriBot back up after it hit something (see Figure 5-49). How about making the robot beep while backing up, like a school bus or large truck?

You know how to use the Sound block to make the robot beep; the challenges are to make it beep while moving and to make sure the robot still backs up the right amount. One way to do this is to start the robot moving using a Move Steering block with the mode set to On and then use a Loop block to stop the robot after it moves at least -300 degrees. Inside the Loop block, you can use a Sound block to make the robot beep. Copy and paste the *BumperBot* program to duplicate it, and then click the newly created *BumperBot2* in the list of programs to open and edit it. In the following instructions, we'll replace the Move Steering block (circled in Figure 5-49) with blocks that will reset the Rotation Sensor, start the TriBot moving, and then beep until the Rotation Sensor reads less than -300 degrees.

1. Add a Motor Rotation block to the left of the Move Steering block highlighted in Figure 5-49.
2. Set the mode to **Reset** and make sure the Port is set to **B**.
3. Change the mode of the Move Steering block highlighted in Figure 5-49 to **On** and the Power parameter to **-15**. This will make the TriBot back up slowly until you stop it.

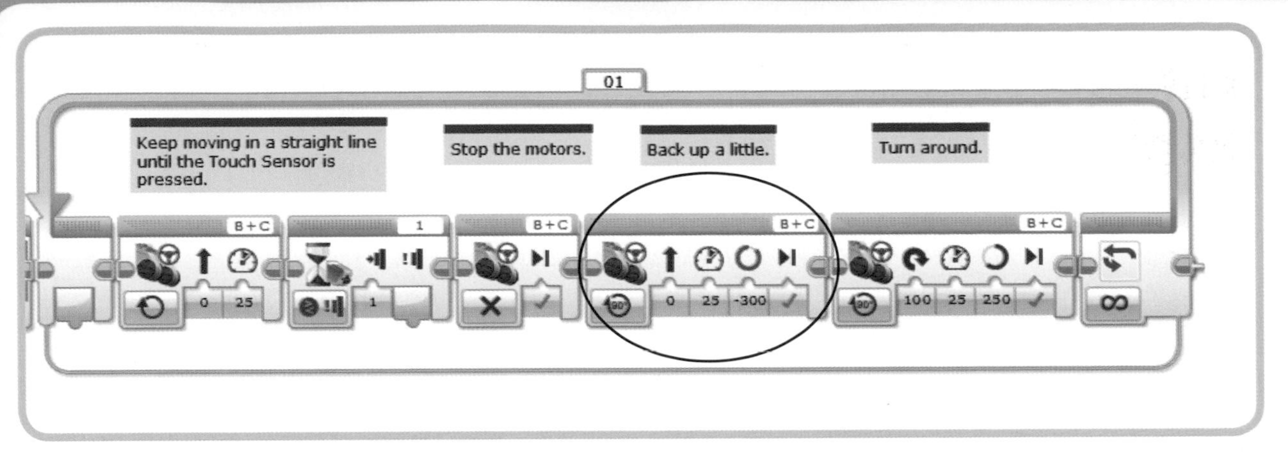

Figure 5-49: The Move Steering block used to back up

Because there is no Duration parameter, we set the Power to a negative value instead to make the TriBot back up. This part of the program should now look like Figure 5-50.

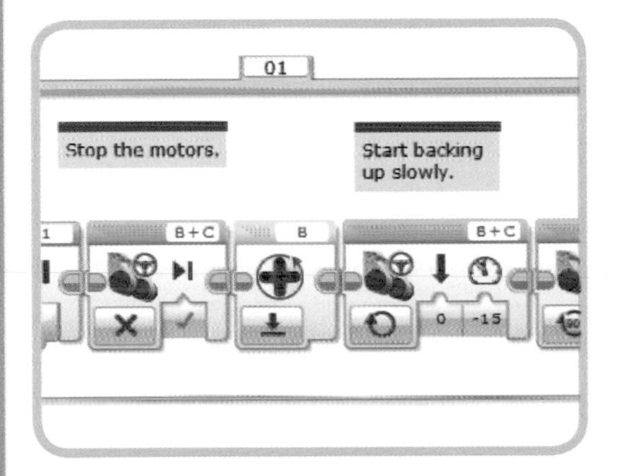

Figure 5-50: Resetting the Motion Rotation Sensor and moving back slowly

4. Add a Loop block to the right of the Move Steering block (the one you just changed to **On** mode).
5. Set the Loop block's mode to **Motor Rotation - Compare - Degrees**.
6. Set the Compare Type to **Less Than** and the Threshold value to **-300**.
7. Drag a Sound block into the Loop block, and set the mode to **Play Tone**.
8. Set the Sound block's Duration parameter to **0.5**.
9. Place a Wait block to the right of the Sound block inside the Loop block. Set the time to wait to **0.25** seconds to pause between the beeps.

The Loop block should now look like Figure 5-51. Run the program. When the TriBot runs into something, it should beep while it's backing up. The robot will back up farther than it did previously (a bit more than -300 degrees) because it only checks the Rotation Sensor after playing the tone and waiting a quarter-second. This isn't a problem because the distance the robot moves back is not critical. (It just has to move far enough to spin around without hitting whatever it ran into.) Experiment with the tone and volume to find a combination that sounds good to you.

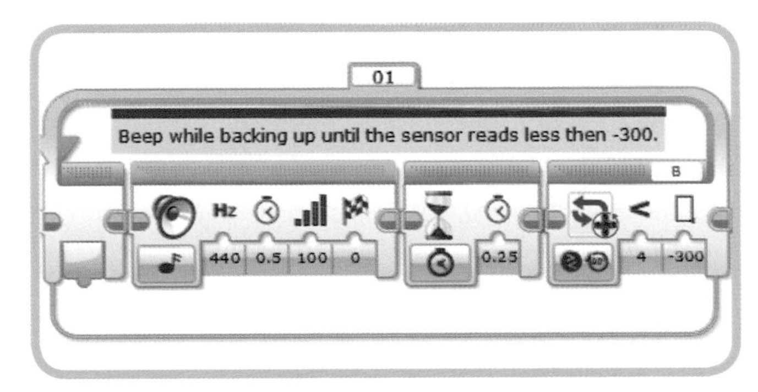

Figure 5-51: Beeping while backing up

further exploration

Here are some activities that will help you explore more possibilities with sensors.

1. Use Port View to experiment with the sensors' modes to see what works well.

    a. What colors and shades work well with the Color Sensor?

    b. How does an object's texture affect the Ultrasonic or Infrared Sensor? Try comparing a tennis ball to a baseball. How does distance affect the reliability?

    c. How does the Gyro Sensor behave when it's turned very fast? What happens when it's turned but not in the plane that it measures, such as when the TriBot is tipped to the side?

    d. How does the color of an object affect the Color Sensor in Reflected Light Intensity mode? How does the distance from the object affect the Reflect Light Intensity reading, and at what distance does this mode become unreliable? Think about how this relates to where you need to place the sensor for a line-finding or line-following program.

2. Use the buttons on the Infrared Remote to control the Lift Arm. Write a program that waits for a button to be pressed, then starts moving the arm and stops the arm when the button is pressed again. Use a slow speed so that you have time to stop the arm before it moves too far. Then, add a Switch block that moves the arm either up or down depending on the button you press.

3. Use the Infrared or Ultrasonic Sensor to create a motion detector. A Wait block in Change mode works well to notice when something has moved toward or away from the sensor. You can decide what the robot should do when it detects an intruder.

conclusion

The EV3 sensors allow you to build a robot that interacts with the world around it. The different sensors (Color, Touch, Ultrasonic, Infrared, Gyro, and Motor Rotation) let your robot perceive its environment in several different ways, allowing you to create robots that exhibit a variety of interesting behaviors. The programs presented in this chapter showed how to use the sensors in combination with Wait, Loop, and Switch blocks to create some interesting EV3 programs, from a simple door chime to the more complex *BumperBot* program.

program flow

Program flow is about controlling the order in which your programming blocks are run. Typically, the blocks run from left to right, but you can control the flow of the program by making blocks wait, repeat, or choose an action based on some condition. The three main blocks used for program flow are the Wait, Switch, and Loop blocks.

You've already seen how the Wait block works. In this chapter, I'll cover the Switch and Loop blocks in depth. The Loop Interrupt block is also used to control Loop blocks, so I'll explain that as well.

NOTE The Switch and Loop blocks also have some features that are used exclusively with data wires, which will be covered in Chapters 9 and 10.

the switch block

The *Switch block* (shown in Figure 6-1) lets your program make a decision about which blocks to run. This type of structure is called a *conditional* because the program flow changes based on whether a condition is met. The Switch block checks the condition in order to choose from two or more groups of blocks; these are called *cases*. This gives your program the ability to make a decision and react to data from the robot's sensors. For example, the *RedOrBlue* program on page 74 uses the reading from the Color Sensor to decide which Sound block to run.

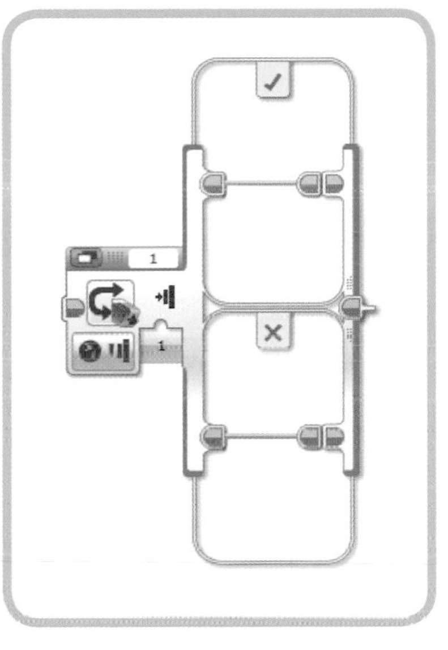

Figure 6-1: The Switch block

setting the condition

To set the condition on a Switch block, first select a sensor and mode from the Mode Selector list. Then enter the Threshold value and any additional parameters, such as the sensor port.

Think of setting the mode as asking a question. The block shown in Figure 6-1 asks the question, "Is the Touch Sensor pressed?" This is a yes-or-no question, so the Switch block can have only two cases: the true case and the false case.

Other questions allow for more than two answers. For example, the question "What color does the Color Sensor detect?" has eight possible answers because the Color Sensor can detect eight values (seven colors and the No Color value). For this question, the Switch block can have up to eight cases.

RESIZING A BLOCK

The Switch and Loop blocks automatically resize when you drag blocks into them. You may want to adjust the size of a Switch or Loop block yourself to shrink a block after removing some of the blocks inside, to see more tabs of a Switch block using Tabbed View, or to make room for comments.

When you click a Loop block or a Switch block case, a set of resizing handles appears, as shown in Figure 6-2. Click and drag a handle to change the size of the block.

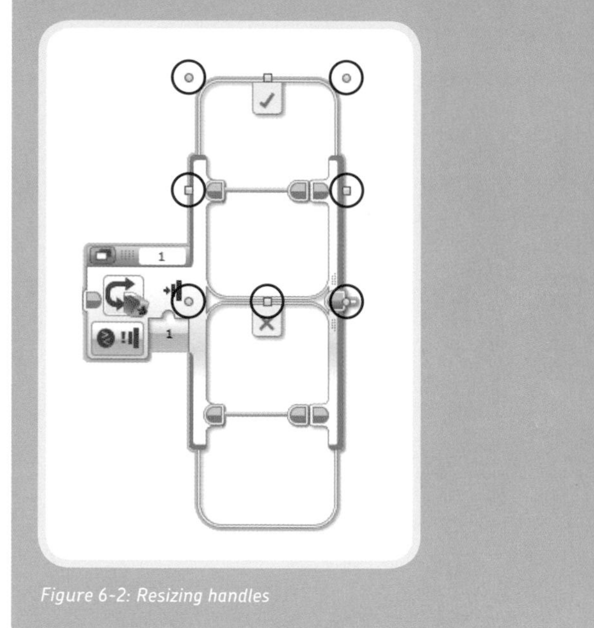

Figure 6-2: Resizing handles

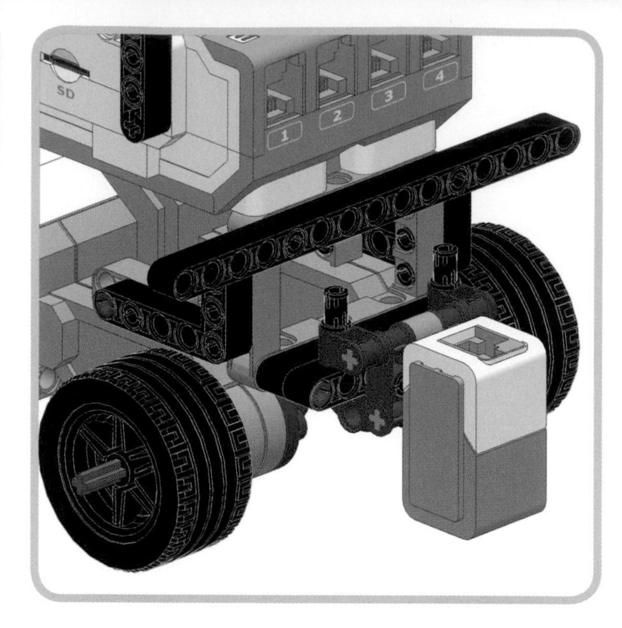

Figure 6-3: Color Sensor position for following a line

To test this program, you need a dark line to follow. You can use black electrical tape or a black marker to mark an oval on a white poster board (see Figure 6-4). Your line should be at least an inch (3 cm) wide and needs to be much darker than the background.

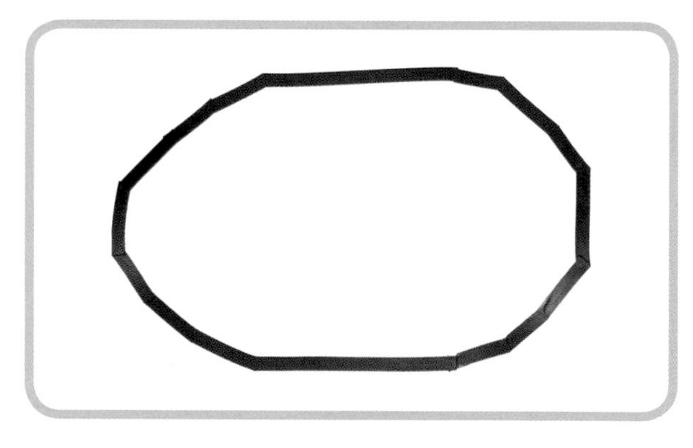

Figure 6-4: A test line made from electrical tape and poster board

the LineFollower program

The *LineFollower* program is a simple line follower, where the robot uses a Switch block to decide the direction to go in. The TriBot uses the Color Sensor in Reflected Light Intensity mode to follow the edge of a line by adjusting the steering based on the Light Sensor reading.

For this program, you need to remove the Touch Sensor bumper from the front of the TriBot and replace it with the Color Sensor. The Color Sensor should be mounted so that it points downward, as shown in Figure 6-3.

NOTE This program works best with smooth turns; sharp corners may cause problems. The final version of the program, presented in Chapter 19, is better at navigating corners.

the basic program

The program makes the robot follow the line by constantly adjusting the direction in which the robot steers in order to keep the Color Sensor over the edge of the line (see Figure 6-5). With the Color Sensor in Reflected Light Intensity mode, the sensor reading tells you how much light is reflected from a small circular area under the sensor. When the sensor is over the white background, it will give a high reading because a large quantity of light is reflected. When the sensor is completely over the dark line, it will give a low reading because a small amount of light is reflected. Between these two extremes, the reading varies based on how much of the line is under the sensor. As the robot moves forward, it will use the Color Sensor reading to tell where it is relative to the edge of the line. If the robot is too far over the line, it will steer to the left, moving it back towards the edge. If the robot moves away from the line, it will steer to the right to get back to the edge.

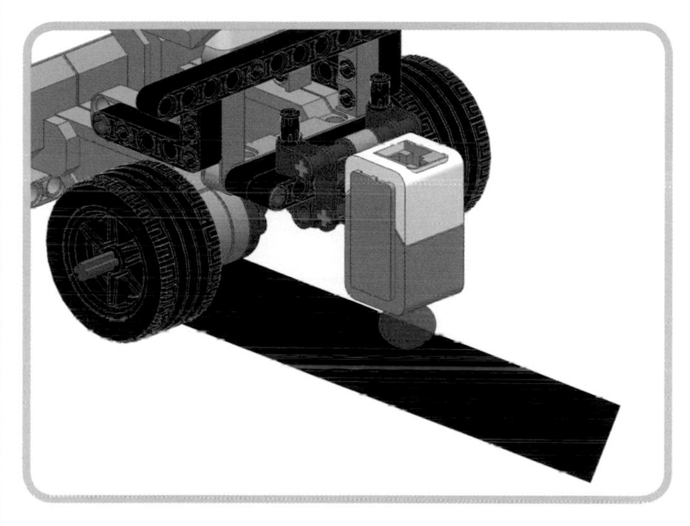

Figure 6-5: The Color Sensor over the edge of the line

The program uses a Switch block to choose between two Move Steering blocks, one of which steers to the left and the other to the right. Figure 6-6 shows the completed program. The entire program is within a Loop block, which keeps the program running until you stop it. The Switch block reads the Color Sensor and decides which Move Steering block to run. The Move Steering block on the upper section (the true case) steers the TriBot to the left, and the one on the lower section (the false case) steers it to the right.

Now let's create the program:

1. Create a new project named *Chapter6*.
2. Create a new program named *LineFollower*.
3. Add the Loop block.
4. Drag a Switch block into the Loop block (the Loop block will grow to fit the Switch block).
5. Add a Move Steering block to each case of the Switch block.
6. Select the Switch block, and set the mode to **Color Sensor - Compare - Reflected Light Intensity**.

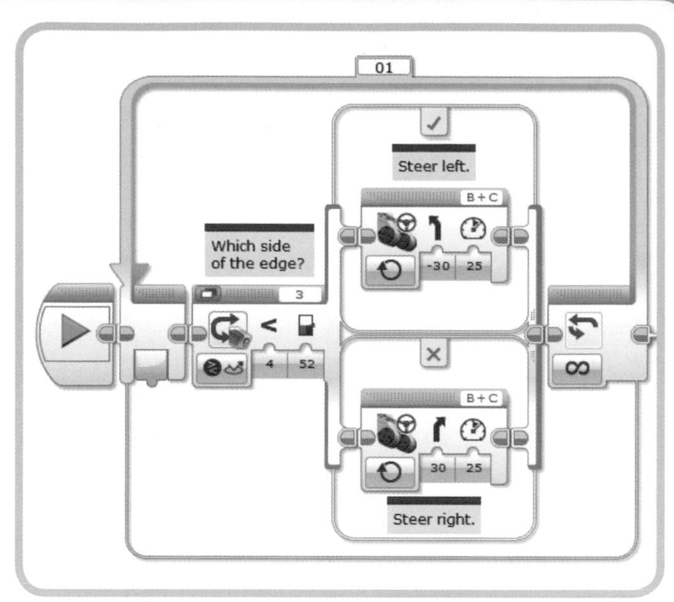

Figure 6-6: The LineFollower *program*

Now you just need to configure the settings for each block. Let's discuss how to decide on a value for each setting.

selecting the color sensor threshold

How should we determine the Threshold value to use for the Switch block? To make the TriBot follow the edge of the line, you'll need to find the value that the Color Sensor reads when it's on the edge of the line. Figure 6-7 shows the Color Sensor readings I get when I place the TriBot at five positions, moving from off the line to over the line. A simple but reliable approach to get a reasonable Threshold value is to take the average of the high (completely off the line) and low (completely over the line) values. Using the values from Figure 6-6 results in a Threshold value of (92 + 13) / 2, which I'll round down to 52 and use that as the Switch block's Threshold value. Note that this is close to the value I got when I placed the TriBot over edge of the line. Your value may be a little different depending on the sensor, the test pad, and the lighting in the room.

7. Set the Switch block's Threshold parameter. The block should now look like this:

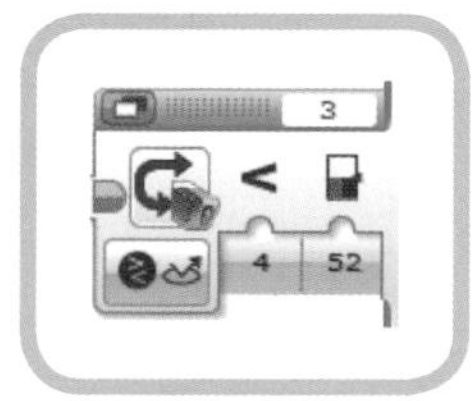

Completely off the line	Slightly over the line	Over the edge of the line	Mostly over the line	Completely over the line
Reading:92	Reading: 74	Reading: 48	Reading: 28	Reading: 13

Figure 6-7: Color Sensor readings at different positions

configuring the move blocks

The two Move Steering blocks have similar settings, except that they steer in opposite directions. The speed of the motors and the steering setting significantly affect how well the TriBot follows the line. If the Steering value is too low, the TriBot won't turn quickly enough to follow a curve in the line. If the Steering value is too high, the TriBot will have a lot of side-to-side motion as it constantly goes from one extreme to the other. Moving too fast makes it harder for the robot to react to changes in the line's direction. Start with a Steering parameter of positive and negative 30 (30 and -30) and a Power parameter of 25:

8. Select the Move Steering block on the upper Sequence Beam.
9. Set the mode to **On**.
10. Set the Power parameter to **25** and the Steering parameter to **-30**. The block should look like this:

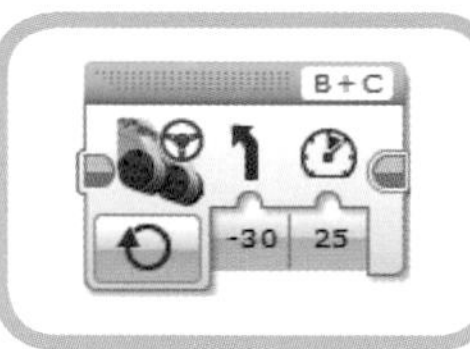

11. Select the Move Steering block on the lower Sequence Beam, and set the mode to **On**.
12. Set the Power parameter to **25** and the Steering parameter to **30**. The block should look like this:

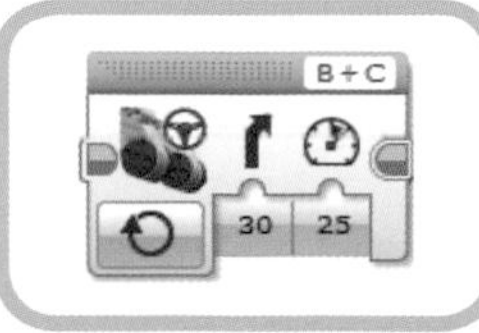

testing the program

Now download and run the program to see how well it works. You might need to adjust how fast the TriBot moves and how sharply it turns. If you do, make sure that you make the same changes to both Move Steering blocks. When it's working, see how much you can increase the speed and still have the robot follow the line reliably.

more than two choices

The first version of the *LineFollower* program makes the TriBot wiggle left and right while following a straight line, because the TriBot is constantly adjusting the steering. You can make the motion smoother with three Move blocks: one to steer left, one to go straight, and one to steer right.

In Color Sensor - Compare - Reflected Light Intensity mode, the Switch block can only choose between two sets of blocks based on the reading from the Color Sensor. To choose between three options, you'll need to use two Switch blocks. The first Switch block decides if the robot should turn to the left, and the second decides whether to go straight or turn to the right. This structure is used often in EV3 programming (and programming in general) to make sophisticated decisions.

Start by making the following changes to the *LineFollower* program:

13. Place a Switch block in the lower section of the existing Switch block, to the right of the Move Steering block.
14. Set the new Switch block's mode to **Color Sensor - Compare - Reflected Light Intensity**.
15. Drag the existing Move Steering block (the one that turns to the right) onto the lower section of the new Switch block.

16. Place a new Move Steering block on the upper section of the new Switch block.
17. Set the mode to **On** and the Power parameter to **25.**

    The Switch block should now look like Figure 6-8.

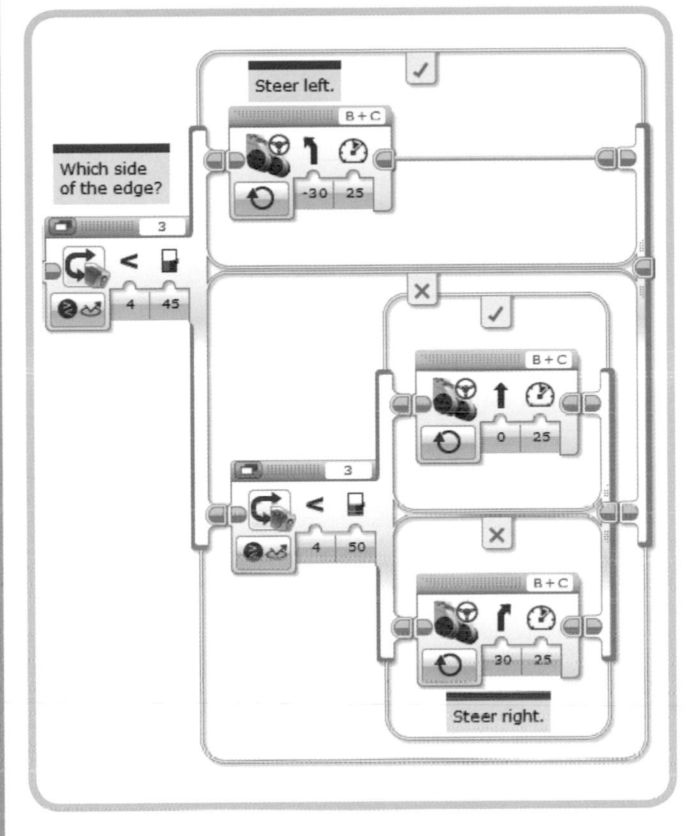

Figure 6-8: Setting the Threshold values for the new blocks

You need to set the Threshold values for the two Switch blocks so that the TriBot will go straight when it's on the edge of the line and turn when it moves away from the edge (by going either too far over the line, or too far away from the line).

Originally, I had a single Threshold value set at 52, which is in the middle of the two extremes, 13 and 92. To get two Threshold values, we can take the values halfway between this middle value and the two extremes. This gives me 32 and 72. Looking back at Figure 6-6, these values make sense because they are close to the readings I got with the sensor mostly on the line and mostly off the line. I'll set the trigger values so that the robot drives straight when the reading is between 32 and 72, and turn left or right when the reading is outside that range. Table 6-1 shows how the program should behave based on the Color Sensor reading. Be sure to calculate your own values, as they will vary depending on the lighting and the materials you used to mark the line.

table 6-1: color sensor ranges and program behavior

Color Sensor Reading	Program Behavior
0–31	Turn left
32–72	Go straight
73–100	Turn right

Follow these steps to complete the program:

18. Select the outer Switch block, and set the Threshold value to the lower limit of the range that should make the robot drive straight (which is 32, using my values).
19. Select the inner Switch block, and set the Threshold value to the lower limit of the range that should make the robot turn right (which is 73, using my values).

    The program should look like Figure 6-9.

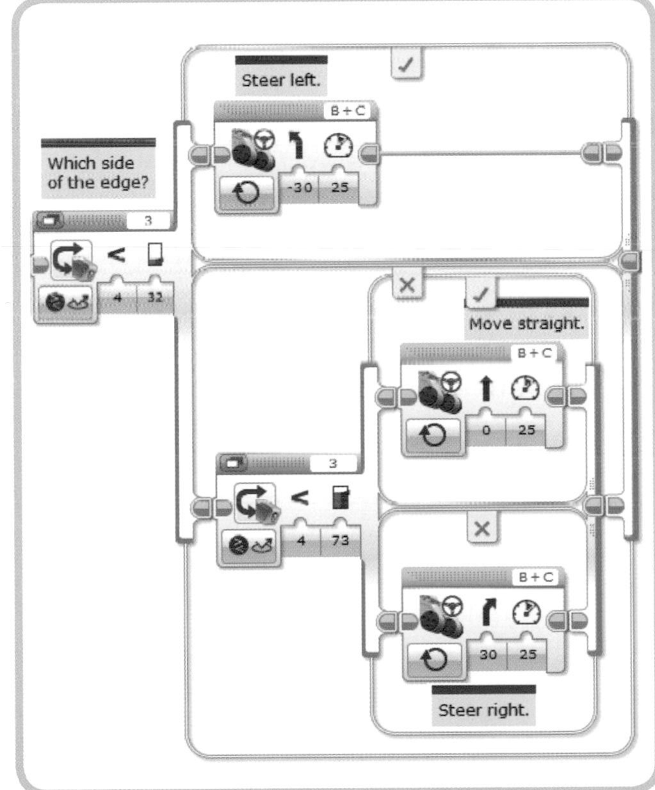

Figure 6-9: Setting the Threshold values

NOTE In this program, each Switch block reads the Color Sensor value to make its decision. This means that the sensor is read twice and the two values may not be exactly the same. But this shouldn't present a problem because the sensor value won't change much in the time it takes the EV3 to run the two Switch blocks.

testing the program

Now when you run the program, you should notice that the robot's motion is much smoother as it follows a straight line. Experiment with different Threshold values and Power and Steering parameters to see how fast you can make the TriBot move without veering off the line and getting lost.

using tabbed view

The *LineFollower* program uses *nested* Switch blocks, where one Switch block is placed within the other. Nested Switch blocks tend to take up a lot of room on the screen, which can make it difficult to work with the rest of the program. To shrink the size of the Switch blocks, switch to Tabbed View (click the **Flat/Tabbed Selector** button). This way, the program takes up much less space on the screen, as you can see in Figure 6-10.

NOTE When using a Switch block in Tabbed View, I recommend putting all comments above the outermost Switch block. Because you only see one tab at a time, comments inside the Switch block might get hidden. Figure 6-10 shows a comment for the code from our line-following program.

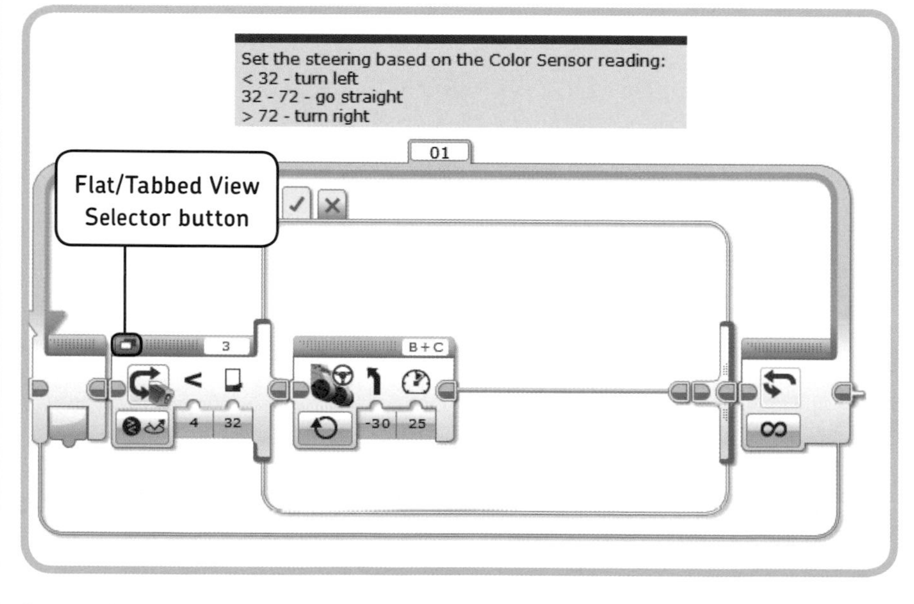

Figure 6-10: The Switch blocks as seen in Tabbed View

CHALLENGE 6-1

The first version of the *LineFollower* program chooses one of two actions each time it loops: turn left or turn right. The second version made the robot drive more smoothly by adding a third option: going straight. Improve the program again by adding two more cases so that the five options are: turn sharply left, turn gently left, go straight, turn gently right, and turn sharply right. You can do this by replacing the Move Steering blocks that turn left and right with Switch blocks that decide to turn sharply or gently based on the reading from the Color Sensor.

the RedOrBlue program

In this section, we'll create the *RedOrBlue* program, which will identify red and blue objects. We'll use the *IsItBlue* program from Chapter 5 as a starting point (Figure 6-11) because it already identifies blue objects. We'll begin by changing the program to recognize red objects, and then we'll update it to do something reasonable if the object is neither red nor blue.

The *IsItBlue* program is saved in the *Chapter5* project, so the first step is to copy the program to the *Chapter6* project and rename it.

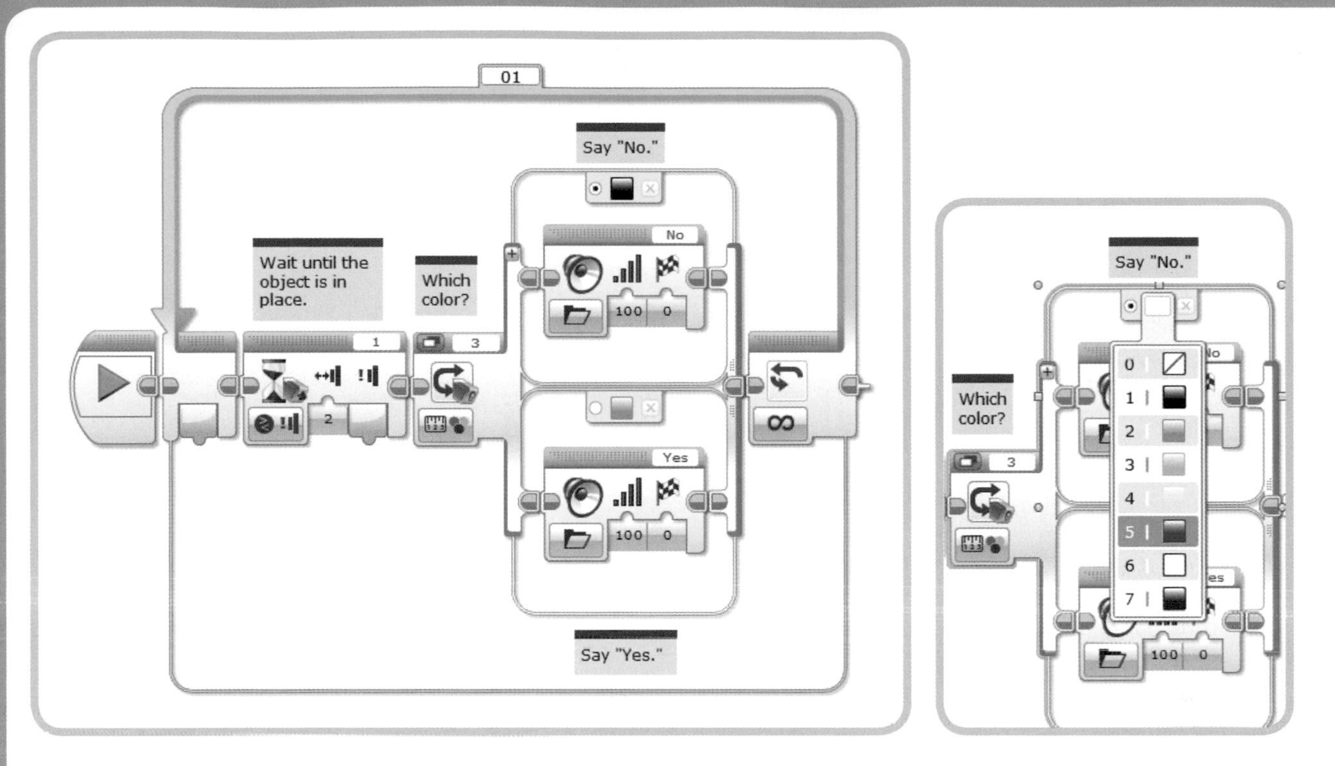

Figure 6-11: The starting point for the RedOrBlue *program*

Figure 6-12: Selecting Red for the upper case

1. Open the *Chapter6* project if it's not already open.
2. Open the *Chapter5* project.
3. Open the Project Properties page for the *Chapter5* project by clicking the small wrench icon at the left side of the program tabs.
4. Select *RedOrBlue* in the list of programs.
5. Click the **Copy** button at the bottom of the window.
6. Select the *Chapter6* project.
7. Open the Project Properties page for the *Chapter6* project.
8. Click the **Paste** button. The *IsItBlue* program will be added to the project.
9. Open the *IsItBlue* program and rename it to *RedOrBlue*.
10. Close the *Chapter5* project.

identifying red objects

The lower case of the Switch block already matches blue objects, so we'll use the upper case to match red objects.

1. Click the black box at the top of the Switch block and choose **Red**, as shown in Figure 6-12.

The yes-or-no response is reasonable for the *IsItBlue* program but isn't quite right for a program that identifies more colors. Change the Sound blocks so that the program says "Red" for red objects and "Blue" for blue objects.

2. Select the Sound block on the upper case and change the sound file to *Red*.
3. Select the Sound block on the lower case and change the sound file to *Blue*.

The program should now look like Figure 6-13. When you run the program, it should correctly identify red or blue objects.

adding a new case

So far, this program only really identifies blue objects, because it will declare any nonblue object to be red (because red is marked as the default case). In this section, you'll modify the program to correctly identify red objects and say "Uh-oh" when it can't determine an object's color. The Switch block currently has two cases: one for red objects and one for blue objects. The *Add Case button*, highlighted in Figure 6-14, will add a new case to the Switch block. If you add a case and then later want to remove it, click the button with the small *x* on the right side of the tab at the top of the case.

NOTE The Add Case button only appears on the Switch block in modes where there are more than two possible cases.

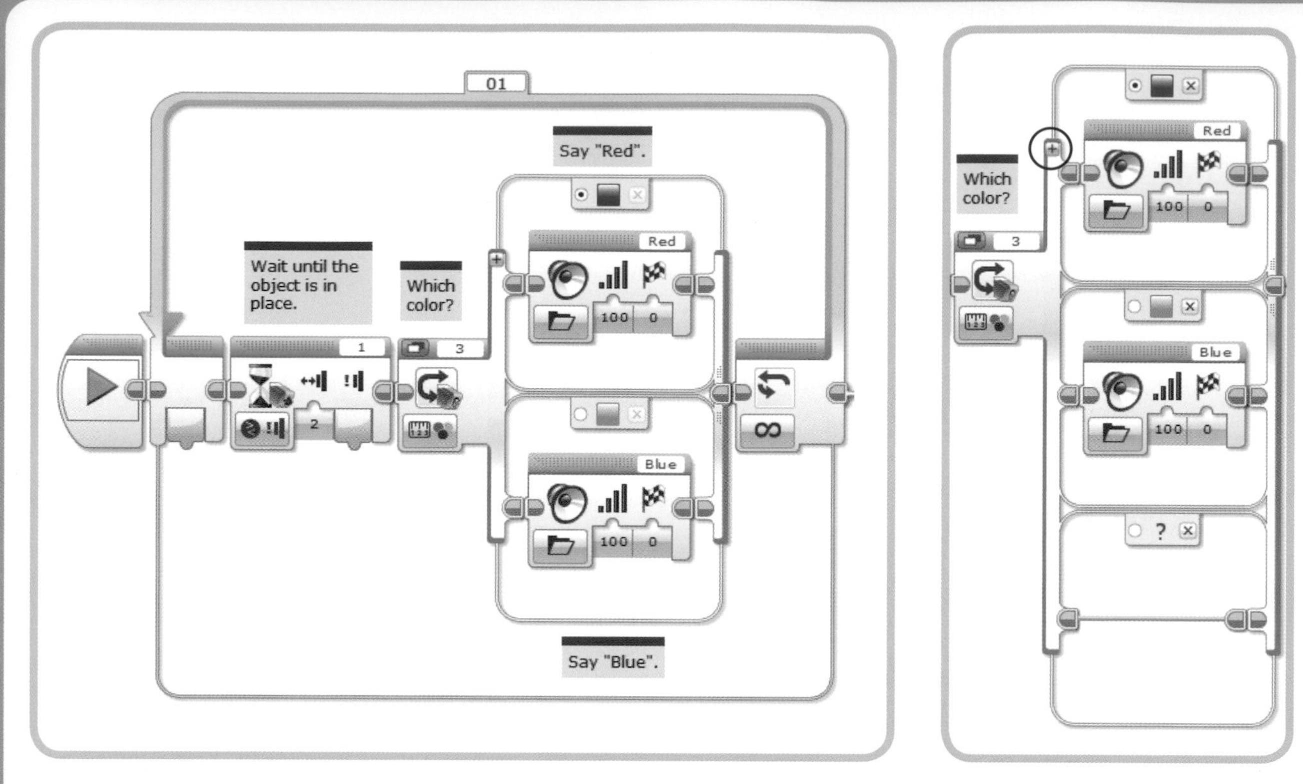

Figure 6-13: Identifying red or blue objects

Figure 6-14: Adding a new case

Now, back to our program:

4. Click the **Add Case** button. The Switch block should look like Figure 6-14.
5. Click the red question mark on the tab at the top of the new case and a menu appears. Select the **No Color** option.
6. Add a Sound block to the No Color case.
7. Click the **Sound File** box and select the *Uh-oh* file. You'll find it under the Expressions group. The program should now look like Figure 6-15.

the default case

Right now, the red case is the default case in our program. This means that when none of the cases are met, the program will run the blocks for the red case, and the program will say "Red" when the Color Sensor reads anything other than red, blue, or No Color (for example, yellow or green). It makes more sense to set the No Color case as the default. This way, the program will say "Uh-oh" whenever it doesn't detect red or blue.

8. Click the **Default Case** button on the No Color case.

Figure 6-16 shows the final configuration of the Switch block.

CHALLENGE 6-2

Extend the *RedOrBlue* program to recognize all seven colors that the Color Sensor can detect.

the loop block

The *Loop block* lets you repeat a group of blocks over and over. The blocks inside a Loop block are called the *loop body*. You set a condition that controls how often the loop body is repeated and determines when the program moves on to the block that appears after the Loop block.

The modes for the Loop block, shown in Figure 6-17, include the same sensor modes as the Switch block, plus four additional modes:

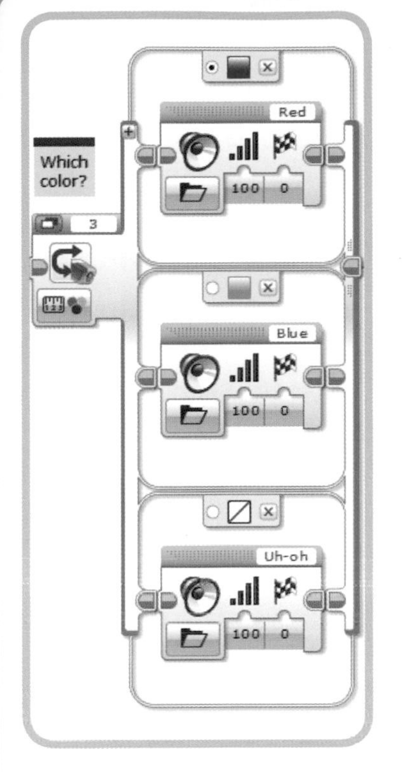

Figure 6-15: The No Color case

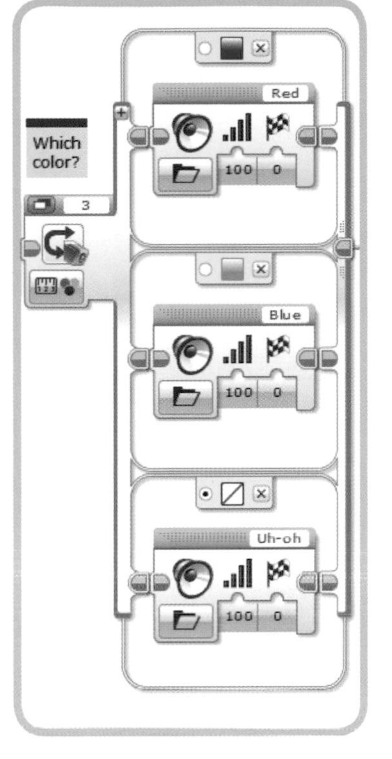

Figure 6-16: Setting the default to No Color

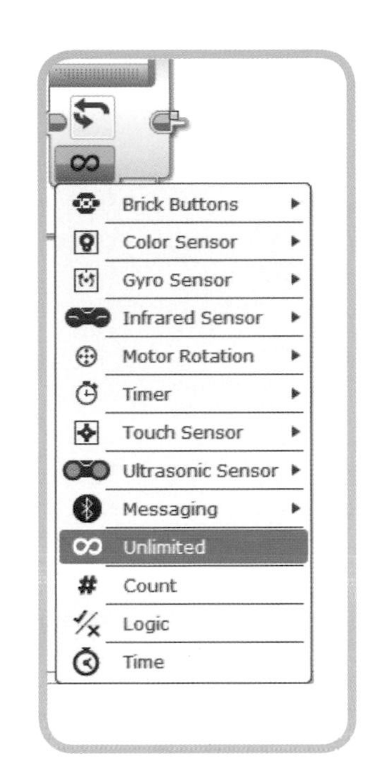

Figure 6-17: Selecting the Loop block mode

Unlimited The Loop keeps repeating until the program ends or the loop is exited using the Loop Interrupt block (discussed below).

Count The Loop repeats for the specified number of times.

Logic A value passed into the Loop using a data wire determines whether the Loop should exit. (Chapter 10 discusses using a Loop block with a data wire.)

Time The Loop continues for the specified number of seconds.

When a Loop block is in a sensor mode, the program checks the sensor measurement after the loop body runs. This means that the loop body is always run at least once. The program then decides whether to continue or leave the loop based on that sensor value. It doesn't matter what value the sensor reads while the loop body is running, only the sensor value measured at the end of the loop affects the program. If you're not careful, this can cause your program to fail. Because the Color Sensor is only checked at the end of the loop, it's very likely that the robot will move past the line without detecting it while the Sound block or Wait block is running. As an example, Figure 6-18 shows a flawed version of the *LineFinder* program that will often fail to find the line.

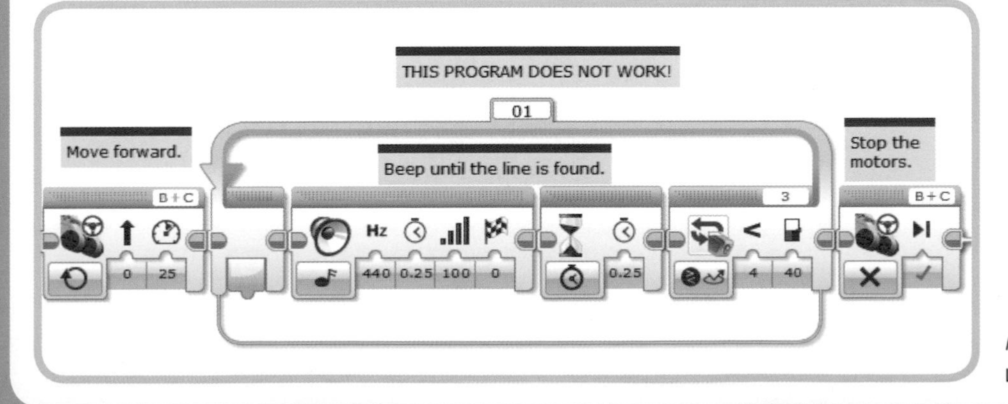

Figure 6-18: A broken LineFinder *program*

the loop interrupt block

The *Loop Interrupt block*, shown in Figure 6-19, gives you another way to exit a loop. The block only has one parameter: the name of the loop to exit. Each Loop block has a tab at the top of the block that holds the name of the loop, shown in Figure 6-20. A loop is named *01* by default; you can click the name to change it.

Figure 6-19: The Loop Interrupt block

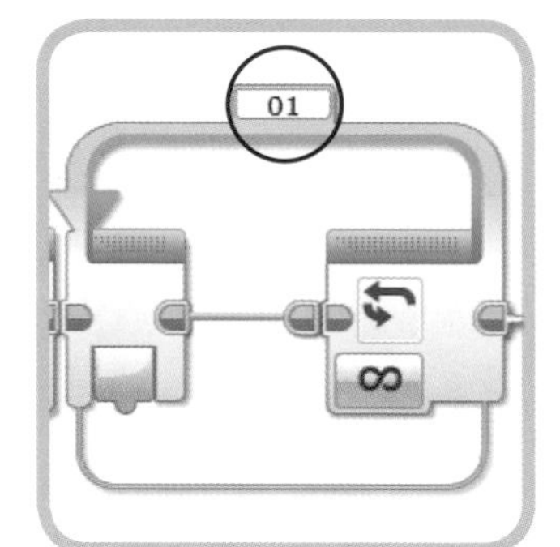

Figure 6-20: The Loop block name

The Loop Interrupt block shows only the first three letters of the name of the Loop block it will interrupt. As a result, it's usually clearer to use numbers or short abbreviations for the loop names; otherwise, it may not be clear which loop a Loop Interrupt block refers to.

the BumperBot3 program

Now you'll make some changes to the *BumperBot2* program to learn how the Loop Interrupt block works. To use this program, return the TriBot to its original configuration, with the Touch Sensor bumper mounted on the front and the Light Color Sensor on the side, as shown in Figure 6-21.

The changes you'll make will exit the loop and stop the program when you turn off the lights in the room. Figure 6-22 shows the part of the original program that we'll change. So far, the program makes the robot move forward and wait for the Touch Sensor to be pressed. When the sensor is pressed, the robot backs up, and the program loops back to the beginning.

Now you'll change the program to make the robot check the Color Sensor while it's waiting. If the Color Sensor reading is very low, indicating that the lights have been turned off, then the Loop Interrupt block exits the loop and the program ends because there are no more blocks after the Loop block.

Figure 6-21: TriBot configuration for the BumperBot3 *program*

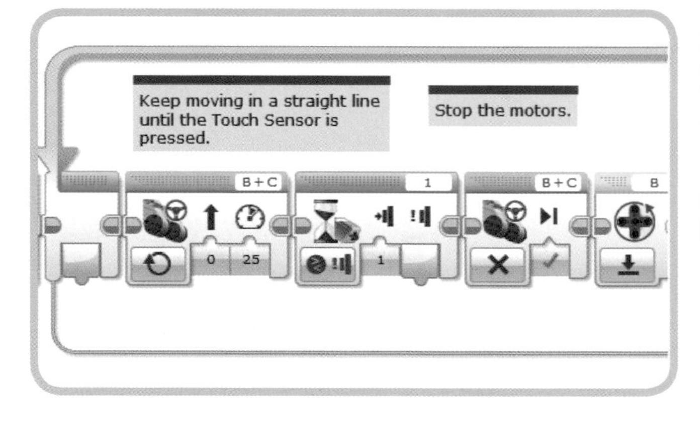

Figure 6-22: The BumperBot2 *program*

Instead of using the Wait block to detect when the bumper is pressed, we'll use a Loop block, as shown in Figure 6-23, that repeats until the Touch Sensor is pressed. Inside the body of the loop, we'll use a Switch block in Ambient Light Intensity mode to check the ambient light level of the room. If the Color Sensor doesn't detect enough light, the program will stop the motors, say "Goodbye," and then exit the loop.

Follow these steps to create the new *BumperBot3* program shown in Figure 6-23:

1. Copy the *BumperBot2* program from the Chapter5 project to the Chapter6 project.
2. Change the name of the program to *BumperBot3*.
3. Change the name of the main Loop block to *02*.

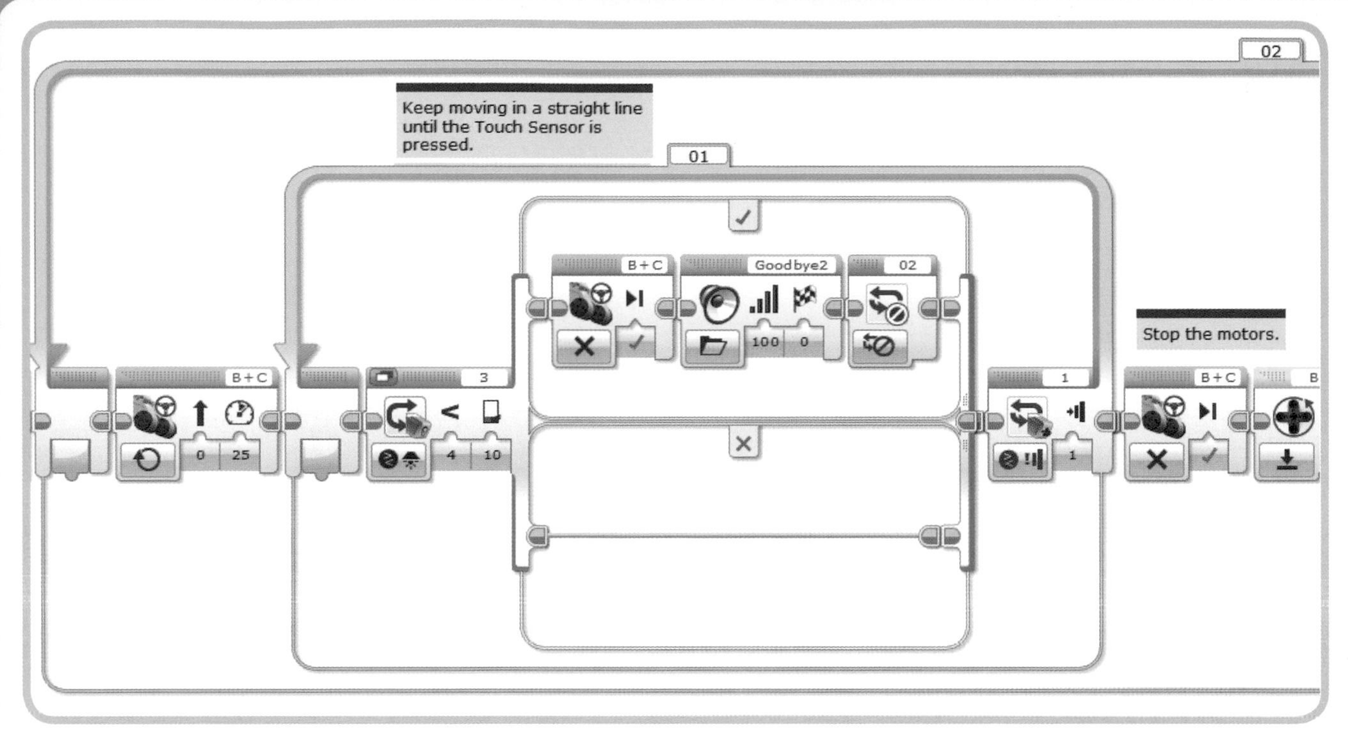

Figure 6-23: The BumperBot3 *program*

4. Delete the Wait block.
5. Add a Loop block where the Wait block was.
6. Set the Loop block's mode to **Touch Sensor - Compare - State**.

   This part of the program should now look like Figure 6-24.

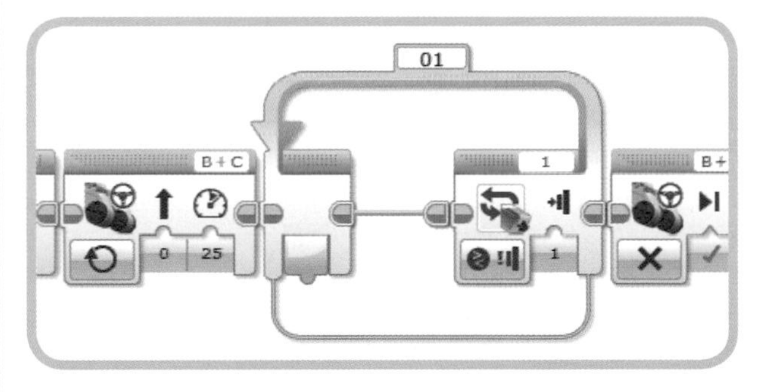

Figure 6-24: Replacing the Wait block with a Loop block

At this point, the empty Loop block does the same thing as the Wait block it replaced: It simply waits for the Touch Sensor to be pressed. But within the Loop block, we can now add some blocks that will run while the program is waiting.

The next step adds a Switch block inside the Loop block to check the Color Sensor:

7. Drag a Switch block into the Loop block.
8. Set the mode of the Switch block to **Color Sensor - Compare - Ambient Light Intensity**.
9. Set the Threshold value to **10**. You can adjust this value after some testing if it's too high or too low. The Loop block should look like Figure 6-25.

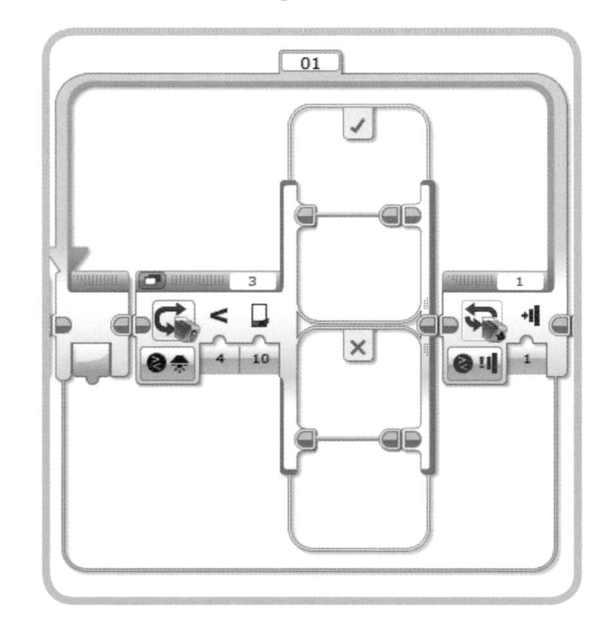

Figure 6-25: Adding the Switch block

When the Switch block runs, it checks the Color Sensor, and if the light intensity is below 10, the blocks on the upper case will run. Now add the following blocks:

10. Drag a Move Steering block to the upper case of the Switch block, and set the mode to **Off**.
11. Add a Sound block after the Move block. Select *Goodbye* from the list of sound files (it's under **LEGO Sound Files ▸ Communication**).
12. Add a Loop Interrupt block after the Sound block. Set the Loop Name to *02*.

Figure 6-26 shows the Loop block with these changes.

Now when you run the program, it should say "Goodbye" and stop if the lights in the room go out. To test the program, you can hold the TriBot in your hand, and after the wheels start moving, cover the Color Sensor with your other hand. When you cover the Color Sensor, the light level drops and the program should say "Goodbye" and exit.

further exploration

Now that you've learned the nuances of the Switch and Loop blocks, here are some activities to practice using them:

1. Write a program that makes the robot follow you—staying close to you but not too close, say between one and two feet. Use the Infrared or Ultrasonic Sensor and a Switch block to make the robot move forward or backward if you are too close or too far away. The robot should not move if you are in the desired range.
2. Use the Infrared Remote to add a pause and resume function to the *BumperBot3* program. Add a new Switch block that checks to see if a button on the remote is pressed. If it's pressed, the program should stop the motors, wait for another button on the remote to be pressed, and then start the motors again. The first button acts as a pause button, and the second button acts as a resume button.

conclusion

In this chapter, you learned to use the Switch block to make decisions, allowing you to choose between two or more groups of blocks. Either by nesting Switch blocks or by adding extra cases to a tabbed Switch block, you can extend the number of choices.

The other major flow control block is the Loop block, which allows you to repeat a group of blocks until a certain condition is reached. Much of the power of EV3 programs comes from the flexibility that's available when configuring blocks. The condition usually compares a sensor reaching a set Threshold value, but you can also configure the Loop block to repeat for a certain number of repetitions or a set amount of time. The Loop Interrupt block provides another way to exit a Loop block, which gives you even more flexibility to control how your program works.

At this point, you've seen how to use the EV3 motors and sensors and learned about the programming blocks you need to build sophisticated programs. In the next chapter, you'll use this knowledge to program the TriBot to find its way out of a maze.

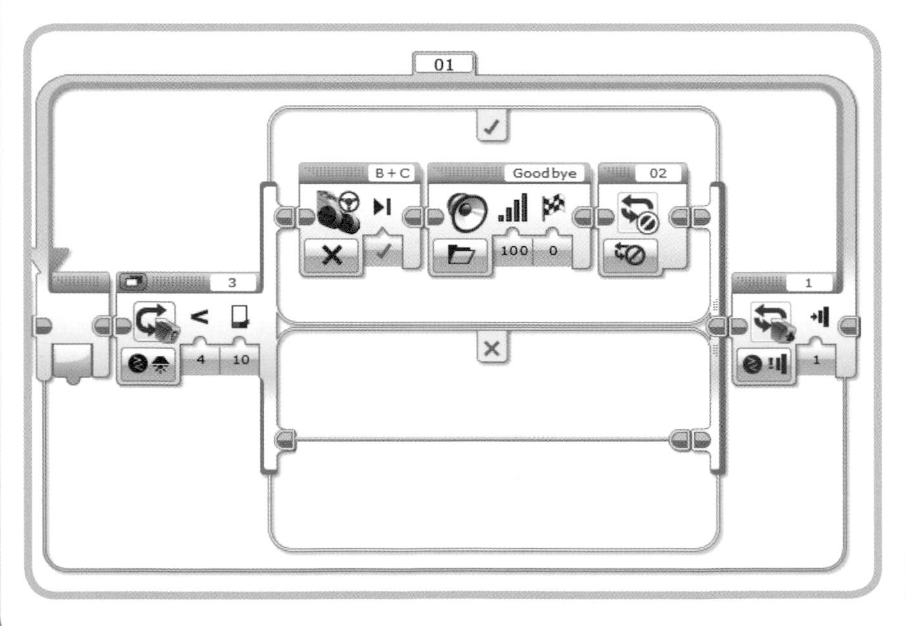

Figure 6-26: Stopping the motors and saying "Goodbye"

7

the WallFollower program: navigating a maze

In this chapter, we'll create the *WallFollower* program that allows the TriBot to find its way out of a simple maze. As you create the program, I'll take you through all the steps required to plan and write a program, from the initial design to the final testing. This will include choosing the initial settings for certain blocks, seeing how well they work, and adjusting those settings as necessary—because programs almost never work exactly right the first time you run them. Let's start by learning how to read and write pseudocode.

pseudocode

The more complex your program is, the more difficult it is to give a short, precise description of the program using normal sentences and paragraphs. There are better ways to describe the logic of your program, and one of the most common is pseudocode.

Pseudocode can be used to describe the most important details about how a program works and the logic behind the program. This gives you an easy way to share your programs with others or to build an EV3 program based on pseudocode written by someone else. Generally, pseudocode resembles traditional text-based programming languages such as Java or C, but it doesn't need to follow strict rules.

For example, Figure 7-1 shows the *RedOrBlue* program from Chapter 6. The program waits for the Touch Sensor to be pressed and then uses the Color Sensor to tell whether the object being tested is blue or red. Listing 7-1 shows the pseudocode for this program.

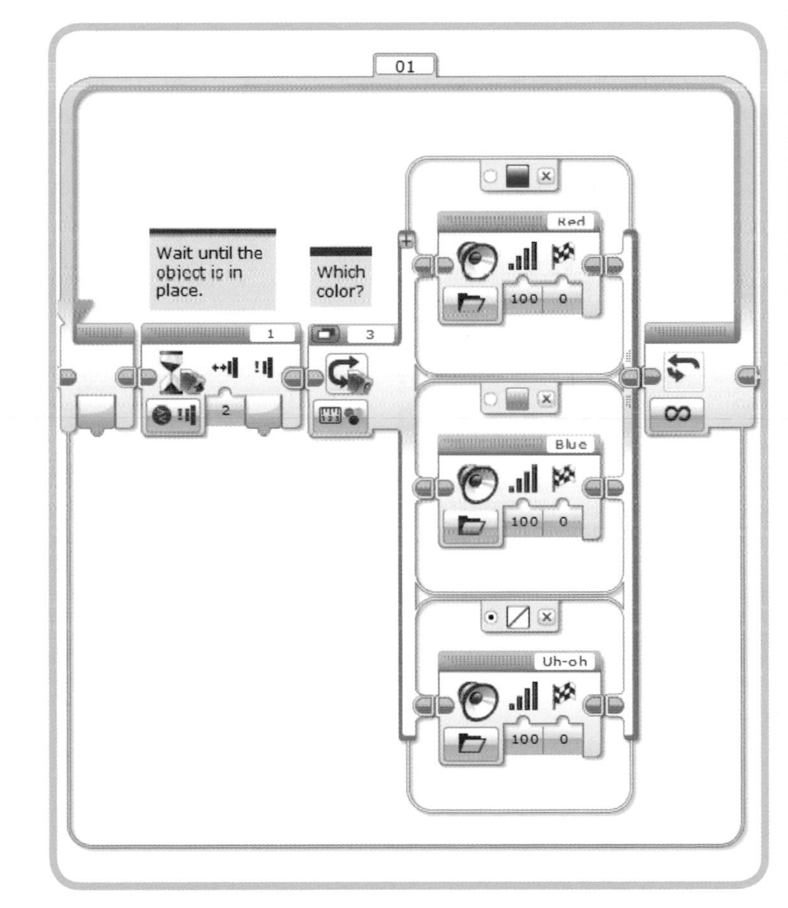

Figure 7-1: The RedOrBlue *program*

```
begin loop
   wait for the Touch Sensor to be pressed
   if Color Sensor detects red then
      use a Sound block to say "Red"
   else if the Color Sensor detects blue then
      use a Sound block to say "Blue"
   else
      use a Sound block to say "Uh-oh"
   end if
loop forever
```

Listing 7-1: Pseudocode for the RedOrBlue *program*

This pseudocode gives a concise, easy-to-understand description of the program. With a little practice, you'll quickly get used to reading pseudocode and turning it into a working EV3 program.

Note the following about Listing 7-1:

* In most cases, a separate line is used for each block.
* The lines are indented to show when blocks are nested (placed inside another block). Indenting makes it easier to see what's happening inside a Loop or Switch block, for example.
* We use the terms `if`, `then`, and `else` to describe the behavior of the Switch block. Many programming languages use `if-then` statements, and you would describe the logic similarly in plain English.
* The indented lines below the first `if` describe the blocks on the Switch block's upper case, and the indented lines after `else if` describe the blocks on the middle case. The lines below the `else` describe the Switch block's default case (the default case should be described last regardless of where it appears in the Switch block). This basically means "If the first case is true, do this set of actions. If the second case is true, do this other set of actions. Otherwise (else), do that set of actions."
* The `end if` line marks where the Switch block ends, after which the program moves on to the next part of the code.

In this example, the pseudocode describes a finished program with all of the details filled in. However, pseudocode is often used when planning and developing a program, in which case the details of the code might not all be filled in yet.

Now let's start planning out the *WallFollower* program. After we figure out what our program needs to accomplish, we can write pseudocode to help us think through the entire program.

solving a maze

There are many well-known approaches to solving a maze. For this program, you will use a method known as the *right-hand rule* algorithm. An *algorithm* is a set of instructions for solving a problem. In this case, the right-hand rule has one fundamental instruction: always follow the wall to the right, and go through any opening on that side.

The right-hand rule algorithm works for mazes without tunnels or bridges, where the start and end points are at the outer edges of the maze. (It won't always work for a maze where the goal is to get to the center.) As long as the maze fits these criteria, the robot is guaranteed to find the exit.

Figure 7-2 shows an example maze with a path drawn using the right-hand rule. To understand the algorithm, imagine that you're walking through this maze, always keeping your hand on the wall to your right. You would follow the path marked in the figure and eventually make your way to the exit. Although this method won't necessarily find the shortest route through the maze, it will eventually find the exit.

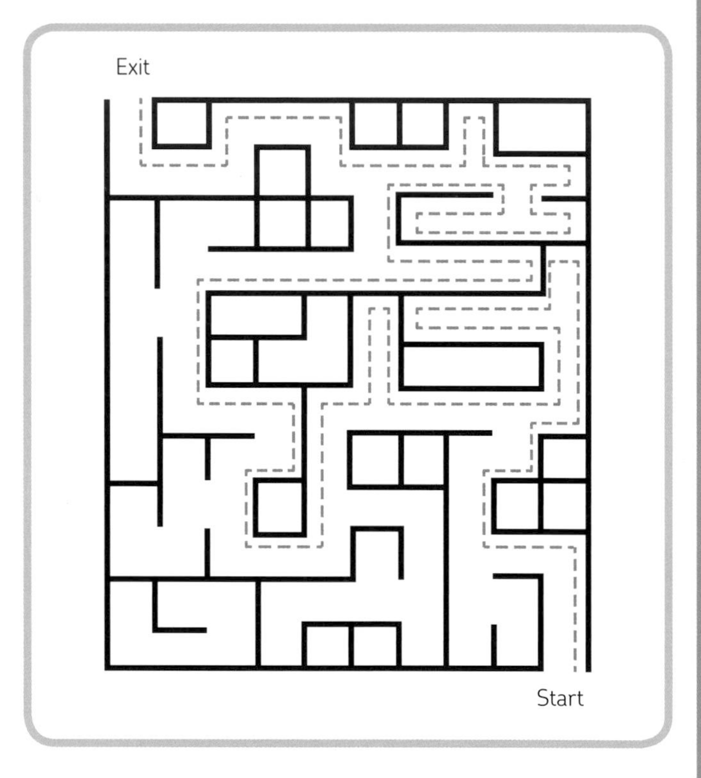

Figure 7-2: The right-hand rule path to the exit of a simple maze

program requirements

The first step in writing this program is to make a simple list of the *program requirements* that describe exactly what the program needs to do. Our program should be successful as long as it meets all the requirements on the list.

To develop the list of requirements, let's think about the different situations the robot will encounter and decide how the program should react in each case. For example, when there is a wall to the robot's right, as shown in Figure 7-3, it should keep moving straight. As it moves forward, the robot needs to keep a consistent distance from the wall so that it doesn't stray too far or veer into the wall.

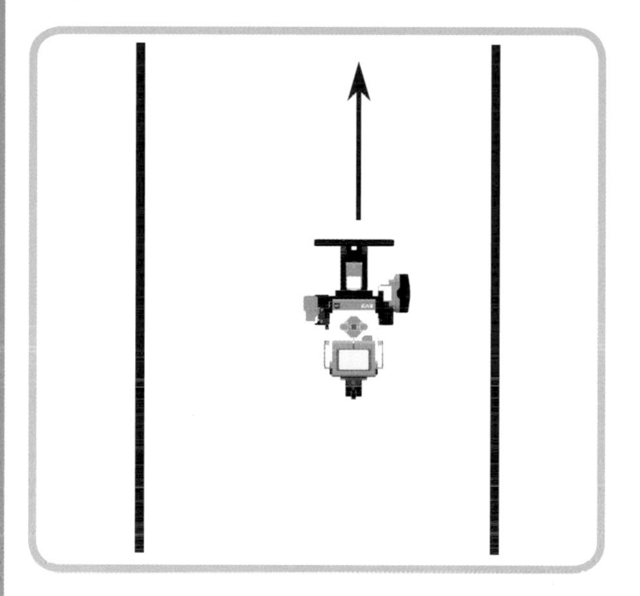

Figure 7-3: Moving parallel to the wall on the right

When the robot meets a corner with a wall to the right and front of it, as shown in Figure 7-4, the robot should turn to the left and continue forward.

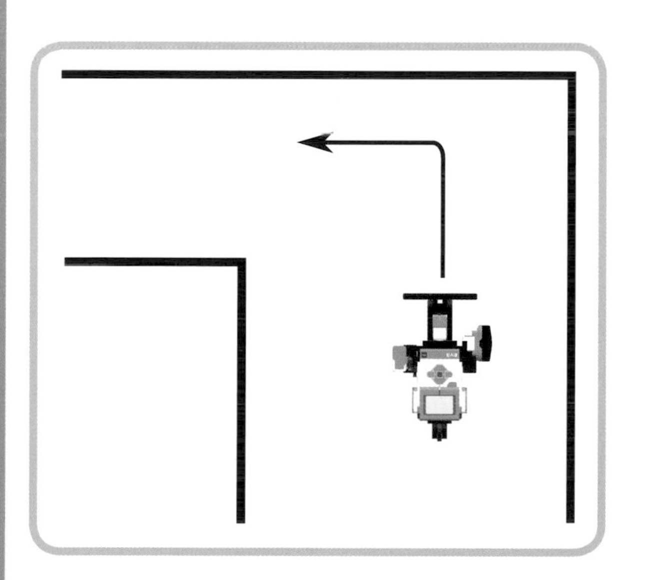

Figure 7-4: Turning left at a corner

When there is an opening in the wall to the right, as shown in Figure 7-5, the robot should turn into the opening. To follow the right-hand rule, the robot should *always* turn into an opening on the right, even when it can go straight as in Figure 7-6, or when it can go left as in Figure 7-7.

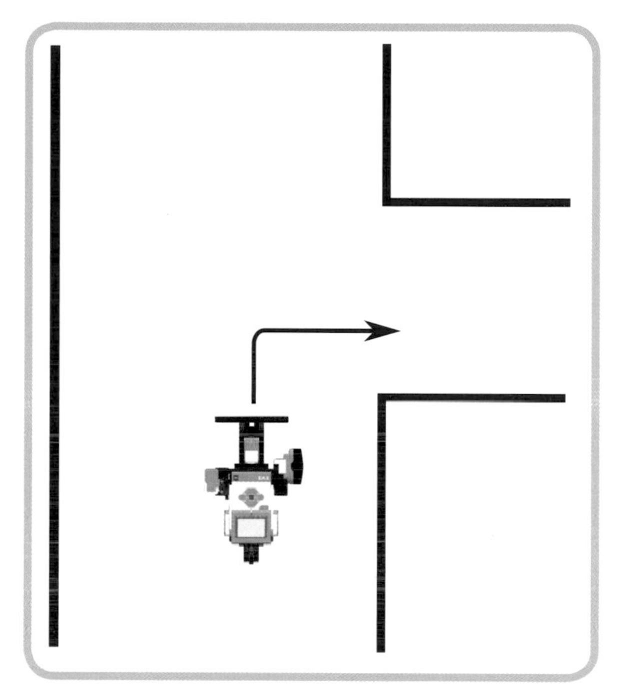

Figure 7-5: Turning into an opening on the right

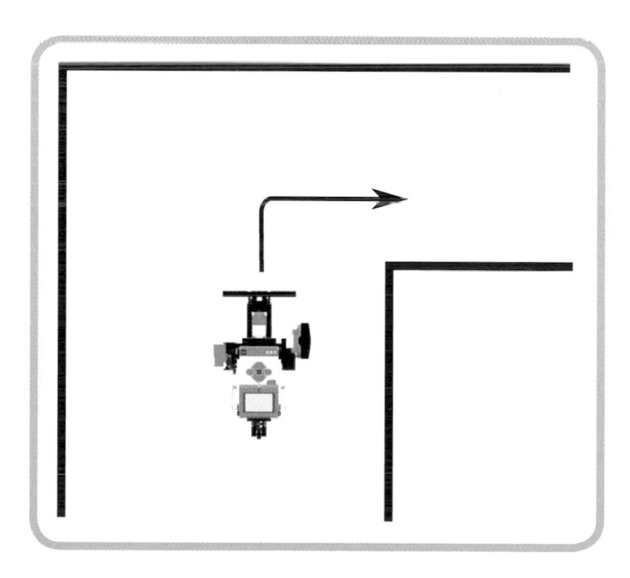

Figure 7-6: Turning right instead of going straight

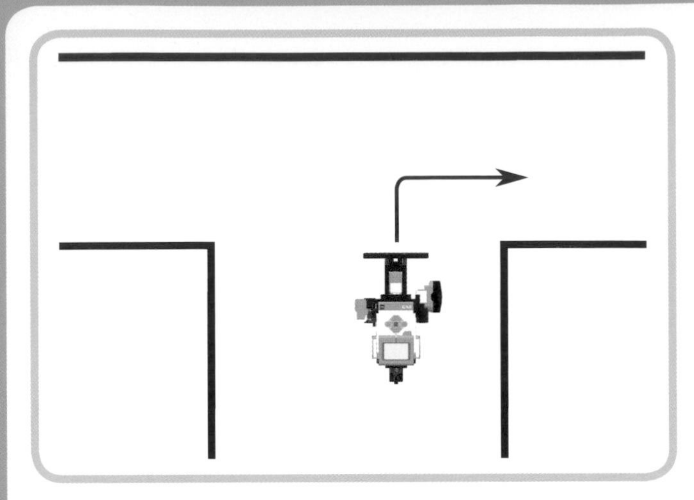

Figure 7-7: Turning right instead of turning left

Based on these situations, you can summarize the program requirements with the following three statements:

* The TriBot should move forward along the wall on its right, staying close to the wall.
* If the TriBot finds a wall in front and to the right, it should turn left 90 degrees and then follow the new wall.
* If the TriBot comes to an opening in the wall on the right, it should turn right 90 degrees and go through the opening.

assumptions

While coming up with the requirements, it's also a good idea to list any assumptions or restrictions about the program. This helps you decide which conditions you need to test and which you can ignore. You can make four assumptions about this program:

* The walls of the maze are straight.
* All openings are big enough for the TriBot to fit through.
* Walls meet at right angles. (This will simplify the code for turning at a corner.)
* At the start of the program, the TriBot will be placed with a wall to its right.

The last item in this list, which describes how the robot is arranged when the program starts, is called an *initial condition*. Starting the program with the robot in place will be much easier than making the robot wander around looking for the maze entrance.

Thinking about these assumptions before you start programming helps you to know which problems you need to solve and which you can ignore. Together with the list of requirements, this will determine what you do and don't expect the program to do.

If your final program ends up being more versatile than you initially planned, that's great! For example, one of the assumptions is that the walls are straight, but you may end up with a program that works with curved walls, too.

initial design

Next, we need to figure out how to make the robot perform each required task. The first task is to make the TriBot travel next to the wall while staying a short distance away from it. You can use the Infrared or Ultrasonic Sensor to detect how far the TriBot is from the wall. Then use Move Steering blocks to command the TriBot to move toward or away from the wall.

Because the wall will be next to the TriBot as it's moving, you need to mount the Infrared Sensor pointing to the side (instead of to the front). Follow the building instructions in "Alternate Placement for the Ultrasonic or Infrared Sensor" on page 31 to mount the sensor so that it points to the side, as shown in Figure 7-8.

Figure 7-8: The TriBot with the Infrared Sensor pointing to the side

Next, we need to make the robot respond correctly when it's following the wall to its right and then runs into a wall in front of it, as shown earlier in Figures 7-4 and 7-5. We'll use the Touch Sensor here. When the TriBot runs into the wall, the Touch Sensor will be pressed, at which point we can make the TriBot back up, make a quarter-turn to the left, and follow the wall it just ran into. (The *BumperBot* program performs similar behaviors, so you should have a good idea of how this will work.)

Finally, we need our program to detect and respond to openings in the wall to the right of the robot. For this, you'll use the Infrared Sensor to detect when the robot passes an opening. If the sensor suddenly reads a large distance while the TriBot is following a wall, you know that it has reached an opening.

We'll place the whole program in a Loop block so that the TriBot will keep moving until you manually end the program.

Now that we've thought through the high-level tasks, we can write pseudocode to describe how the program will work (see Listing 7-2). This listing gives a concise summary of the steps expressed in the previous few paragraphs and depicted in Figures 7-5 through 7-7. Because we are in an early stage of developing the program, the listing covers only the main points. In the following sections, we'll develop the details of each part of the EV3 program.

```
begin loop
  if too close to the wall (use Infrared Sensor) then
     drive forward, steering away from the wall
  else
     drive forward, steering toward the wall
  end if
  if Touch Sensor is pressed then
     back up a little to get room to turn around
     spin one quarter-turn to the left
  end if
  if an opening is detected (by the Infrared Sensor) on
     the right then
     spin one quarter-turn toward the opening
  end if
loop forever
```

Listing 7-2: Initial design for the WallFollower *program in pseudocode*

USING THE EDUCATION EDITION

The Infrared Sensor (from the Home Edition) and the Ultrasonic Sensor (from the Education Edition) can both measure the distance from the robot to the wall, so you can use either one for this program. The only difference will be the Threshold values we use in the program. Throughout the rest of this chapter, I'll refer to the Infrared Sensor in the figures and instructions, and I'll only mention the Ultrasonic sensor when you need to do something different for it.

The other notable difference between the Home and Education Editions is the size of the wheels, which impacts the Distance parameters used for the Move Steering blocks. I'll use the wheels from the Home Edition in the instructions and note any changes needed for using the wheels from the Education Edition.

following a straight wall

The first section of EV3 code will make the TriBot travel along the wall. You'll use a Switch block to choose between two Move Steering blocks: one that goes forward while steering toward the wall and one that goes forward while steering away from the wall. This block should keep the TriBot moving forward at a consistent distance from the wall. This approach is similar to the first version of the *LineFollower* program in Chapter 6, and it's a pattern that you'll often see when using a sensor to control a motor.

writing the code

To write this part of the code, first we need to decide what the TriBot's distance from the wall should be. The TriBot should stay close to the wall but still have enough room to turn when it gets to a corner. You can get a reasonable starting value by using the Port View and following these steps:

1. Put the TriBot on the floor next to the wall, with the Infrared Sensor facing the wall. Make sure there is enough room to spin the TriBot all the way around (see Figure 7-9).
2. Use the Port View window in the EV3 software or on the EV3 Brick, and note the reading from the Infrared Sensor.

I got a reading of 7, so that's what I'll use in the following instructions. Remember that the Infrared Sensor does not measure an exact distance, so your value may be different from mine.

Now we can start writing the program. We'll begin by putting the blocks together according to the instructions below, and then we'll fine-tune the settings after some testing:

1. Create a new project named *Chapter7*.
2. Create a new program named *WallFollower*.
3. Drag a Loop block onto the program. Use the default **Unlimited** mode.
4. Drag a Switch block into the Loop block.
5. Set the mode to **Infrared Sensor - Compare - Proximity**.
6. Set the Threshold parameter to the value you determined earlier (my value was 7).

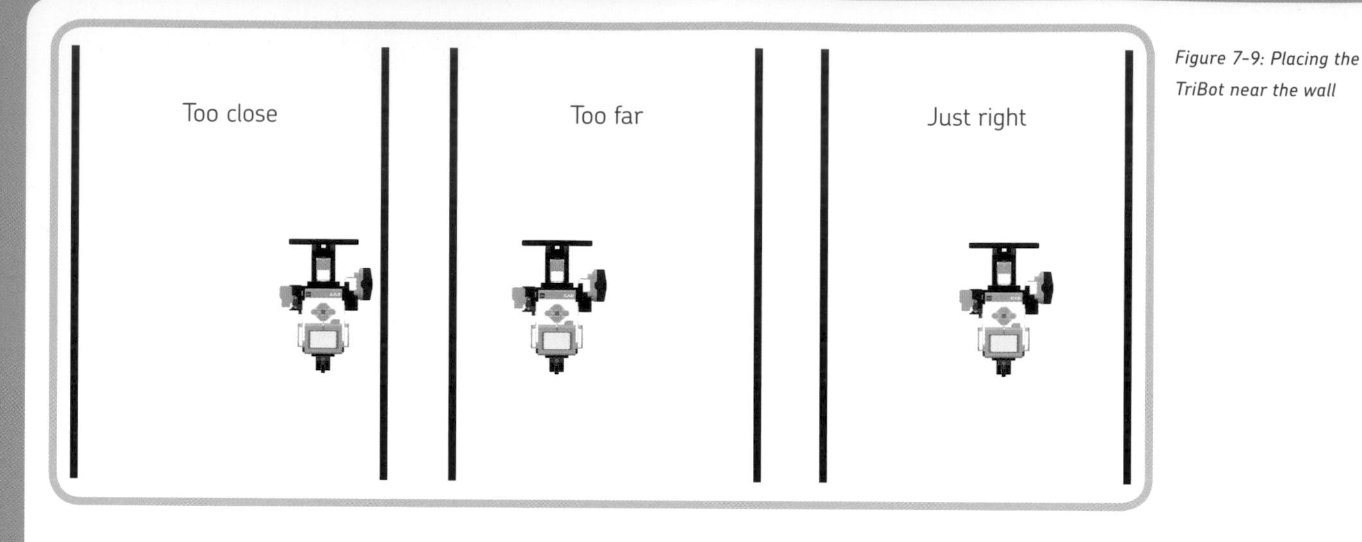

Figure 7-9: Placing the TriBot near the wall

NOTE If you're using the Ultrasonic Sensor, set the mode on the Switch block to Ultrasonic Sensor - Compare - Distance Inches (or Centimeters). Set the Threshold value to 5 inches (about 13 cm).

The Switch block should look like Figure 7-10.

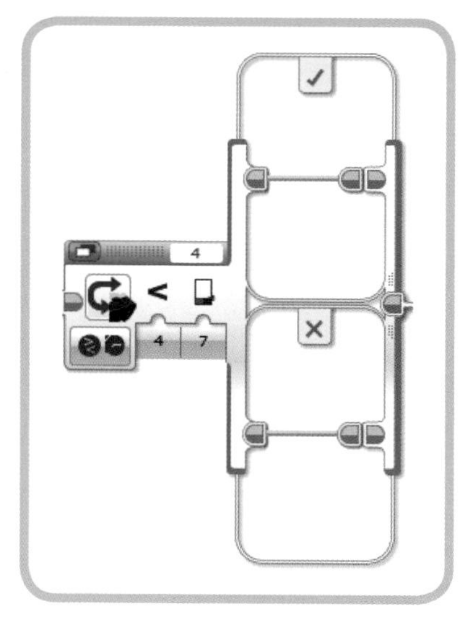

Figure 7-10: Switch block based on the distance from the wall

7. Drag a Move Steering block onto the Switch block's upper case.
8. Set the mode to **On** to keep the motors moving while the steering is adjusted each time through the loop.

The Move Steering block will run when the Infrared Sensor reads less than the Threshold value, meaning the TriBot is too close to the wall. In this case, the robot should steer to the left, away from the wall. Start with a Steering value of -10 to get a slight turn.

9. Set the Steering value to **-10**.

Add a similar block to the bottom case to move the robot toward the wall by using a positive Steering value.

10. Drag a Move Steering block onto the Switch block's lower case.
11. Set the mode to **On** and the Steering value to **10**.

At this point, the Switch block (Figure 7-11) should enable the TriBot to travel along a wall. The next step is to do a little testing and modify the program as needed.

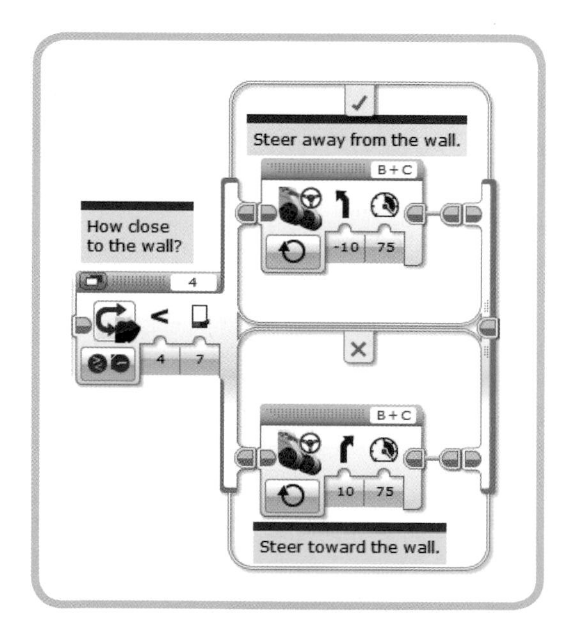

Figure 7-11: Following a wall

testing

Now let's do some testing to see how well this first part of the program works. Then we can make some adjustments to improve the program. It's difficult to get the settings just right on the first try, so it's important to test them out and adjust the program as necessary.

To test the *WallFollower* program, you need a wall with a corner and an opening, or you can use a full maze (which is a little more fun). To build a maze, you just need walls that are tall enough for the Infrared Sensor to detect. Boxes, piles of books, or large blocks of wood work well for the maze walls. You could even build a maze out of LEGO blocks!

When you test the program, you want to see if the robot runs into the wall or wanders too far away from it. If this happens, the Power parameter might be too high, so the robot moves too fast to adjust its path in time. Another problem could be that the steering is not turning sharply enough to keep the robot at the right distance from the wall.

Let's adjust both of these settings in order to make this part of our program more reliable. First, slow the robot down to make adjusting the steering a little easier.

12. Set the Power parameter on both Move Steering blocks to **25**.

Slowing the TriBot down in this way should help a lot, but we can improve the program even more by adjusting the Steering parameter to control the sharpness of the TriBot's turns. Test different Steering values to see how they affect the robot's motion, and remember to change the setting on both Move Steering blocks. Table 7-1 shows my results.

table 7-1: steering test results

Steering Value	Result
10	The TriBot does fine for a while but eventually runs into the wall because it's too close and doesn't respond quickly enough.
20	The TriBot stays close to the wall without hitting it. The motion is not perfectly smooth, but it's not bad.
30	The TriBot stays close to the wall without hitting it. The movement is very choppy, with a lot of side to side motion.

Based on my results, I suggest setting the Steering value to 20.

13. Set the Steering parameter of the block on the upper case to **-20**.
14. Set the Steering parameter of the block on the lower case to **20**.

Figure 7-12 shows the changes to the program. At this point, the TriBot should do a good job following a straight wall with no corners or openings.

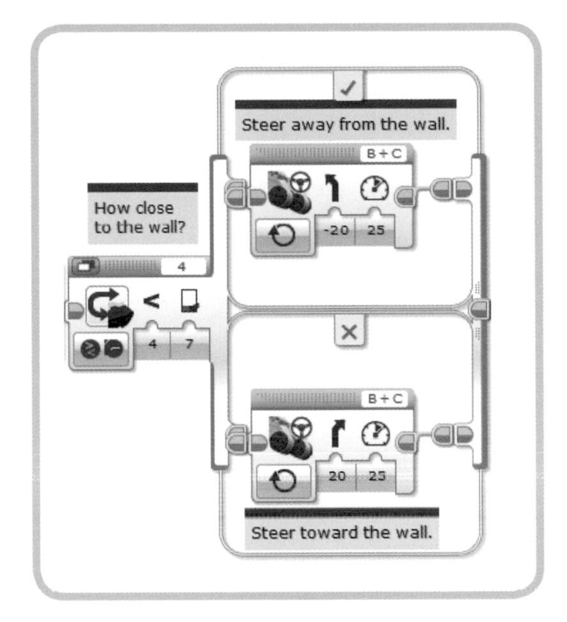

Figure 7-12: Settings for following a wall after testing

turning a corner

The next part of the program uses the Touch Sensor to detect when the robot reaches a corner, at which point it turns the TriBot to the left so that it can follow the new wall. This is similar to the *BumperBot* program, which also makes the TriBot back up and turn when it runs into something.

Listing 7-3 shows the pseudocode for this section of the program.

```
if the Touch Sensor is pressed then
  stop the motors
  back up far enough to turn the robot
  spin a quarter-turn to the left
end if
```

Listing 7-3: Pseudocode for turning a corner

After the TriBot backs up and turns, it needs to be positioned the correct distance from the wall so that it doesn't run into it or stray from it. To get the right positioning, you have to determine the correct Duration settings for the two Move Steering blocks. You can start with the values from the *BumperBot* program and adjust the values as needed after some testing.

writing the code

Follow these steps to add this section to the program:

15. Drag a Switch block into the Loop block, placing it to the right of the existing Switch block. The program should look like Figure 7-13. Keep the default settings (Touch Sensor in **Compare - State** mode). The blocks on the upper case of the Switch block will be used if the robot runs into a wall and the Touch Sensor is pressed.
16. Drag a Move Steering block to the upper case of the new Switch block. Set the mode to **Off**.
17. Add another Move Steering block to the upper Sequence Beam. Set the mode to **On for Degrees** and the Degrees parameter to **-300**; this block will make the TriBot back away from the corner.
18. Set the Power parameter to **25** to match the other Move Steering blocks. (The robot moves more smoothly when all the Move Steering blocks use the same Power parameter.)
19. Add another Move Steering block to the upper case. Set the Steering parameter to **-100**, the Degrees parameter to **250 degrees**, and the Power parameter to **25**. This block spins the TriBot so that the Infrared Sensor faces the new wall.

NOTE The Education Edition tires are bigger than the Home Edition ones, so you'll need to change some of the settings if you're using them. Use 185 degrees instead of 250 degrees for the Degrees parameter.

20. Set the Steering parameter to **-100** to make the robot spin to the left.
21. This Switch block uses only the upper case because the program doesn't need to do anything if the Touch Sensor is not pressed. Click the **Flat/Tabbed View** button on the Switch block to use Tabbed View, and hide the empty bottom case.

    Figure 7-14 shows the completed Switch block.

testing

To test the new code, start the TriBot close to a corner and see how it reacts when it bumps into the wall. Figure 7-15 shows roughly how the robot should move as it backs up and turns around.

My initial testing revealed a couple of flaws: The TriBot moves back too far, and it spins a little too far. After trying a few values, I settled on backing up for 150 degrees and then spinning for 210 degrees. Figure 7-16 shows the two Move Steering blocks with the new settings.

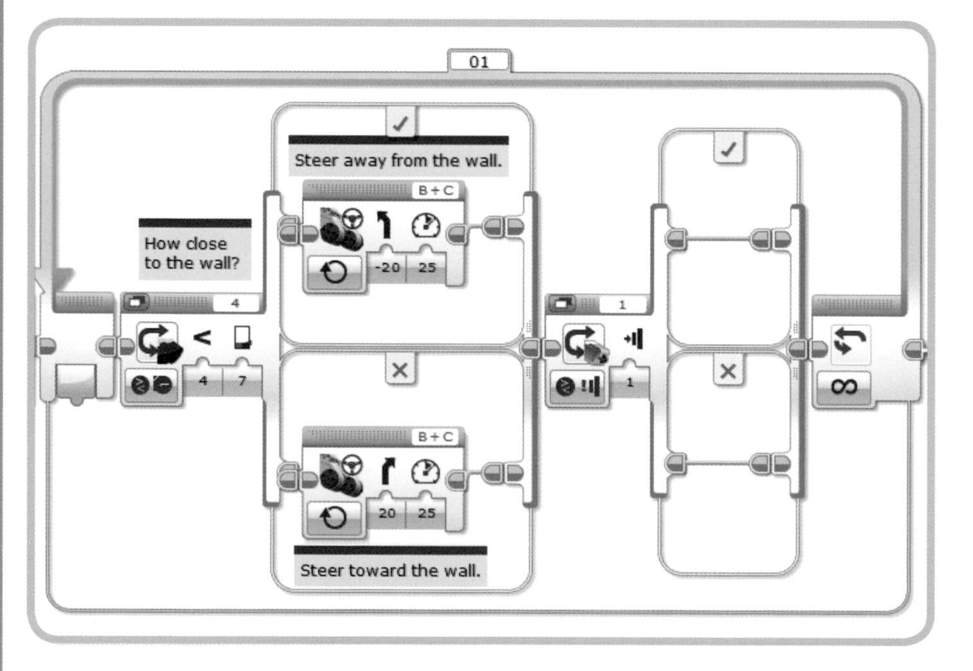

Figure 7-13: Adding another Switch block

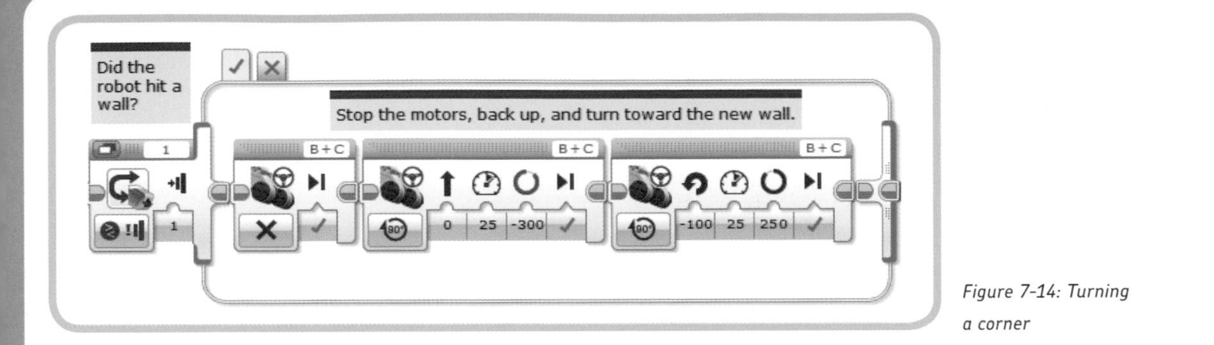

Figure 7-14: Turning a corner

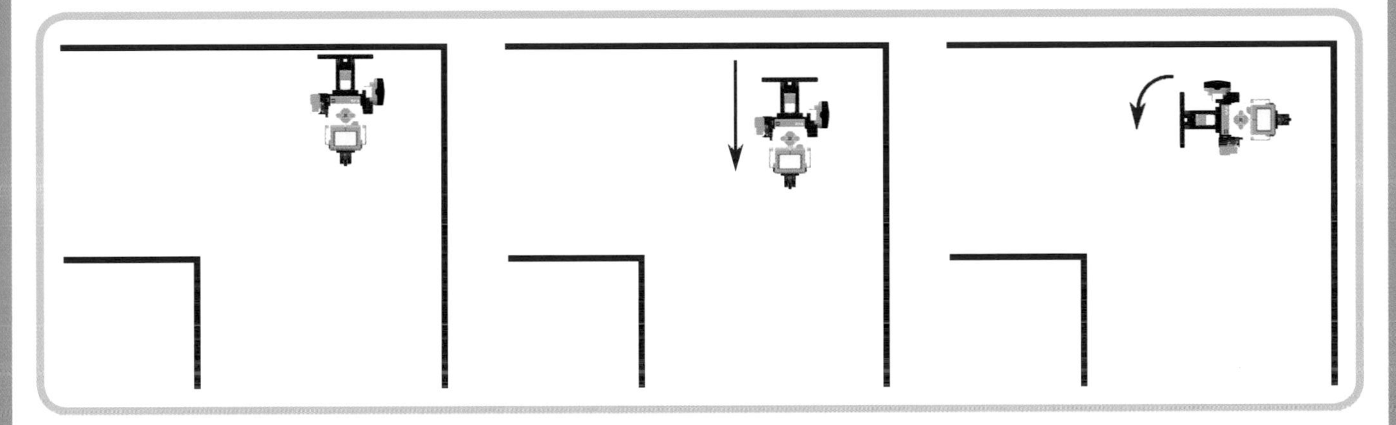

Figure 7-15: Backing away from the wall and turning to the left

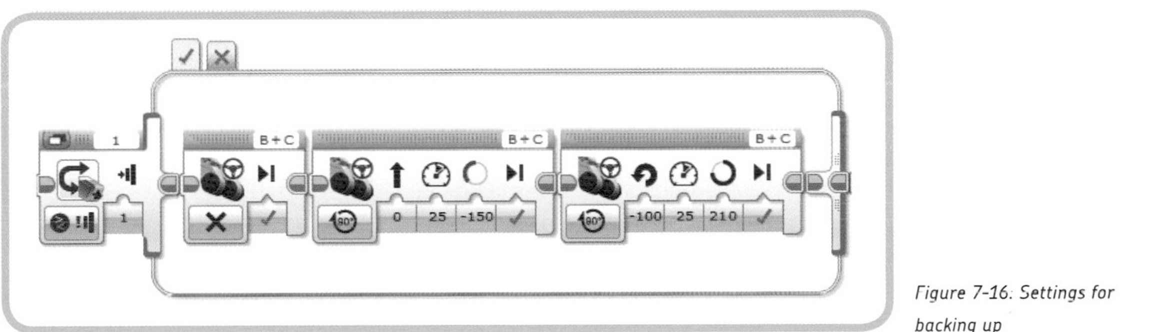

Figure 7-16: Settings for backing up

NOTE For the Education Edition tires, use 110 degrees for backing up and 160 degrees for turning.

Before moving on to the next section of the program, retest the wall-following code from earlier to make sure it still works as expected.

NOTE Whenever you add new code, it's a good idea to test the parts of the program that worked before you made changes. This makes it easier and quicker to find any bugs that may have accidentally been introduced by the new code.

going through an opening

When the TriBot comes to an opening in the wall on its right, it should turn to go through the opening and continue to follow the wall on its right.

When the Infrared Sensor faces an opening, it will suddenly read a much greater distance than it did when following the wall. The Infrared Sensor will reach the opening before the back half of the robot, so the robot needs to move forward a

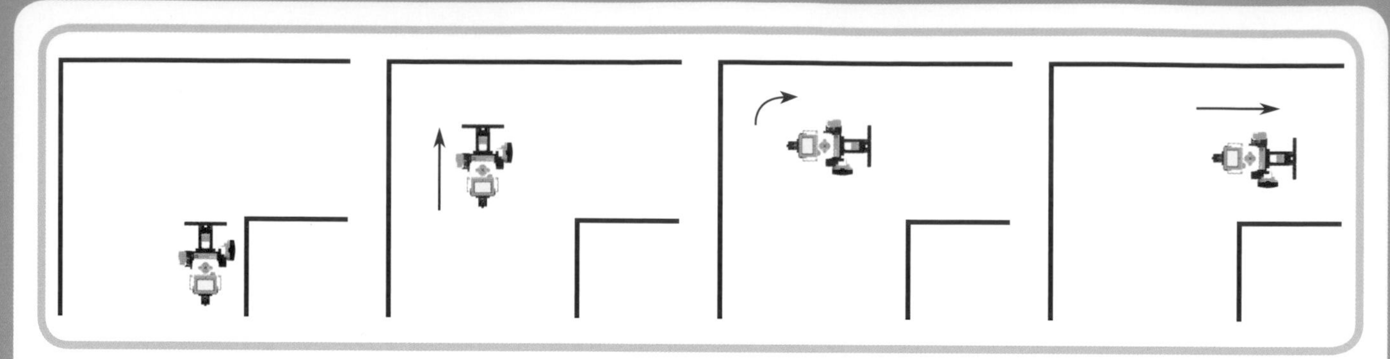

Figure 7-17: Turning and moving through the opening

little more before turning. After the TriBot has spun a quarter-turn toward the opening, the robot needs to move forward a little more before the Infrared Sensor will be next to the wall. Figure 7-17 shows how the TriBot should move, and Listing 7-4 shows the pseudocode for this section of the program.

```
if the Ultrasonic Sensor detects an opening then
  stop the motors
  move forward a little
  spin a quarter-turn to the right
  move forward into the opening
end if
```

Listing 7-4: Pseudocode for moving through an opening

Before adding any blocks to the program, you need to determine the values to use for the Switch block and the Move Steering blocks. For the Switch block, begin with a Trigger value of 15 inches (10 inches for the Ultrasonic Sensor), which should be big enough to make sure the TriBot has found a real opening and not just a small bump in the wall. For the Move Steering blocks, begin with the same value you used to back the TriBot away from the corner, and then adjust the value after some testing.

writing the code

Follow these steps to make the TriBot move through an opening:

22. Add a Switch block to the end of the code inside the Loop block.
23. Set the mode to **Infrared Sensor - Compare - Proximity**. Set the Threshold parameter to **15** and the comparison to **>** (greater than).
24. Click the **Flat/Tabbed View** button. You only need to add blocks to the true case.
25. Add a Move Steering block to the true case of the Switch block. Set the mode to **Off**.
26. Add another Move Steering block. Set the mode to **On for Degrees**, the Power to **25**, and the Degrees to **150**. This block will move the robot forward so that the whole robot is next to the opening.
27. Add a third Move Steering block. Set the mode to **On for Degrees**, the Steering parameter to **100**, the Power to **25**, and the Degrees to **210**. This block will spin the robot to the right so that it's facing the opening.
28. Add a fourth Move Steering block. Set the mode to **On for Degrees**, the Power to **25**, and the Degrees to **150**. This block will move the TriBot forward into the opening.

NOTE For the Education Edition tires, use 110 degrees for moving forward and 160 degrees for turning.

Figure 7-18 shows this part of the program.

testing

Test the new code by placing the TriBot along the wall near an opening and running the program. Carefully observe how the robot moves into the opening and starts following the new wall. Here are some things to look for:

* If the Threshold value for the Infrared or Ultrasonic Sensor is too large, the robot may not notice an opening.
* If the Threshold value for the Infrared or Ultrasonic Sensor is too small, the robot will turn toward the wall when there isn't an opening.
* If the Infrared or Ultrasonic Sensor has trouble detecting the wall because it's not a good surface for the sensor (too soft, too dark, etc.), the robot will turn toward the wall (as if it sees an opening).
* If the robot does not move forward far enough, it will turn too soon and hit the corner of the wall.
* If the robot moves too far forward before turning, it will miss the opening or end up too far from the wall after turning.

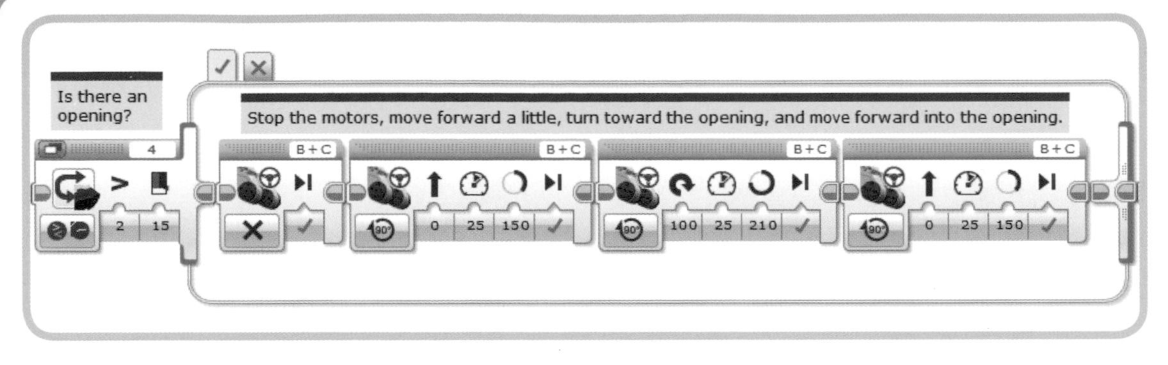

Figure 7-18: Turning into an opening

USING SOUND BLOCKS FOR DEBUGGING

Turning into an opening requires four Move Steering blocks. When you're testing and changing this code, it can be useful to know when one block ends and the next begins so that you'll have a better idea of which block to adjust if necessary.

It can be helpful to add Sound blocks playing different notes before each block so you can tell where each Move Steering block begins and ends. Be sure that the Sound block's Play Type is set to **Play Once** (instead of Wait for Completion) so that the TriBot doesn't stop while it plays the sound. Figure 7-19 shows a Sound block added between the first and second Move Steering blocks.

Whenever you add code for debugging purposes, such as a Sound block, be sure that the added code has as little influence as possible on the program's timing; otherwise, when you remove the debugging code, the program may act differently.

When you're done testing, you can remove the Sound blocks, but be sure to test the program again after removing them to make sure that it still works.

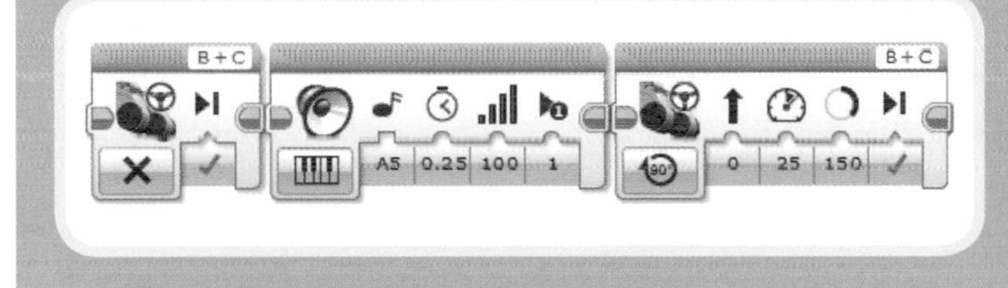

Figure 7-19: Making a sound before moving forward

* If the turn is too short, it will make the TriBot move too far from the wall.
* If the turn is too long, the TriBot will run into the wall.
* If the final move is too short, the TriBot won't enter the opening, and the Infrared Sensor won't detect the new wall.
* If the final move is too long, the TriBot might move past a tight corner or run into another wall.

My testing showed that the durations for all three moves were a little off. After some experimentation, I settled on Duration parameters of 300 degrees for the move forward, 190 degrees for the turn, and 350 degrees for the move into the opening. I also noticed that sometimes the TriBot acts as if there is an opening when there isn't one. Setting the Threshold value for the Infrared Sensor in the Switch block to 20 yields a big improvement.

NOTE If you're using the Education Edition, try setting the Durations to 225, 140, and 260 degrees and the Threshold value for the Ultrasonic Sensor to 13 inches (about 33 cm).

Figure 7-20 shows this section of the program with the final values.

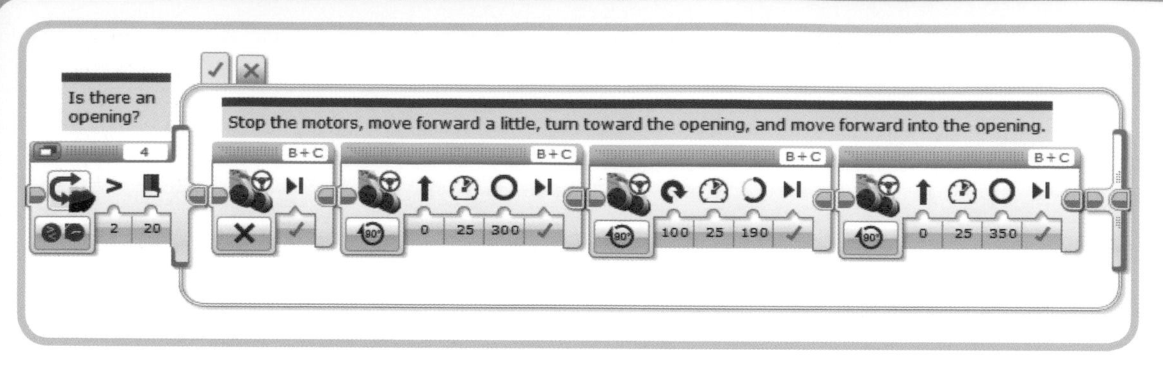

Figure 7-20: The settings for turning into an opening, after testing

final test

When the TriBot can move through an opening correctly, it's time to test the entire program. The requirements for this program (as listed in "Program Requirements" on page 82) state that the TriBot should be able to follow a straight wall and negotiate corners and openings. If the TriBot can do these things, then it should be able to navigate a simple maze. Try a variety of test mazes to see whether the program correctly handles several turns and corners in a row.

When you are sure that the program fulfills its requirements, see how it works in other situations. For example, adjust the spacing of the walls (make the corridors narrower or wider) to see how that influences the robot's behavior. You can also try adding some curved or slanted walls to see how the robot responds. Even though the program wasn't designed to handle these situations, it might work just fine. If not, think about how you might adjust the program to make it more versatile.

further exploration

After you have the program working well, try these activities:

1. If you can keep the TriBot at just the right distance from the wall, then you don't necessarily need a separate section of code for recognizing an opening. The robot will automatically move around the corner of a wall on the right using the code in the first section, which sets the Steering value based on the Infrared or Ultrasonic Sensor. The challenge is to find a Threshold value that keeps the TriBot at the right distance so that it can complete the turn without straying or running into the wall. Try different values to see if you can get this to work for your test area.
2. This program currently makes 90-degree turns because we said the maze would contain only right angles. But neither of these turns has to be perfect because the robot will soon straighten out based on the Infrared or Ultrasonic Sensor readings. Try a maze with corridors that meet at odd angles, and experiment with different turning angles to let the TriBot handle varied situations. Is it better to have the TriBot spin too far or not far enough?
3. The *LineFollower* program from Chapter 6 used nested Switch blocks to reduce side-to-side motion. The same technique can be applied to the code that makes the TriBot follow the wall. Make the necessary changes, see how smooth you can make the robot's motion while following the wall, and then see what effect, if any, this has on how well the program recognizes an opening on the right.
4. Instead of using the Touch Sensor to detect a wall, try using the Color Sensor and a brightly colored area on the wall at the intersection. Use different colors to give the program special directions, such as playing a certain sound when it sees red or changing speed or direction when it sees blue.

conclusion

In this chapter, I've taken you through the steps involved in developing a typical EV3 maze-navigating program. At each step, you added a logical piece of the program, did some testing, and then made necessary modifications to adjust the move durations and fix bugs.

The next chapter starts your exploration of data wires—one of the most powerful features of EV3 programming.

data wires

In this chapter, you'll learn how to use data wires to pass information from one block to another. Using data wires, you can change a block's settings while a program is running (by using data from a new sensor reading, for example). Data wires are one of the most powerful EV3 programming features, and learning how to use them will open up many more possibilities for your programs.

I'll begin with a simple example to show you what a data wire is and how it works. We'll then spend the rest of the chapter developing a more complex program that turns the TriBot into a sound generator. Along the way, I'll cover all the basic concepts you need to successfully use data wires in your own programs. I'll also introduce some new blocks that are particularly useful when working with data wires.

what is a data wire?

Most blocks require information—or *data*—to perform an action. For example, the Move Steering block needs to know which motors to use, how fast to move them, and for how long. This is called the block's *input data*. Up until this point, you've manually configured the input data for each block.

Some blocks create data that can be used by other blocks. This is called *output data*. For example, the Motor Rotation block reads data from the Motor Rotation Sensor and can provide that information to other blocks.

A *data wire* takes output data from one block and uses it as the input data for another block. This gives you a lot more flexibility than typing values for the parameters because it allows a block's settings to change while the program is running. For example, you could use the output from a sensor to control another block.

the GentleStop program

The *GentleStop* program is designed to show you how to use a data wire. The program moves the TriBot forward and then makes it gradually slow down and stop when it reaches a wall.

The key to this program is that the speed of the TriBot depends on how far it is from the wall, as shown in Figure 8-1. As the robot moves closer to the wall, its speed smoothly decreases until it stops. An Infrared Sensor block measures how far the robot is from the wall, and a Move Steering block uses this value for its Power parameter to control the robot's speed. Figure 8-2 shows how these two blocks connect in the program.

NOTE If you have the Education Edition, use the Ultrasonic Sensor and Ultrasonic Sensor block in place of the Infrared Sensor and block. This program works equally well with either sensor.

As the program runs, the Move Steering block's Power parameter adjusts according to the robot's distance from the wall. For example, when the sensor reads 80, the TriBot is relatively far from the wall, so the Power parameter is also 80 and the robot moves very fast. But when the sensor reads 20, the TriBot is getting close to the wall, so the Power parameter decreases to 20 and the robot moves much more slowly. The TriBot eventually bumps into the wall, pressing the Touch Sensor and stopping the program.

NOTE To measure the distance to a wall, the Infrared Sensor must be pointing forward. If the Infrared Sensor isn't pointing forward, move it to the position shown in Figure 8-3.

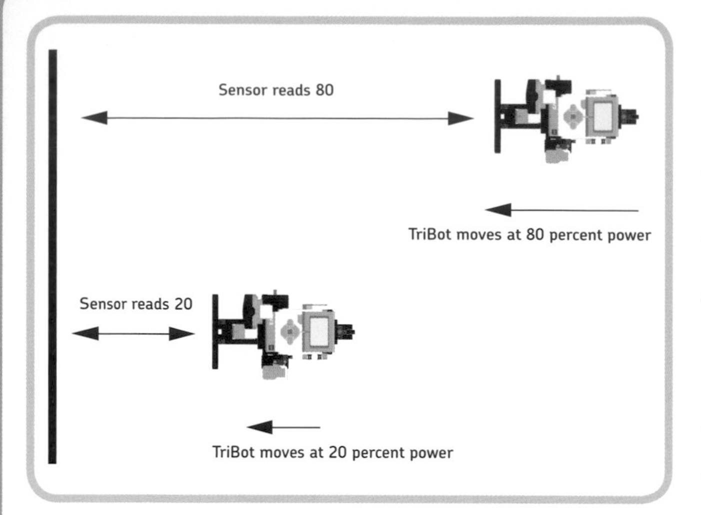

Figure 8-1: The TriBot slows down as it approaches the wall.

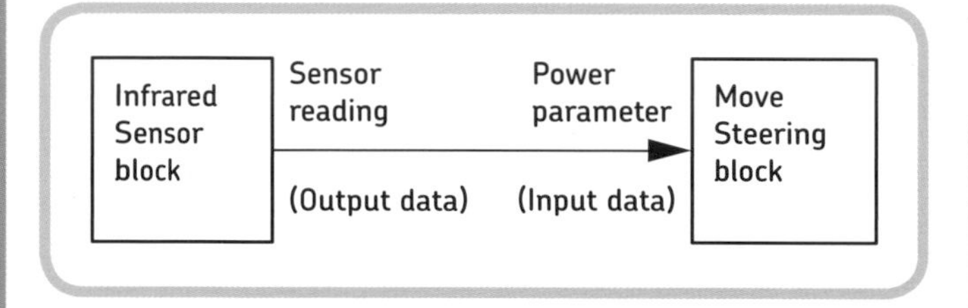

Figure 8-2: Combining the Infrared Sensor block and the Move Steering block

Figure 8-3: The Infrared Sensor pointing in front of the TriBot

writing the program

For this program, you'll place an Infrared Sensor block and a Move Steering block inside a Loop block. The loop will exit when the Touch Sensor is pressed so that the program stops when the TriBot gently bumps into the wall. Then you'll use a data wire to connect the Infrared Sensor output to the Power parameter of the Move Steering block.

The Sensor blocks (such as the Infrared Sensor block, Touch Sensor block, Color Sensor block, and so on) have the same modes and controls as the Wait, Switch, and Loop blocks. Each Sensor block has one or more output plugs, which you can use (with data wires) to control something other than a Wait, Switch, or Loop block. For example, the Infrared Sensor block shown in Figure 8-4 uses Measure - Beacon mode and has output plugs that report whether the beacon has been detected, the distance to the beacon, and the heading (the relative direction to the beacon). The Ultrasonic Sensor block shown in Figure 8-5 has a single output plug that measures the distance to the nearest detected object.

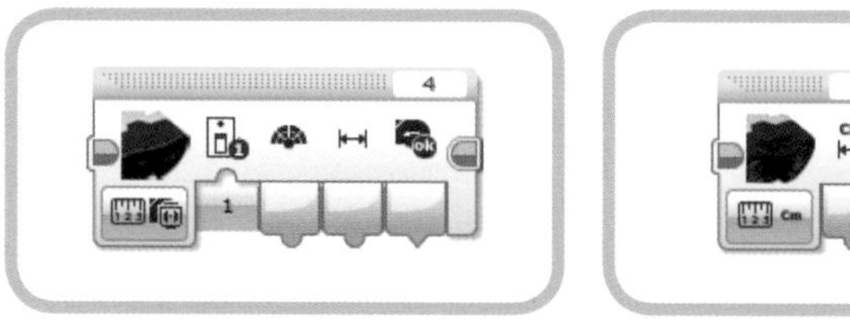

Figure 8-4: The Infrared Sensor block

Figure 8-5: The Ultrasonic Sensor block

Follow these steps to create the program:

1. Create a new project named *Chapter8*.
2. Create a new program named *GentleStop*.
3. Copy the blocks in Figure 8-6. The Move Steering block uses the **On** mode, and the Infrared Sensor block uses **Measure - Proximity** mode.

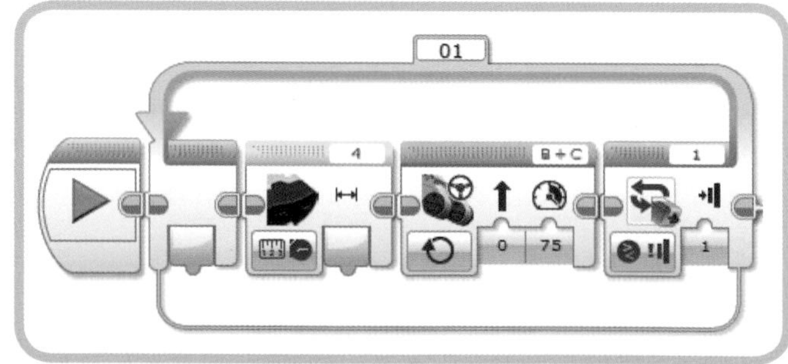

Figure 8-6: The GentleStop *program, before connecting the data wire*

NOTE **If you're using the Ultrasonic Sensor block, select the Measure - Distance Centimeters mode.**

The three blocks are in place, so you're ready to connect the data wire between the Infrared Sensor block and the Move Steering block.

In Figure 8-6, the Move Steering block has two block inputs: the Steering and Power parameters. The Infrared Sensor block as configured in Figure 8-6 doesn't have any inputs; the Proximity measurement is a *block output*. You can tell that this is an output because it has a half-circle on the bottom of the gray square (called an *output plug*). Notice that block inputs, like the ones on the Move Steering block, have the half-circle on the top of the square (called an *input plug*).

When the Infrared Sensor block runs, it will measure the proximity to the wall and make that value available in the Proximity block output. Next, we'll use a data wire to connect this output to the input plug for the Power parameter of the Move Steering block.

4. Move your cursor over the block output of the Infrared Sensor block. The cursor should change to a spool of wire, as shown in Figure 8-7.

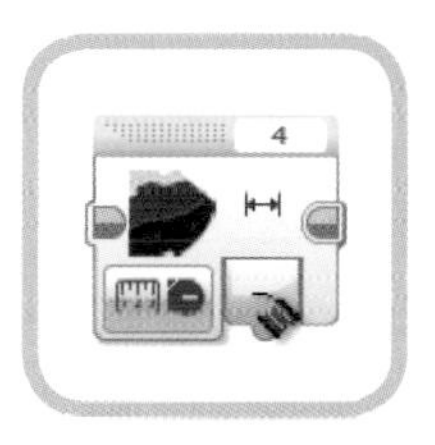

Figure 8-7: The cursor for drawing a data wire

5. Click the output plug. When you drag the output plug, a yellow wire is drawn between the cursor and the Infrared Sensor block, as shown in Figure 8-8.

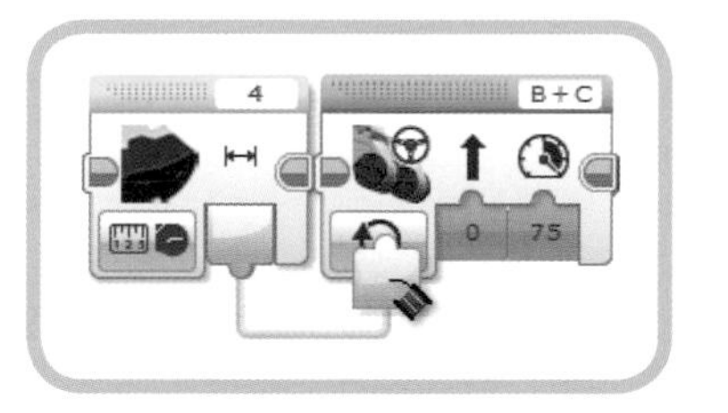

Figure 8-8: Drawing the data wire

6. Drag the output plug (and wire) to the Move Steering block's Power parameter to attach the wire. The program should look like Figure 8-9.

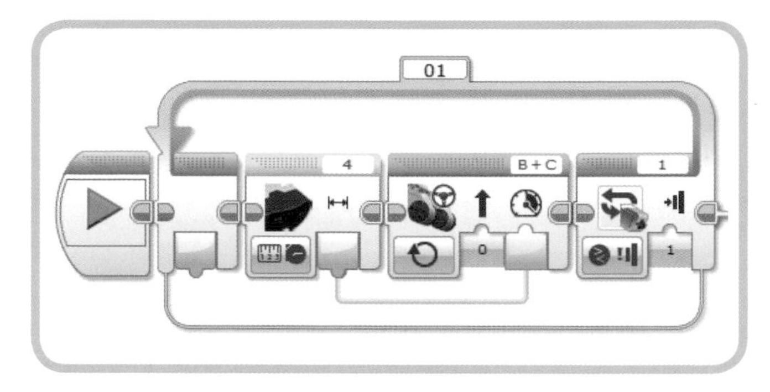

Figure 8-9: The GentleStop *program*

NOTE **If you accidentally connect the data wire to the wrong plug (which is easy to do), select Edit ▸ Undo (or press CTRL-Z) to erase the data wire and start again.**

Now it's time to download and test the program. Place the TriBot in the middle of the room, and point it straight at a wall. The TriBot should start off quickly and then slow down as it gets closer to the wall. It should come to a gentle stop when it hits the wall.

NOTE **Don't start with the TriBot very close to the wall. The motors may not move at all if the sensor reading is less than the minimum Power parameter needed to move the robot (which is 5 for my TriBot).**

tips for using data wires

Usually, connecting two blocks with a data wire is simple, as in our *GentleStop* program. However, occasionally you might want to delete a data wire or have more control over how the wire is drawn. Here are some tips to make life a little easier when working with data wires:

* Wait for the cursor to change to the wire spool before drawing the data wire.
* To erase a wire while drawing it, press ESC.
* You can erase a wire immediately after drawing it by selecting Edit ▸ Undo.
* To delete a data wire, drag it off the block input plug. (In the *GentleStop* program, this would be the Power parameter of the Move Steering block.)

* After a data wire is drawn, you can click and drag the wire up and down. This keeps the data wires organized when you are using several of them in the same part of a program.
* Double-click a data wire to have the EV3 software automatically choose the placement for the data wire.
* Deleting a block will also delete all data wires connected to it.
* While your program is running, you can move your mouse over a data wire to display the current value in the data wire, as shown in Figure 8-10. If there is no value (for example, if the block supplying the value has not run yet), it will display "----". This feature is incredibly helpful when debugging a program, but it only works when the EV3 Brick is connected to the software and the program is started by EV3 software (rather than by using the buttons on the Brick).

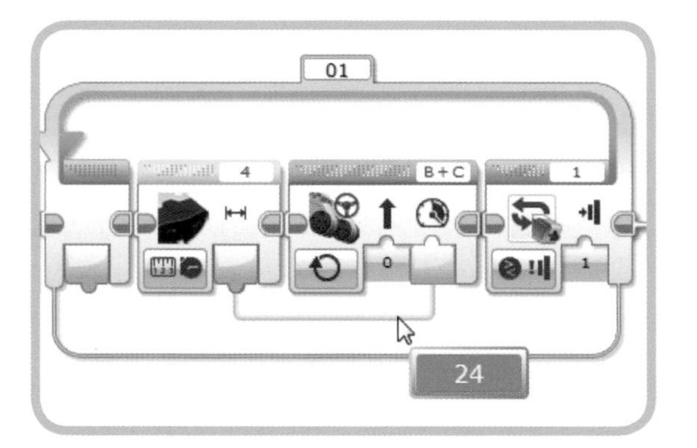

Figure 8-10: Showing the value on the data wire

* Each input plug can be connected to only one output plug (otherwise the block receiving the input wouldn't know which value to use). An output plug can be connected to more than one input plug because there's no problem with passing the same value to multiple blocks.

CHALLENGE 8-1

Mount the Color Sensor on the front of the TriBot as you did for the *LineFollower* program. Make a copy of the *GentleStop* program and use the Color Sensor in **Ambient Light Intensity** mode in place of the Infrared Sensor block. Using a flashlight, you should be able to guide the TriBot across the floor. The robot won't steer to follow the light, but it will move more quickly or more slowly depending on how close the light is to the sensor.

the SoundMachine program

The next program, *SoundMachine*, turns the TriBot into a simple sound generator. The wheel attached to motor B controls the volume, and the wheel attached to motor C controls the tone (or pitch). Turn motor B's wheel to make the sound louder or softer, and turn C's to make the sound higher or lower, as shown in Figure 8-11.

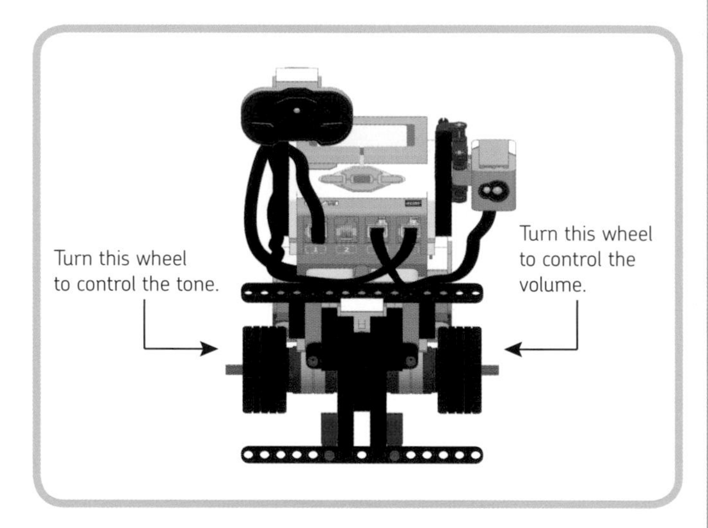

Figure 8-11: Using wheels to control the Sound block

The program uses a Sound block to create the sound and two Motor Rotation blocks to measure how far each wheel has turned. You'll use data wires to connect the output from the Motor Rotation blocks to inputs on the Sound block. I'll present the program in three parts. First you'll control only the volume, then you'll add some code to control the tone, and finally you'll display the Volume and Tone values on the EV3 screen.

controlling the volume

The first part of the program looks similar to the *GentleStop* program, except that it uses Rotation Sensor and Sound blocks instead of Infrared Sensor and Move Steering blocks. You configure the Sound block to play a tone, and use the value from the Rotation Sensor to control the volume. The entire program is contained in a loop.

Running a Sound block over and over too quickly can make the sound distorted. A small pause of 0.02 seconds at the end of the loop body fixes this problem. Listing 8-1 shows the pseudocode for this part of the *SoundMachine* program.

```
begin loop
   read the Rotation Sensor for motor B
   use a Sound block to play a tone; use the
      Rotation Sensor value for Volume
   use a Wait block to pause for 0.02 seconds
loop forever
```

Listing 8-1: Pseudocode for controlling the volume

Figure 8-12 shows the program before connecting the data wire. The Motor Rotation block is configured to use motor B, and the Sound block has the mode set to **Play Tone**. The Sound block's Play Type is set to **Play Once** so that the program won't pause while it plays the sound. As the loop repeats, the sound continues playing, the Rotation Sensor is continually checked, and the volume adjusts accordingly.

Now connect the Motor Rotation block's Degrees output to the Sound block's Volume input to complete this part of the program (Figure 8-13).

When the program starts, you won't hear anything because the Rotation Sensor will read 0, but turn motor B forward, and the sound should get louder.

The Sound block's Volume parameter uses values from 0 (no sound) to 100 (the loudest setting). The value from the Motor Rotation block is measured in degrees, and 100 degrees is a little more than a quarter-turn of the wheel. This means you can adjust the volume from the quietest to the loudest setting by turning the wheel just past a quarter of a rotation.

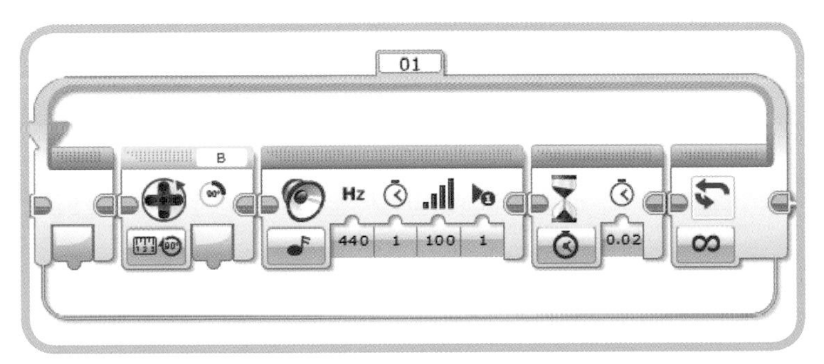

Figure 8-12: The SoundMachine *program before connecting the data wire*

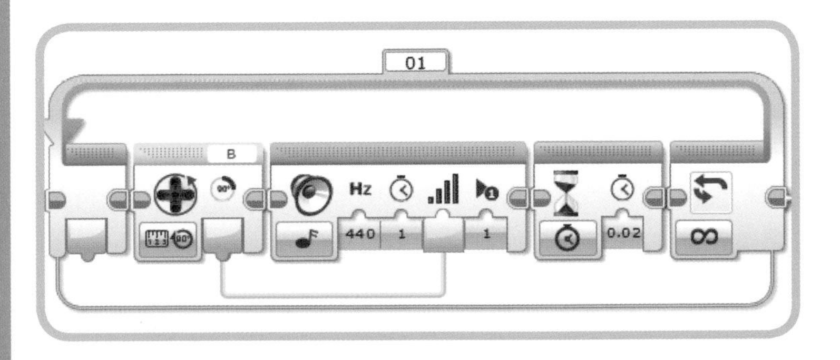

Figure 8-13: The wheel rotation controls the volume.

The Motor Rotation block reports a positive value when the motor is rotated forward and a negative value when a motor is rotated backward. Therefore, you must rotate the wheel forward for this program to work. Rotating the wheel backward will cause the Motor Rotation block to generate a negative number, and as a result, the Sound block won't make any noise because the Volume parameter will be 0. The range of the Volume parameter is 0 to 100, and any value less than 0 is treated as 0, while any value greater than 100 is treated as 100.

using the math block

For the next part of the *SoundMachine* program, you'll need to do a little math, which in EV3 programming is accomplished using the Math block, shown in Figure 8-14. You'll find the Math block on the red Data Operations palette.

The Math block takes one or two numbers as input and lets you configure an operation to perform on the numbers. You can enter the numbers yourself or supply them using data wires. The result of the operation is available in an output plug.

The mode of the Math block determines what operation is performed (Figure 8-15). The Absolute Value and Square Root modes only require one number, so when you select either of these, the second parameter (labeled *b*) will disappear. The Advanced mode allows you to evaluate more complex equations and is covered in Chapter 13.

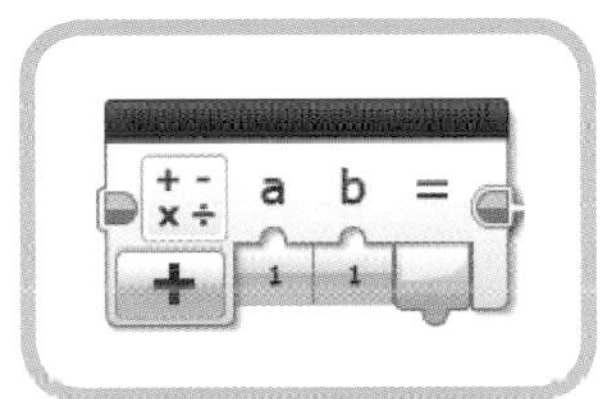

Figure 8-14: The Math block

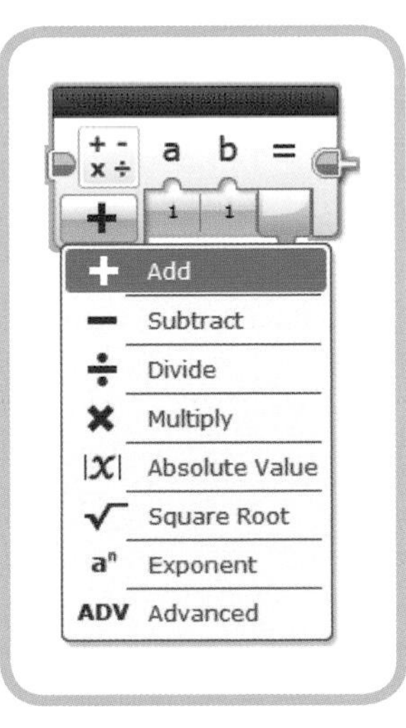

Figure 8-15: Selecting an operation for the Math block

adding tone control

You add tone control to the *SoundMachine* program using motor C as a dial. Then you use another Motor Rotation block and connect its output to the Sound block's Frequency input. The Frequency parameter takes a value measured in *hertz (Hz)*, with a range from 300 Hz (the lowest pitch) to 10,000 Hz (the highest pitch).

NOTE You can use the Context Help window or the full help file to learn about the values that a block input can accept. The Context Help window will show the range of values and a short description. The help file contains more detail and has a section for each block that includes a table describing all the block inputs and outputs. To open the Context Help window, press CTRL-H. To open the help file, press F1.

The range of values used for the Tone Frequency parameter is quite large, which makes controlling the Sound block's Frequency parameter more complicated than controlling the Volume parameter. If you connect the Motor Rotation block directly to the Frequency parameter as you did for the volume, then you'll need to turn the wheel about 27 full rotations to produce the highest pitch, which isn't very convenient.

You can solve this problem by using a Math block to multiply the value from the Motor Rotation block by 100 before passing the value to the Tone Frequency data plug. This converts Rotation Sensor values between 0 and 100 (the same range you used for the volume) into Tone Frequency values between 0 and 10,000. (Rotation values less than 3 produce Tone Frequency values that are less than the minimum value of 300, so the EV3 will use 300 instead. You can ignore this to keep things simple.)

Listing 8-2 shows the pseudocode for the program, with the new parts in bold:

```
begin loop
   read the Rotation Sensor for motor B
   read the Rotation Sensor for motor C
   use a Math block to multiply the motor C
      rotation by 100
   use a Sound block to play a tone; use the Rotation
      Sensor value for Volume; use the Math block
      result for the Tone Frequency
   use a Wait block to pause for 0.02 seconds
loop forever
```

Listing 8-2: Pseudocode for adding tone control

Figure 8-16 shows the program with the changes. The output from the Motor Rotation block is first passed to the Math block, where the value is multiplied by 100 and then passed to the Sound block's Frequency parameter. To make the program easier to read, I enlarged the Loop block by dragging the lower border down and pulled down the Volume data wire. It's much easier to understand a program if the data wires are not all jumbled together.

Now when you run the program, you should be able to control both the tone and the volume using the TriBot's wheels.

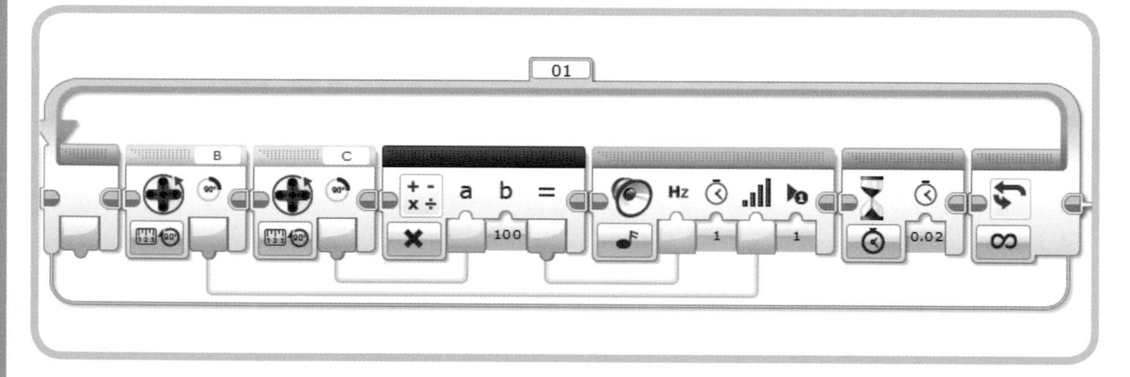

Figure 8-16: The SoundMachine *program with tone control added*

understanding data types

Before you finish the *SoundMachine* program, you need to learn about one more aspect of data wires. So far, all the data you've worked with (the Infrared and Rotation Sensor readings and the Power, Volume, and Frequency parameters) have been numbers. But numbers aren't the only type of information we can use with EV3; there are also Text values and Logic values. Think about how you'd answer the following three questions:

* What is your name?
* How old are you?
* Are you the oldest child in your family?

Each question asks for a different kind of information. The first answer is a word, the second is a number, and the third is either "yes" or "no." In computer programming lingo, different kinds of information are called *data types*. Your answers to the previous three questions correspond to the three data types that you'll use in EV3 programming:

* *Text values* are groups of characters that may include letters, numbers, and punctuation. For example, in the first program you created, you used the Display block to print the Text value "Hello." In EV3 programs, Text values are used mainly for displaying information on the EV3 screen.
* *Number values* are used to represent readings from sensors and to set Threshold values. Numbers are also used for many other block settings, such as the Move Steering block's Power and Steering parameters.
* *Logic values* can be either true or false. For example, you can use the Infrared Sensor block to see if the Distance value is greater than a Threshold value. The result of this comparison is a Logic value; it's true if the reading is greater than the threshold, and it's false if it isn't. Depending on how a Logic value is used, you may see it labeled as either True/False or Yes/No. This type of value is often called a *binary value* because it can have only one of two possible values.

There are actually two more EV3 data types: Number Array and Logic Array. An array holds a group of values. We'll discuss arrays in Chapter 15. Until then, we'll work with only Number, Text, and Logic values.

You can tell which data type a block input or output uses from its shape. An item that uses a Number value will have a half-circle, one that uses a Logic value will have a triangle, and one that uses a Text value will have a square. Data wires use different colors to indicate the data type: yellow for numbers, green for logic, and orange for text. Table 8-1 shows how the block outputs and data wires look for each data type. (Block inputs look the same as block outputs, except that the shape is on the top of the gray block).

table 8-1: block outputs and data wire colors by data type

Data Type	Block Output	Data Wire
Text		
Number		
Logic		

Each block input is meant for one particular data type, but there are a few automatic data-type conversions that the EV3 can perform. This lets you connect an output of one type to an input of another type in some situations. A Logic value output can be connected to a Number input, with true or false being converted to 1 or 0, respectively. Both Logic and Number values can be connected to a Text input, and the Number value will be converted to text. For example, the Number value 5.3 would automatically become the Text value "5.3". Conversions don't work in the opposite direction, though, so you can't use a Text value where a number is expected.

displaying the frequency and volume values

The next set of changes to the *SoundMachine* program displays the Frequency and Volume values on the EV3 screen. First, use a Display block to show the Frequency value.

1. Add a Display block after the Sound block. Set the mode to **Text Grid** (Figure 8-17).

   In the past when you used the Display block, you typed the text into the box at the top of the block. To tell the Display block that you want to use a data wire to supply the text instead, use the Wired option.

2. Click the box and select **Wired** (Figure 8-18). The block will show a new input plug for connecting a data wire (Figure 8-19).

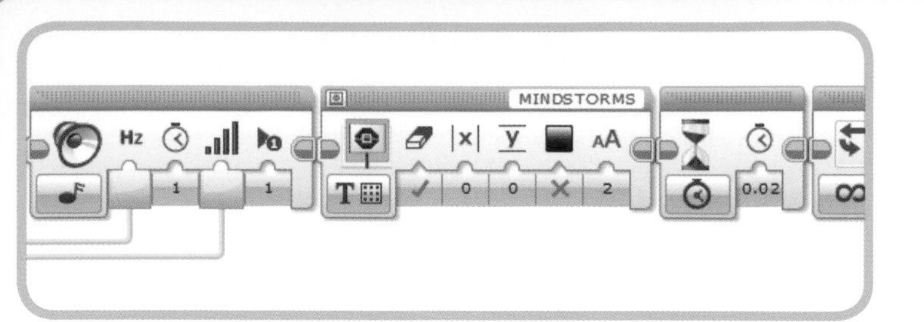

Figure 8-17: Adding the Display block

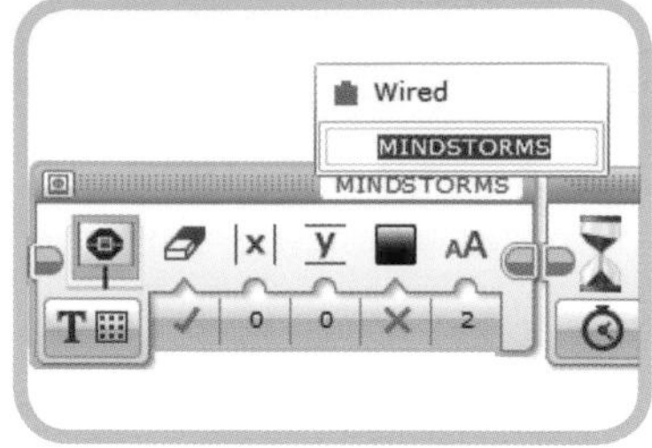

Figure 8-18: Selecting Wired for the text input

Figure 8-19: The Text input plug

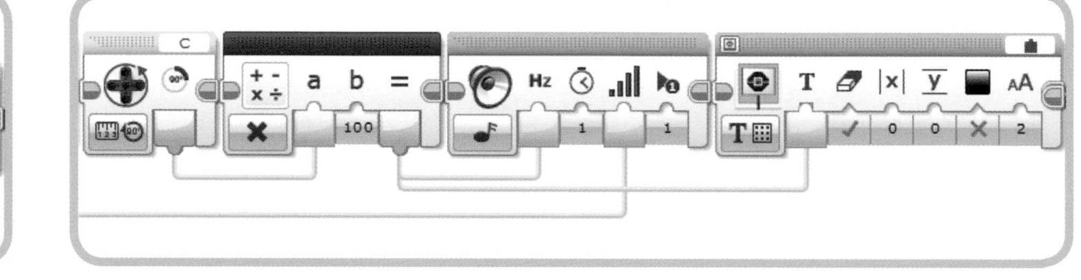

Figure 8-20: Connecting the Math block to the Display block

We want to display the Sound block's Frequency value, which comes out of the Math block. Connect the data wire from the Math block to the Display block. Even though the Text input normally expects a Text value, we can use the number from the Math block because, as mentioned earlier, the EV3 software can automatically convert numbers to text.

3. Connect the Math block's output to the Display block's Text input.

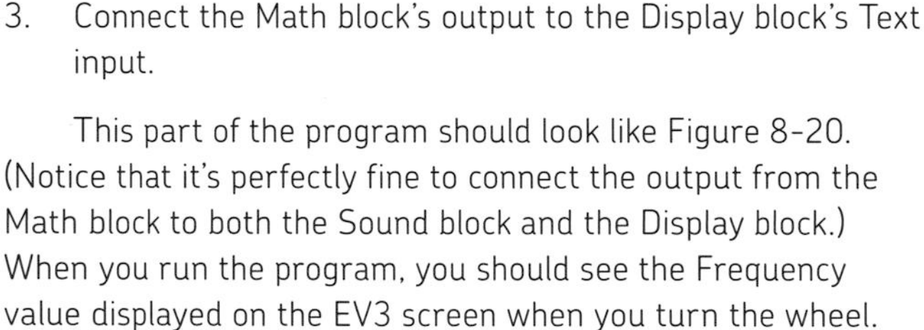

This part of the program should look like Figure 8-20. (Notice that it's perfectly fine to connect the output from the Math block to both the Sound block and the Display block.) When you run the program, you should see the Frequency value displayed on the EV3 screen when you turn the wheel.

using the text block

For the final set of changes to the program, we'll use the Text block shown in Figure 8-21, which lets you join up to three pieces of text. This can be useful for adding labels to values you display on the EV3 screen. You'll find the Text block on the red Data Operations palette.

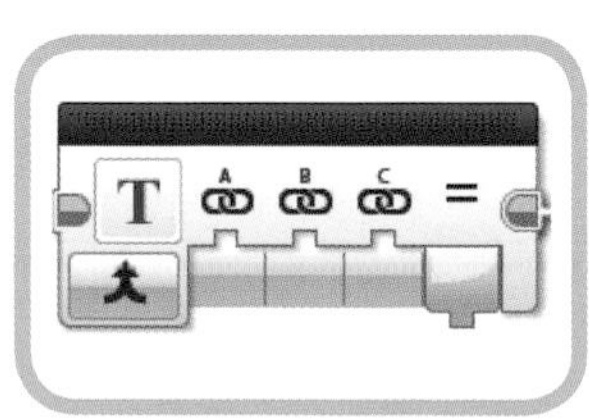

Figure 8-21: The Text block

The Text block has three input parameters that can be set to Text values. In most cases, you'll type into one or two of the boxes and supply the other input using a data wire. The output from the Text block is created by merging (or joining together) the three pieces of text. The Text block won't add a space between each item, so if you want a space between a label and a value, you need to add it yourself (as shown in the next section).

adding labels to the displayed values

The previous version of the *SoundMachine* program displayed the Frequency value on the EV3 screen. You can improve this by using a Text block to add more information. Instead of just displaying a number, you can display a label and the unit for the value. Instead of displaying "2500", the program displays "Tone: 2500 Hz".

Figure 8-22 shows the changes to the program. The output from the Math block is passed into the Text block as the *b* value, and the *a* and *c* values set the label and unit text, respectively. Although you can't see it in Figure 8-22, there is a space after the label. The value entered is "Tone: " not just "Tone:". Likewise, there is a space before Hz in the *c* parameter.

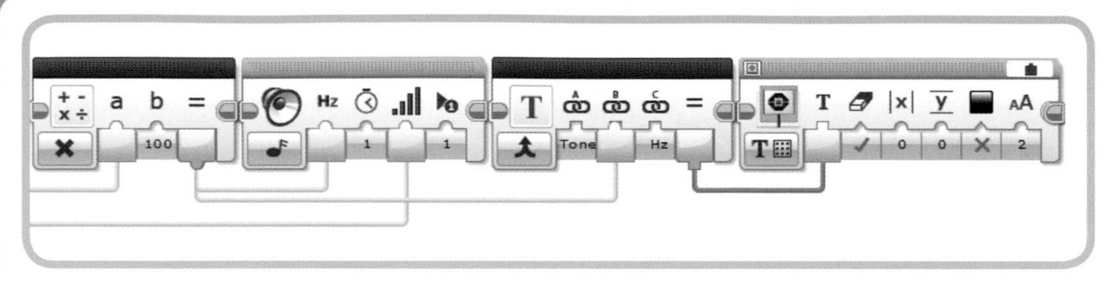

Figure 8-22: Adding a label to the Frequency value

Make the changes shown here and then test the program. You should notice that it works fine at the beginning, when the Frequency value is 0. However, as you increase the frequency, the full text will be too long to fit on the display, and the value and unit will get cut off. Luckily, this is easy to fix with the Font parameter.

The Display block's Font parameter gives you three choices for how the letters are drawn: Normal, Bold, and Large, which are set using the numbers 0, 1, and 2, respectively. The default setting is Large, which is easier to read but takes up more space. The Normal and Bold choices use letters about half the size of the default setting, so more text will fit. To make the Frequency value fit on the screen, configure the Display block to use smaller letters by setting the Font parameter to 1 (Figure 8-23).

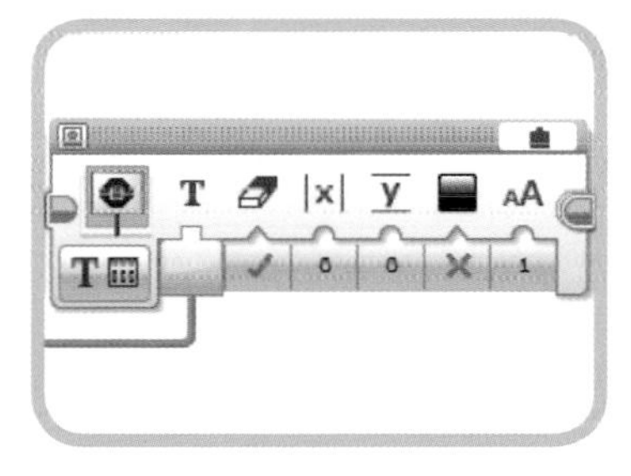

Figure 8-23: Selecting a smaller font

Now when you run the program, you should be able to see the Frequency value even when the number is large.

displaying the volume

The final change to this program adds the volume level to the display. The new code is similar to the code used to display the Frequency value, taking the output from the Motor Rotation block and displaying it as "Volume: 50%".

Figure 8-24 shows the new Text and Display blocks. Just as before, be sure to enter a space after the volume label (though you don't need a space before the percent sign). For the new Display block, after setting the mode to **Text Grid**, change the Row parameter so that the volume is shown below the frequency (instead of writing over it), and set the Clear Screen parameter to **False** so that the tone display is not erased.

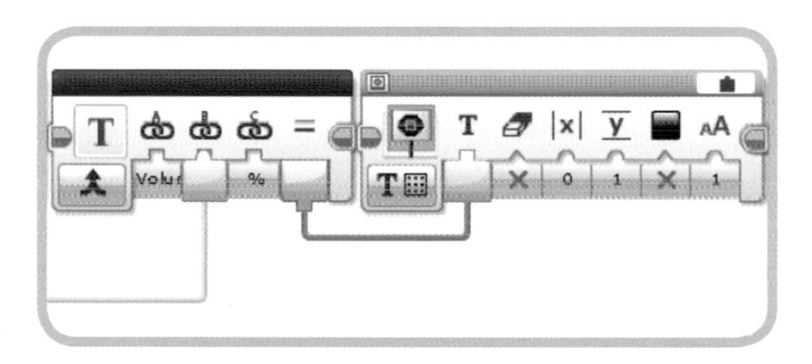

Figure 8-24: Displaying the volume

The completed program is shown in Figure 8-25. When you run the program, both the Frequency and Volume values should be displayed with appropriate labels and units.

CHALLENGE 8-2

Modify the *SoundMachine* program so that you can rotate the wheels in either direction. A simple way to accomplish this is to pass the value from the Rotation Sensor through a Math block with the mode set to Absolute Value.

CHALLENGE 8-3

The tone doesn't start changing until you move the wheel a little because the minimum Tone value is 300. Fix this issue by adding a Math block to the program that ensures the Tone parameter is always at least 300.

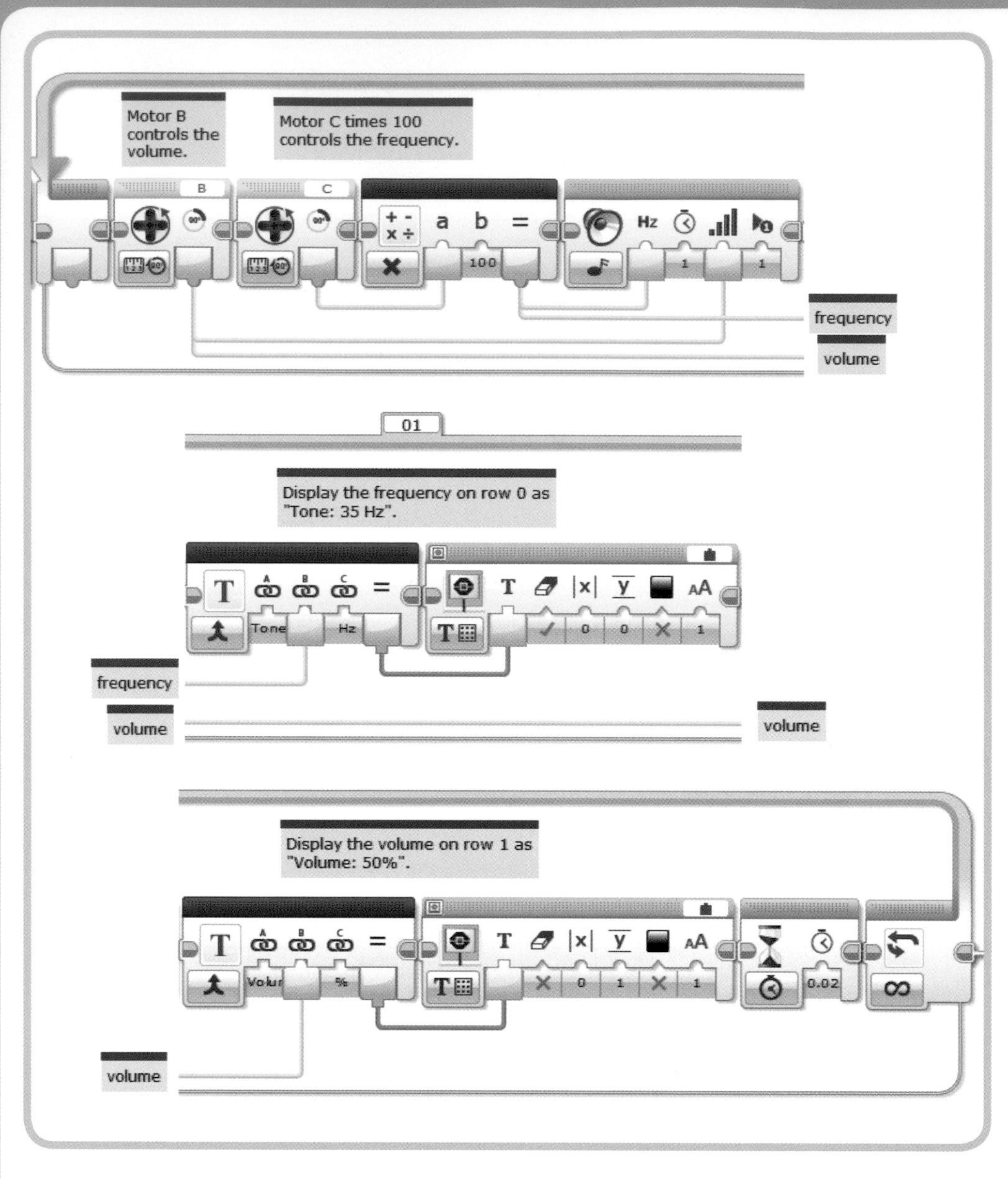

Figure 8-25: The complete SoundMachine *program*

CHALLENGE 8-4

Rotating the wheel 100 degrees (a little more than a quarter-turn) will cover the entire range of Tone values. This makes it easy to make large changes but difficult to make small changes. You can achieve a finer degree of control by changing the multiplier used in the Math block. A smaller value there will make it so that rotating the motor adjusts the tone frequency more slowly, which lets you make more precise changes. Experiment with different values to see what works best.

further exploration

Try these activities to gain more experience using data wires:

1. Run the *SoundMachine* program with the EV3 connected to the software. Hover the mouse over the data wires to display the values while the program is running. Think about how you could use this feature while testing and debugging.
2. The Motor Rotation block's Current Power motor can measure how fast a motor is moving. Try connecting the Current Power value from one motor to the Power para meter for a second motor to make the second motor follow the motion of the first motor. For example, connect the Lift Arm to the EV3's A port, and write a program that uses the Current Power from the B motor to control the Power parameter of a Medium Motor block. When you move the wheel attached to motor B, the Lift Arm should follow the same movement, like a remote-controlled tollgate.
3. Create a new program that makes the TriBot follow the Infrared Beacon (available in the Home Edition). You can do this by modifying the *GentleStop* program to use the Infrared Sensor in Measure - Beacon mode and connecting the Beacon Heading to the Move Steering block's Steering control. (Think of how the Color Sensor is used in the *LineFollower* program.)

conclusion

Data wires move information between blocks, allowing you to change a block's settings while your program is running. The programs presented in this chapter have shown you some simple ways to use data wires and have introduced some of the blocks that are designed to work with data wires. The Sensor blocks use data wires to make sensor readings available to other blocks in your program. The Math and Text blocks are used almost exclusively with data wires to transform data or convert between data types.

The next two chapters will show you how to use data wires with the Switch and Loop blocks. You'll get lots of practice using data wires because you'll use them extensively for the programs in the rest of this book.

9

data wires and the switch block

You can use the Switch block to make decisions about which blocks to run by choosing between two or more alternatives. For example, the *WallFollower* program in Chapter 7 uses a Switch block that reads the Infrared Sensor and decides whether to move the TriBot toward or away from the wall. In this chapter, you'll learn how to use the Switch block to make decisions based on a value supplied by a data wire. You'll also learn how to pass data between the blocks inside the Switch block and the ones that come before or after the Switch block.

the switch block's value modes

Up until now, you've used the Switch block to take a sensor measurement and make a decision based on that measurement. The Switch block can also make a decision based on a value supplied by a data wire. The *Text*, *Logic*, and *Numeric modes* (Figure 9-1) let you create cases to match different possible values passed in to the Switch block from a data wire. The only difference between these three modes is the data type of the value used.

In each of these modes, the Switch block will have a single parameter, which is usually supplied by a data wire. When the block runs, it will choose which case to use based on the value from the data wire. For example, Figure 9-2 shows a Switch block in Numeric mode, with cases for the values 5, 7, and 9. When this block runs, it looks at the value on the data wire. If the value is 5, it chooses the top case; if the value is 7, it chooses the middle case; and if the value is 9, it chooses the bottom case. For other values, the top case is used because it's marked as the default.

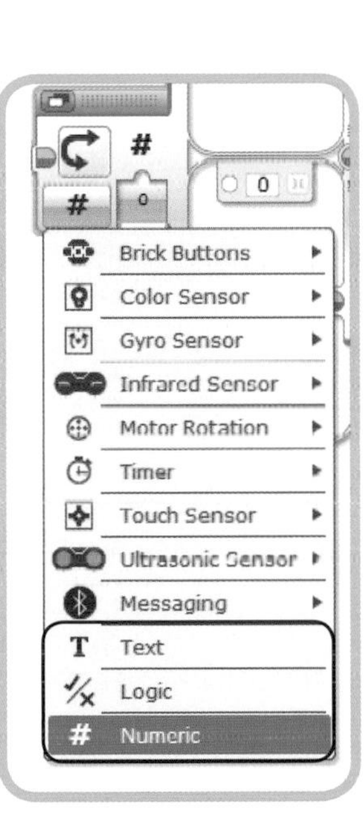

Figure 9-1: The Switch block's value modes

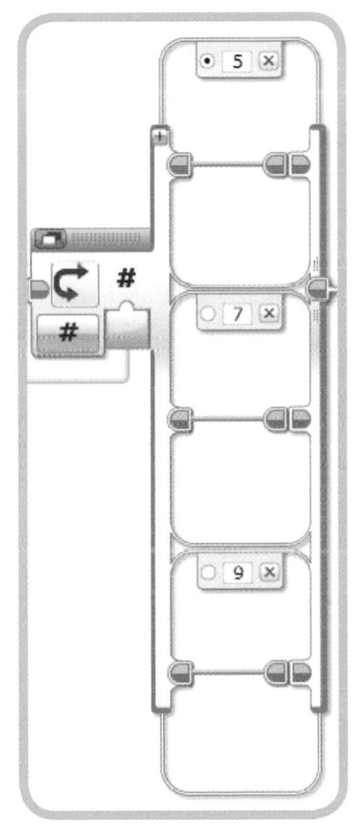

Figure 9-2: Selecting the case using a number

The values in Numeric mode must be integers (whole numbers). For example, if you type in 5.25 to identify a case, the value will be changed to 5. When the Switch block runs, it takes the number on the data wire and rounds it down to the nearest whole number. So in the example above, the middle case will match any value between 7 and 8, including 7 but not including 8.

Text mode works the same way as Numeric mode, except that the values you enter for each case are letters or words instead of numbers. For Numeric or Text mode, you can create as many unique cases as you need. There are only two Logic values, True and False, so Logic mode always has exactly two cases—no more and no less.

rewriting the GentleStop program

Recall that the *GentleStop* program makes the TriBot slow down and gently stop as it approaches and bumps into a wall. In our first version of the program, the robot stops when the Touch Sensor is pressed, but we could improve the program by making the motors stop when the robot is very close to the wall, before the Touch Sensor is bumped. In this section, you'll rewrite the *GentleStop* program to do just that, using the output from the Infrared Sensor block as the trigger for a Switch block.

NOTE For the Education Edition, use the Ultrasonic Sensor and Ultrasonic Sensor block in place of the Infrared Sensor and block. The program works equally well with either sensor.

We'll start by making the TriBot move forward with the Power set to 75 and then stop when the reading from the Infrared Sensor is less than 20. Then we'll add a data wire so that the Infrared Sensor block controls the Power parameter, as in the original program. This makes the robot move forward (with a Power parameter matching its distance to the wall) until the Infrared Sensor measures a distance of 20 or less. At this point, the robot stops.

Listing 9-1 shows the pseudocode for the program, which keeps the TriBot moving forward until the Infrared Sensor reads less than 20, at which point the motors are stopped:

```
begin loop
   read the value from the Infrared Sensor
   if the distance > 20 then
      move forward, use the Infrared Sensor measurement
      for the Power parameter
   else
      stop moving
   end if
loop until the Touch Sensor is pressed
```

Listing 9-1: Pseudocode for rewriting the GentleStop *program*

Start by adding and configuring the blocks as shown in Figure 9-3. The Infrared Sensor block uses the Compare - Proximity mode to read the sensor and compare the proximity reading to the Threshold value to determine if it's greater than 20. The result of this comparison is a Logic value, so set the Switch block to Logic mode to make its decision based on this value. The comparison result is passed to the Switch block using a data wire connected to the Infrared Sensor block's Compare Result block output. If the value is true, meaning the distance to the wall is greater than 20, the Switch block executes the Move Steering block on the upper case, moving the robot forward. If the distance is 20 or less, the Switch block uses the Move Steering block on the lower case, and the robot stops moving.

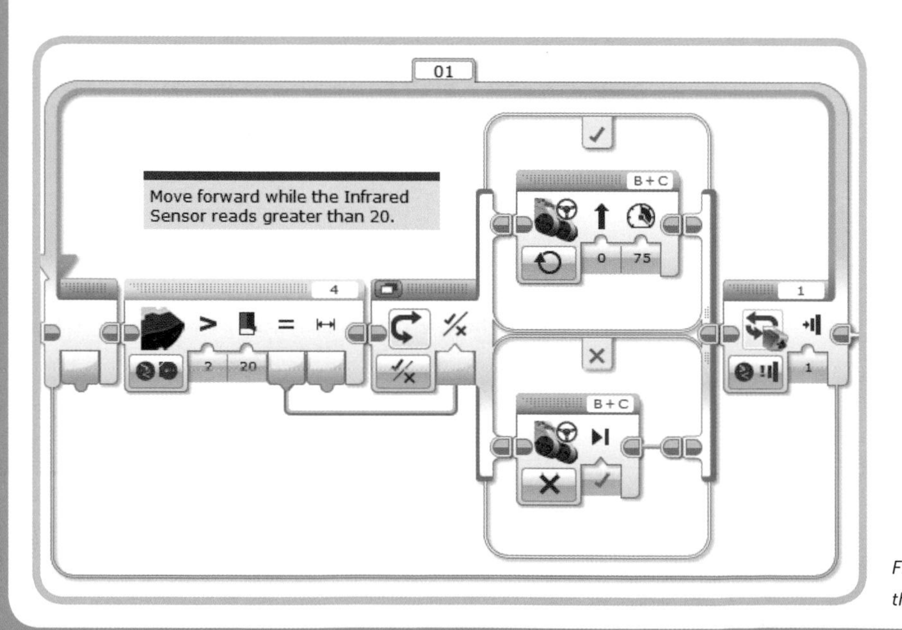

Figure 9-3: Moving forward while the distance is greater than 20

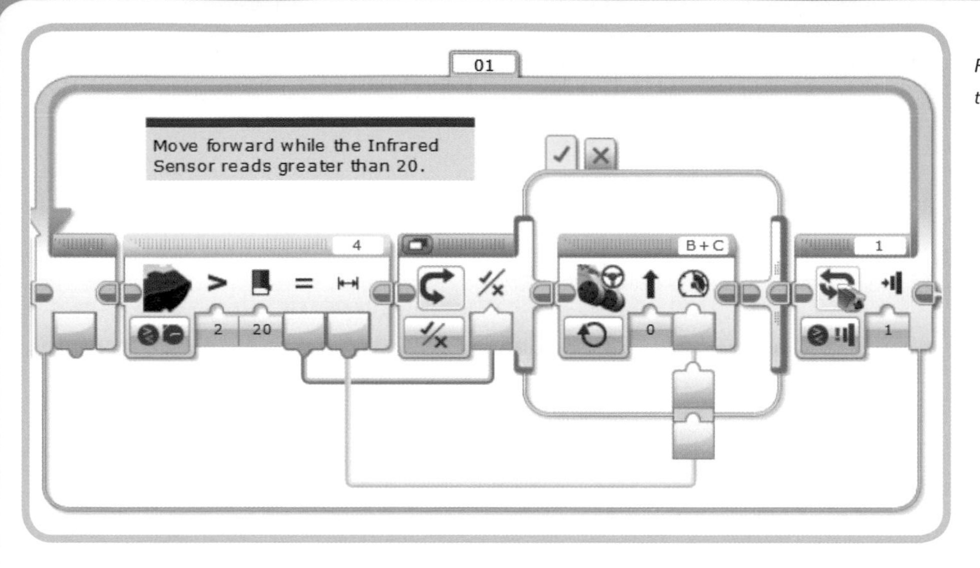

Figure 9-4: Connecting the Proximity to the Power parameter

NOTE For the Ultrasonic Sensor block, select the *Measure - Distance Centimeters* mode, and set the Threshold value to 15.

Next, we connect another data wire to one of the cases in the Switch block, so we can use the Proximity measurement to control the Power parameter.

passing data into a switch block

Now we're going to use a data wire to make the TriBot's speed depend on its distance from the wall. To connect a data wire from outside the Switch block to a case inside the Switch block, you must put the Switch block in Tabbed View by clicking the Flat/Tabbed View button. When the Switch block is in Tabbed View, you can draw a data wire between the Infrared Sensor block's Proximity output and the Motor Steering block's Power input, as shown in Figure 9-4.

When you drag the data wire across the Switch block's boundary, two block input boxes—called a *tunnel*—are created: one outside the boundary and one inside. The block input appears on the inside of the Switch block for each tab, and you can choose whether or not to use it in each case. For the *GentleStop* program, the proximity reading is not used in the case that stops the motor (Figure 9-5), so the block input remains unconnected there.

When you run this version of the program, it should act like the original except that the TriBot should stop the motors when it gets close to the wall instead of bumping into it. One other difference is that the original program ended when the robot hit the wall and pressed the Touch Sensor. In this version, the program keeps running after the robot has stopped moving because the Touch Sensor never gets pressed. We'll address this issue in the next chapter.

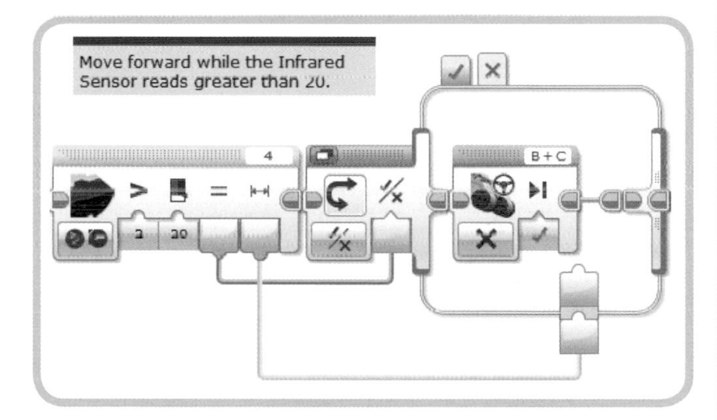

Figure 9-5: The data wire is not used for the false case.

advantages of using a sensor block

In this version of the *GentleStop* program, we used an Infrared Sensor block to take a measurement, compare it to a threshold, and pass the result to the Switch block instead of using just a Switch block in Infrared Sensor-Compare-Proximity mode to make the measurement. There are several advantages to using a Sensor block with data wires like this rather than just using a Switch block:

* A Sensor block gives you access to the sensor reading as well as the result of the comparison. We just took advantage of this in the *GentleStop* program, where we used the Proximity value to control the Move block's Power parameter, in addition to using the Comparison Result to stop the robot.
* You may want to use a condition that's more complex than a greater than or less than comparison. Using data wires and the Math, Logic, and Comparison blocks (covered in Chapter 13), you can test for just about any condition you can think of.
* The value used by the Switch block can also be passed to other blocks in the program and used to control other behavior.

passing data out of a switch block

Now let's build the *LogicToText* program shown in Figure 9-6 so we can see a simple example of passing data out of a Switch block. This program reads the state of the Touch Sensor and writes "True" on the display if the button is pressed and "False" if it's not. Both cases of the Switch block contain a Text block with an output plug connected to the Display block. Note that the Text blocks in both cases need to be connected to the Display block.

To build this program, start by configuring the blocks as shown in Figure 9-7. The Touch Sensor block uses the default Measure State mode, and the result is used as the trigger value for the Switch block. The Text block on the Switch block's upper case generates the text "True", and the one on the lower case generates "False". To show the text, set the Display block's mode to **Text Grid**, and select **Wired** in the box where the text normally goes.

Now connect the output from the Text blocks to the Display block by following these steps:

1. Click the **Flat/Tabbed View** button of the Switch block. The Switch and Display blocks should look like Figure 9-8.
2. Draw a data wire from the Text block's Result output to the Display block's Text input (Figure 9-9). Notice that two block output boxes are created where the data wire crosses the Switch block boundary.
3. Click the **X** tab at the top of the Switch block to display the other case.
4. Connect the Text block's Result to the block output box on the boundary of the Switch block. This part of the program should now look like Figure 9-10.

When the Switch block runs, it uses only one of the tabs (either the true case or the false case), and the corresponding output of the Text block is sent to the Display block. Try running the program; you should see the display print "True" when the Touch Sensor button is pressed and "False" when it's released.

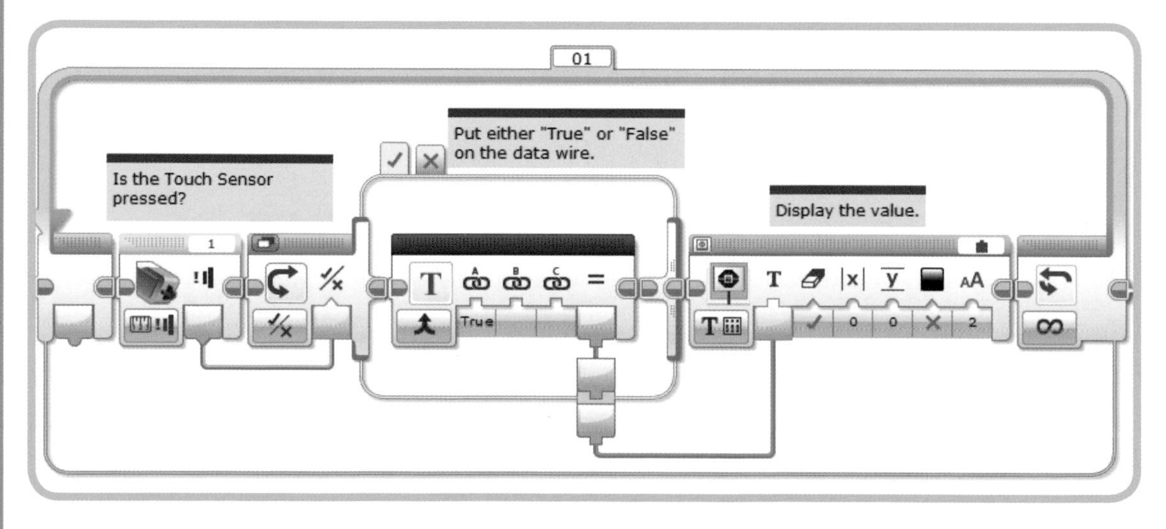

Figure 9-6: The LogicToText *program*

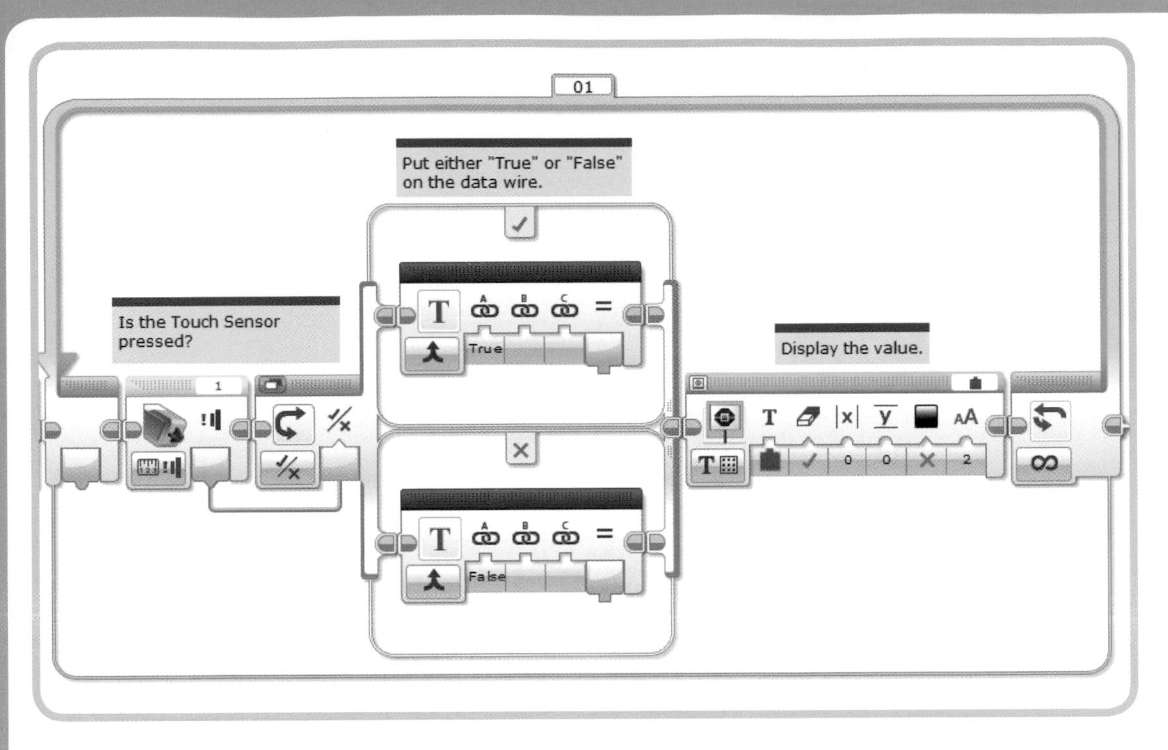

Figure 9-7: The LogicToText program starting point

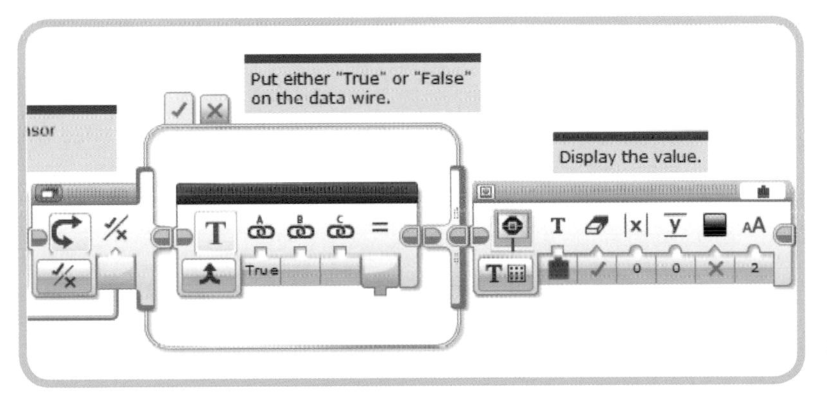

Figure 9-8: The Switch block in Tabbed View

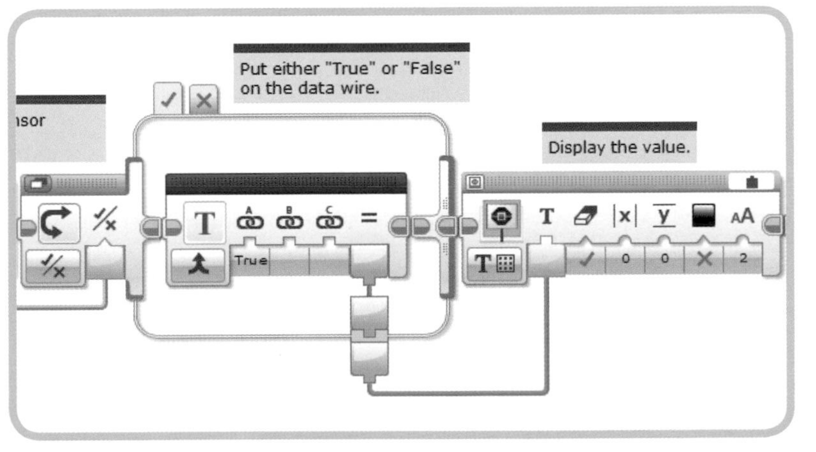

Figure 9-9: Connecting the first data wire

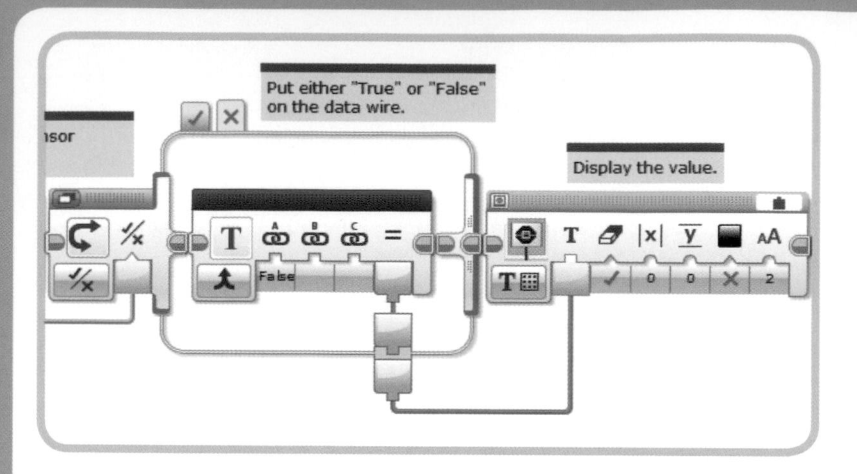

Figure 9-10: Connecting the second Text block to the Display block

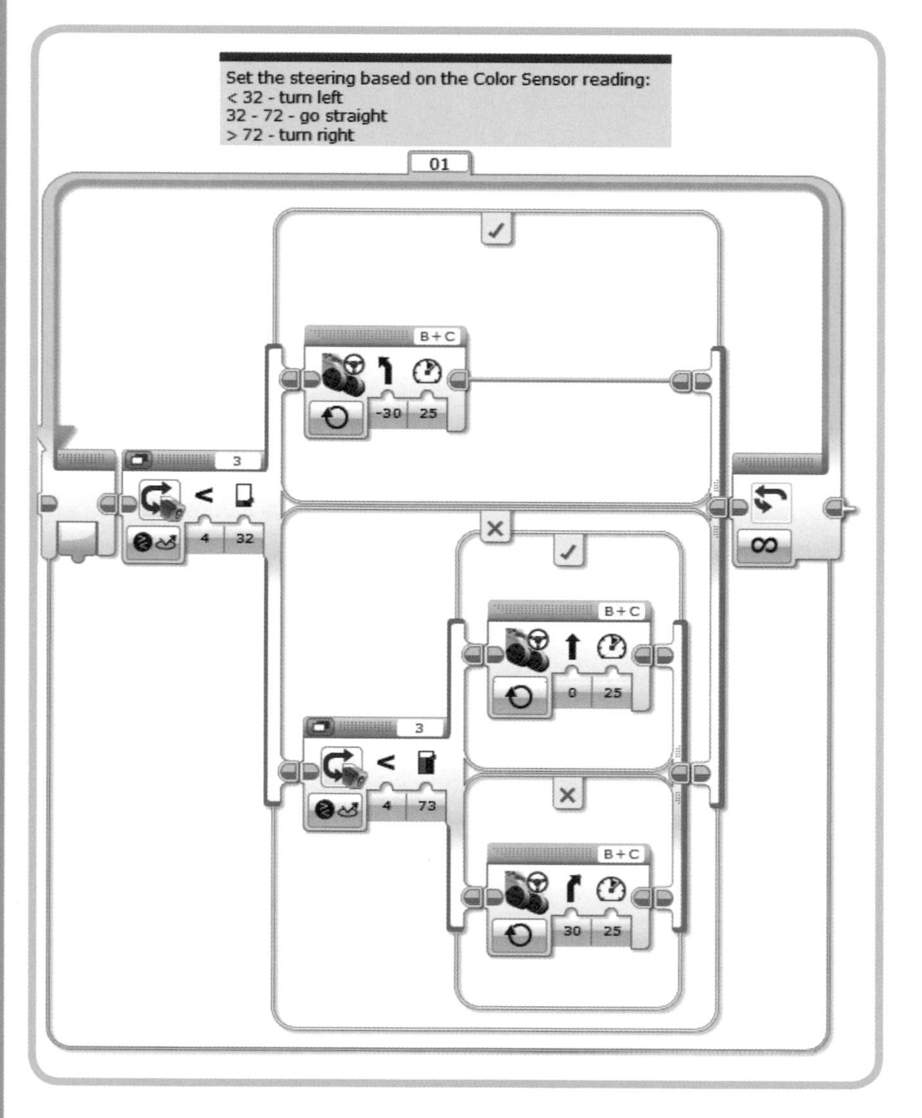

Figure 9-11: The LineFollower *program using nested Switch blocks*

CHALLENGE 9-1

Write a *ColorToText* program based on the *LogicToText* program that converts the color detected by the Color Sensor to the appropriate text string ("No Color", "White", "Green", and so on).

simplifying the LineFollower program

The *LineFollower* program from Chapter 6 (see Figure 9-11) uses nested Switch blocks to choose between three Move Steering blocks. One block turns the robot to the left, one moves the robot straight, and one turns the robot to the right.

A Switch block in Sensor Comparison mode can only have two cases; either the sensor reading meets the criteria or it doesn't. To choose from three cases (as in our program above), you need two nested Switch blocks. If you wanted to choose among five cases, you would need four nested Switch blocks. You can see how this can get out of hand quickly.

Because a Switch block in Numeric mode can have any number of cases, you can often replace a group of nested Switch blocks that use a single sensor mode with one Switch block in Numeric mode. Numeric mode lets you add as many cases as you want, without having to nest. At the same time, however, you don't want to create a different case for each possible sensor reading (from 1 to 100, say). The key is to group all the expected sensor readings into a small set of numbers. This approach, often used in programming, is called *binning*, which uses a little arithmetic to map the sensor reading ranges you are interested in to a group of small numbers. It's called binning because it takes a large group of numbers and puts related values into groups or bins. In this version of the *LineFollower* program, you'll take the range of expected sensor values and put them into three bins: one set of values that causes the robot to turn to the left, one set that causes the robot to go straight, and one set that causes the robot to turn to the right.

To calculate our bins, we need three numbers: the lowest and highest value you expect from the sensor, and the number of bins. Recall from Chapter 6 that testing the Color Sensor on the line and background revealed that the smallest reading I can expect is 13 and the largest is 92. Figure 9-12 shows a graphical representation of this range of values, out of the possible values from the sensor. When the *LineFollower* program runs, I expect that all the readings from the Color Sensor will fall in the area marked by the gray box.

Figure 9-12: The range of expected values

The first step in the binning process is to slide the range to the left so that it starts at 0, as shown in Figure 9-13. In the program, you can do this by subtracting 13 (the lowest expected value) from the sensor reading. Once you do that, you'll have a value in the range of 0 to 79. The range of values has to start at 0 in order for the next step in the process to work correctly.

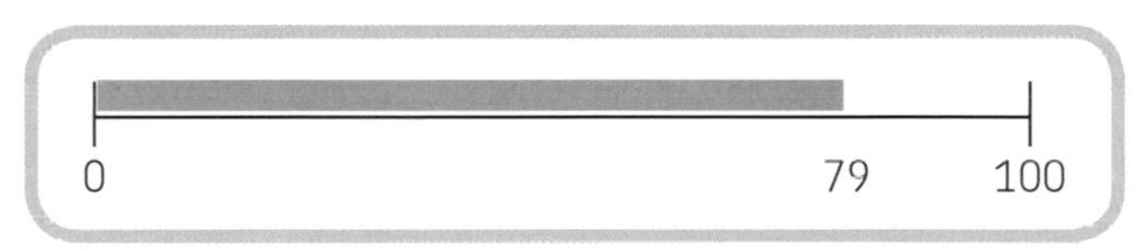

Figure 9-13: The range starting at zero

The second step is to divide the range into three bins (Figure 9-14). I've numbered the bins 0, 1, and 2 because that will make the math work out easier. The size of the total range of values I expect is 79, so we can divide that by 3 to find out the size of each bin: 79 divided by 3 is 26.33333, but I find it easier to deal in whole numbers than decimals, and the boundary between each bin doesn't have to be exact, so I'll round up to 27 here instead. To go from a sensor value to the bin number, all you need to do is subtract 13 to get a value in the range of 0 to 79, and then divide by 27. The result of the division is a decimal number, but when you pass the result to the Switch block, it rounds the value down to the next lowest whole number, which gives us the bin number. For example, if you divide 60 by 27, you get 2.22, which the Switch block rounds down to 2 to give you the right bin number.

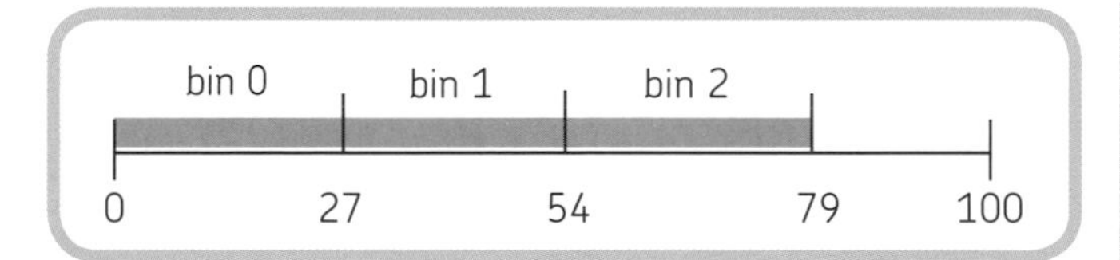

Figure 9-14: Dividing the range into three bins

The new version of the program contains one Switch block with three cases—one for each Move Steering block. The value for each case corresponds to the bin numbers (0, 1, and 2). The program uses the binning process to convert the Color

Sensor reading into one of the three cases by first subtracting 13 and then dividing by 27 (and then rounding down the result). You can write this as a formula:

Case number = (Sensor reading – 13) / 27

Table 9-1 shows the range of values for each case and the associated program behavior. The ranges shown here are a little different from those used in Chapter 6 because we used a different method for determining the limits of each range.

table 9-1: program behavior based on color sensor reading

Color Sensor Reading	Case Number	Program Behavior
13–39	0	Turn left
40–66	1	Go straight
67–92	2	Turn right

Now you're ready to rewrite the program. The first part of the program uses the Color Sensor block to read the sensor and two Math blocks to subtract 13 and divide by 27 (Figure 9-15). The Color Sensor block uses Measure-Reflected Light Intensity mode, as in the original program.

The second part of the program uses a Switch block in Numeric mode to match the bin number to the appropriate Move Steering block (Figure 9-16). I've kept the Switch block in Flat View so you can see all three cases; in your own program you may want to use Tabbed View to conserve space. The bottom case (the one for bin 2) is selected as the default so that the program acts reasonably if the sensor value is above 92.

What happens if the sensor value is less than 13? In that case, the result from the second Math block is a small negative number, which is rounded to 0 by the Switch block, and the correct case is selected. It turns out that as long as the sensor reading is within half a bin-width (27 / 2, or 13.5, in this example) of the lowest expected value, the program works correctly.

In this case, all the possible values below our lowest value (0–13) fall within half a bin width, so the binning process works even for those readings, but keep this in mind if you use a binning technique with different ranges in other programs. Selecting the right values for the highest and lowest expected sensor readings is very important. You don't want to end up with unexpected behavior in cases where a reading is outside that range.

When you run the program, it should behave much like the previous version. Essentially, you simplified the Switch block by adding the complexity of the Math blocks. On balance, the program is now more elegant and easier to enhance (such as with the following challenge).

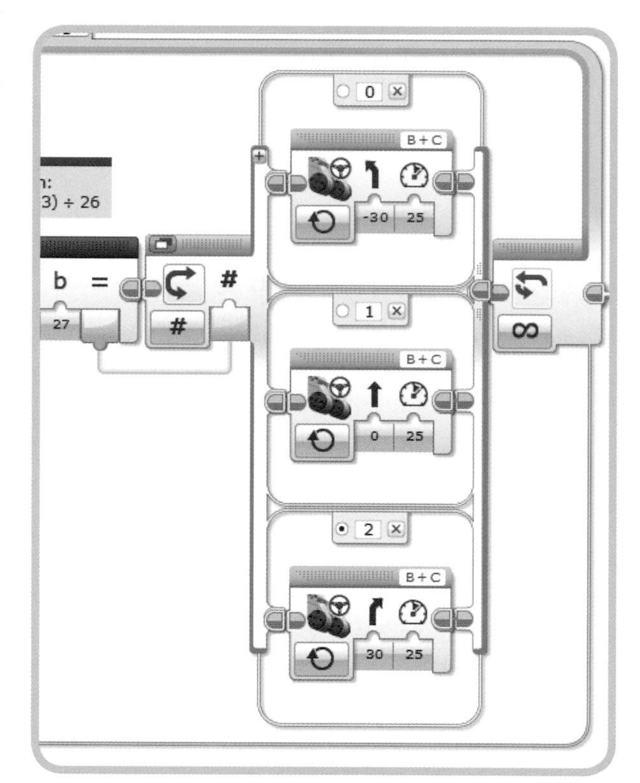

Figure 9-16: Moving the robot based on the bin number

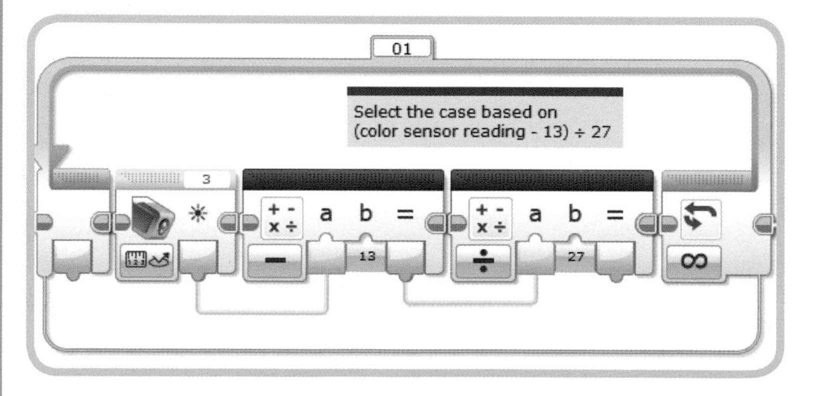

Figure 9-15: Converting the sensor reading to a bin number

CHALLENGE 9-2

Expand the *LineFollower* program to turn slightly when the robot is near the line and more sharply when it's farther from the line. This gives you five cases, with two types of turn for each side. All you need to do is change the value of the second Math block from 27 to 16 and add two more cases to the Switch block.

This challenge is very similar to Challenge 6-1, which expanded the program to five cases using nested Switch blocks. Compare how much effort it took to expand the program using each approach (nested Switch blocks or bins) and how the size and visual complexity of the program increased.

further exploration

Here are some activities you can try that involve using data wires with the Switch block:

1. Experiment with drawing data wires into and out of a Switch block. Try the following:
   a. Put the Switch block in **Flat View** and try to draw a data wire into and out of the block.
   b. Put the Switch block in **Tabbed View** and draw a wire between a block outside and a block inside the Switch block. Then click the **Flat/Tabbed View** button.
   c. Draw a few data wires into the Switch block. Practice moving the data wires to become familiar with how the data wires can be arranged both inside and outside the Switch block. In addition to moving the data wires, you can drag the tunnels to the place where the wire crosses into or out of the Switch block.
2. Change the *SoundMachine* program so that it displays the volume as "Soft", "Medium", "Loud", or "Very Loud" instead of a percentage. Choose the text to display by dividing the volume into four ranges using the binning process.
3. You can use a Switch block with data wires to place a limit on a value. For example, a Sensor block can compare a measurement with a Threshold value and pass the result to a Switch block in Logic mode. One case of the Switch block would pass the value from the Sensor block through the Switch block, and out through a data wire, with no blocks in between (Figure 9-17). The other case would pass on the maximum value instead.

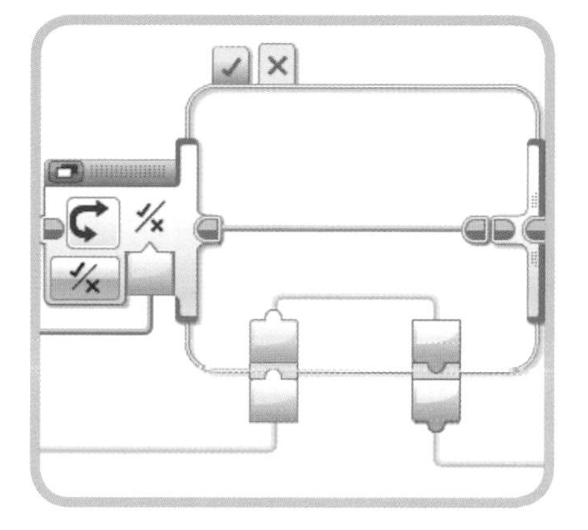

Figure 9-17: Passing a value through a Switch block unchanged

4. Write a program that moves the TriBot using the ambient light level for the Power parameter of a Move Steering block, but limit the value to a maximum of 75.

conclusion

Using a data wire to supply the input to a Switch block gives you a lot of flexibility in the types of decisions your programs can make. Using a Sensor block, you can perform a comparison outside the Switch block, which allows you to make more complex decisions than the Switch block alone supports. Data wires also allow you to pass data between the blocks inside the Switch block and those before or after, so you can easily customize each choice or make decisions that affect the rest of the program.

Using a Number or Text value as input, you can make the Switch block choose from more than two possible alternatives. This allows your program to more easily make complex decisions and can help avoid deeply nested Switch blocks. The binning process used in the *LineFollower* program is a very common method for accomplishing this.

10

data wires and the loop block

In this chapter, you'll learn how to use two special features of the Loop block that were designed to be used with data wires. The Loop block's Logic mode gives you flexible control over when the Loop block finishes, and the Loop Index output tells you how many times the loop completed.

logic mode

The Loop block's Logic mode lets you choose when to exit the loop using a Logic value from a data wire. Figure 10-1 shows how a Loop block looks with Logic mode selected and a data wire attached. After the blocks in the loop body run, the value on the data wire is checked. If the value is false, the loop continues and repeats the blocks in the body. If the value is true, the loop exits. The Loop condition is always checked after the body runs, so those blocks will run at least once even if the value on the data wire is true from the beginning.

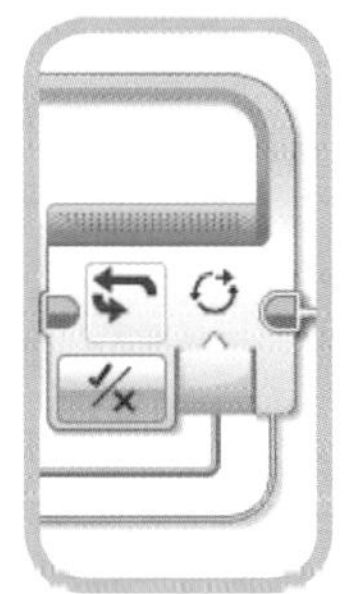

Figure 10-1: The Loop block's Logic mode

The Loop block's Sensor modes are flexible enough for most programs, but there are some situations where Logic mode is a better option. For example, when you already have a block that can perform the comparison you're interested in, you can simply pass the result of that comparison to the Loop block. The *GentleStop* program uses an Infrared Sensor block to read the proximity to the wall, so you can use this same block to decide when to exit a loop. Another example is when you want to make a decision based on more than one sensor, such as stopping the loop if the Touch Sensor is pressed *or* if the Infrared Sensor detects an object closer than a certain distance. You'll learn how to make this type of decision in Chapter 13.

Using Logic mode, you can rewrite the *GentleStop* program from Chapter 8 (see Figure 10-2) and make it a little simpler. Instead of using a Switch block to determine when to stop the motors, you can put the Loop block in Logic mode and exit the loop when the TriBot gets close to the wall. This means that the program will no longer stop when the TriBot runs into something. But because the program has been changed to stop the motors before the robot gets to the wall, this shouldn't be a problem as long as the path between the robot and the wall is clear of obstacles.

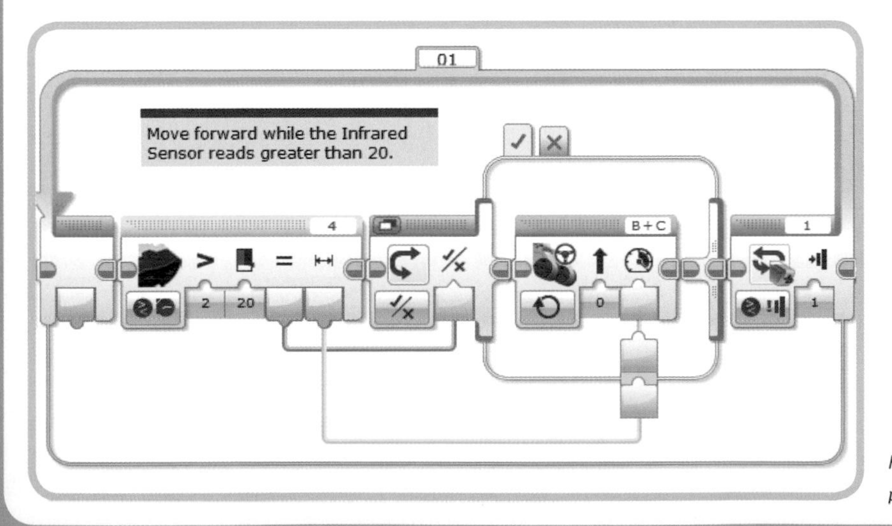

Figure 10-2: The GentleStop *program from Chapter 8*

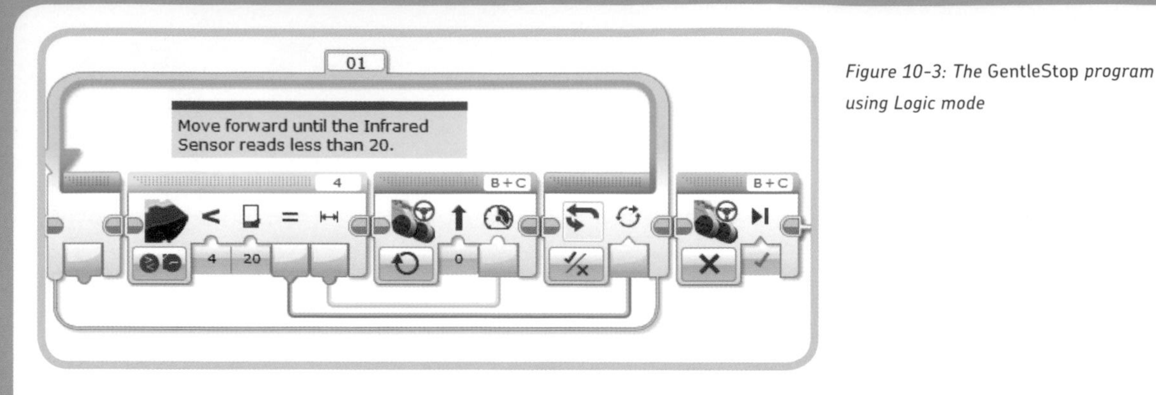

Figure 10-3: The GentleStop *program using Logic mode*

The updated version of the program is shown in Figure 10-3. In addition to removing the Switch block and moving the Move Steering blocks, you'll need to set the Loop block's mode to **Logic** and change the comparison used by the Infrared Sensor block. Because the Loop block exits when the value on the data wire is true, you need the comparison to be **Less Than** so that the value is false until the TriBot gets near the wall. Notice that the Power parameter of the Move Steering block is still controlled by the Infrared Sensor's Proximity reading, so the robot should still slow down as it approaches the wall.

Download and run the program; it should behave almost the same as the previous version, moving the TriBot forward quickly at first and then slowing down and stopping close to the wall. This version actually behaves a little better than the previous one because the program exits after stopping. The Switch block version keeps running even though the TriBot stays still.

CHALLENGE 10-1

Add a sound effect to the *GentleStop* program using the proximity from the wall to control the volume. The sound should be loud when the program starts and gradually become softer as the TriBot approaches the wall.

the loop index

The *Loop Index* tracks the number of times the loop body has been repeated. The output plug for the Loop Index is on the left side of the Loop block (Figure 10-4). The first time through the loop, this value starts at 0. Each time the program goes back to the beginning of the loop body, the index value increases by one. When the loop finishes, the Loop Index value is one fewer than the number of times the loop completed because it isn't updated after the last time the body runs.

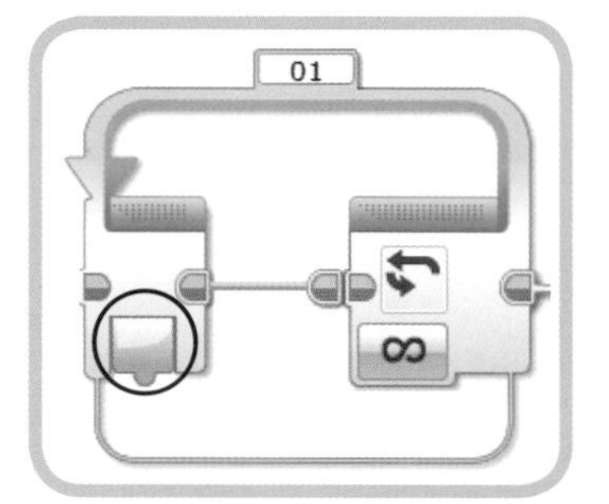

Figure 10-4: The Loop Index output plug

the LoopIndexTest program

The *LoopIndexTest* program (shown in Figure 10-5) demonstrates how the Loop Index behaves. The Loop block uses Count mode and is set to repeat five times. Each time through the loop, the Display block shows the Loop Index, and the Wait block adds a short pause to give you time to read the value. When you run this program, the display should show "0", "1", "2", "3", and "4".

This program and the two that follow aren't very interesting to watch (you won't impress your friends by building a robot that counts to four), but they demonstrate how the Loop block works. When you begin working with a block or feature you haven't used before, it's helpful to write small programs like these to explore how they work so you know how to use them in more complex programs.

restarting a loop

The *LoopIndexTest2* program shown in Figure 10-6 shows how a Loop Index works when it's nested inside another Loop block. The outer Loop block is set to run twice.

The first time the inner loop runs, the display should show "0", "1", "2", "3", and "4", just as it did in *LoopIndexTest*. But what happens the second time the inner loop runs? Will it display "0", "1", "2", "3", and "4" again or continue counting and show "5", "6", "7", "8", and "9"?

When you run the program, the display repeats "0", "1", "2", "3", and "4" twice. This tells you that the Loop Index resets to 0 each time you reach the nested Loop block.

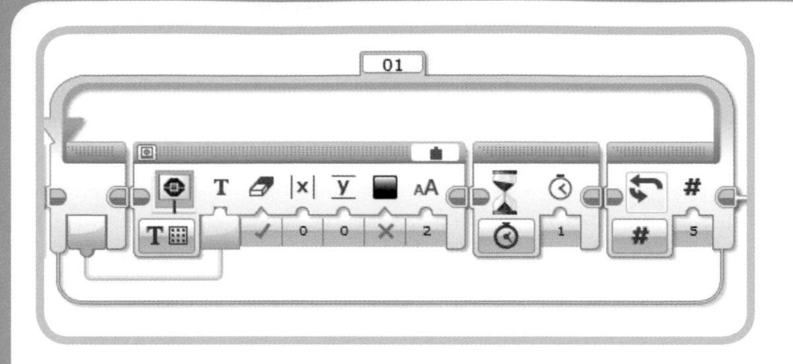

Figure 10-5: The LoopIndexTest *program*

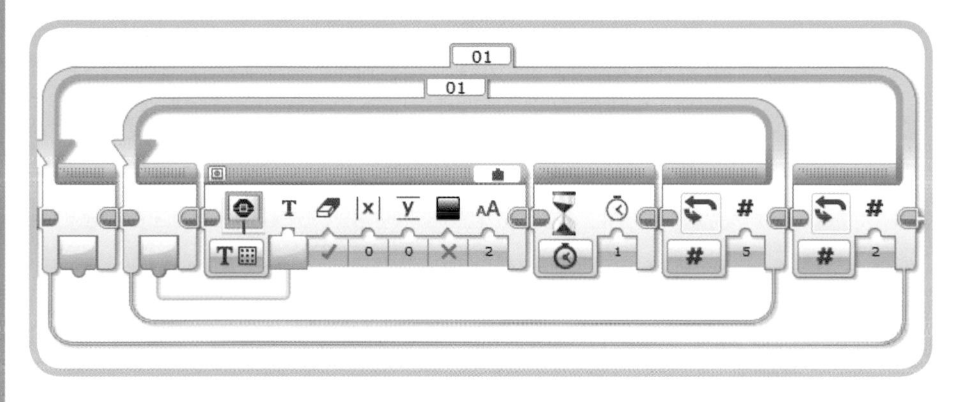

Figure 10-6: The LoopIndexTest2 *program*

the final loop index value

The *LoopIndexTest* and *LoopIndexTest2* programs use the Loop Index within the loop body. This same value can also be used by blocks that follow the Loop block, as demonstrated by the *LoopIndexTest3* program (Figure 10-7). *LoopIndexTest3* runs the Loop block five times and then prints the final Loop Index value on the EV3 screen.

This program displays the Loop Index value that's sent from the data wire after the last run through the loop. This number will be one fewer than the total number of times the loop repeats. (Remember: after the loop runs for the last time, it doesn't go back to update the Loop Index; it just moves on to the next block.) Because the Loop block is set to run five times, when you run the program, it should print "4" on the display.

NOTE It's common in computer programming to begin counting from zero instead of one. This doesn't usually pose a big problem, but it's easy to get *off-by-one errors* in which a loop repeats one too few or one too many times.

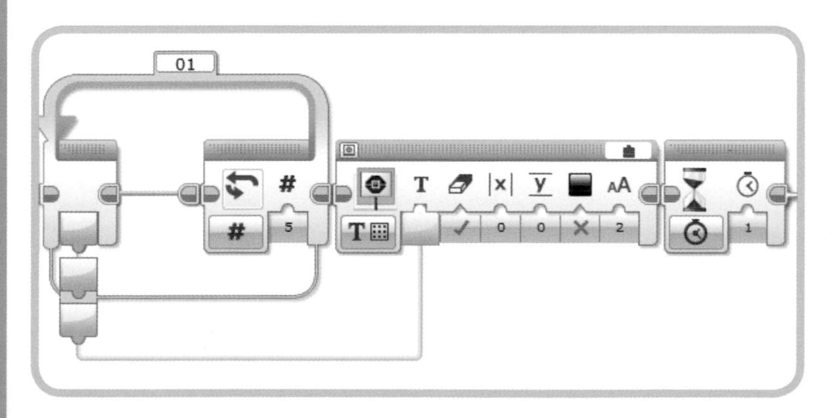

Figure 10-7: The LoopIndexTest3 *program*

the Spiral-LineFinder program

The *LineFinder* program from Chapter 5 (Figure 10-8) moves the TriBot forward in search of a dark line. That program works fine as long as you start with the robot pointing the right direction, but you can make the program more effective by making the TriBot search in a spiral pattern instead of a simple straight line.

In a rectangular spiral (Figure 10-9), each segment is longer than the previous one, creating an ever-expanding path from a central starting point. The *SpiralLineFinder* program makes the TriBot follow a rectangular spiral path and stop when the Color Sensor finds a line.

following a spiral

To follow a spiral, the TriBot needs to repeat the following steps: move forward, make a quarter-turn, move forward a little farther than the first move, make a quarter-turn, and so on. By now, you've seen that the Loop block is the best way to repeat an action like this. Using Move Steering blocks to move the TriBot forward and make the turn should be familiar as well. The only new idea in this program is changing how far the robot moves forward each time the program goes through the loop by using the Loop Index.

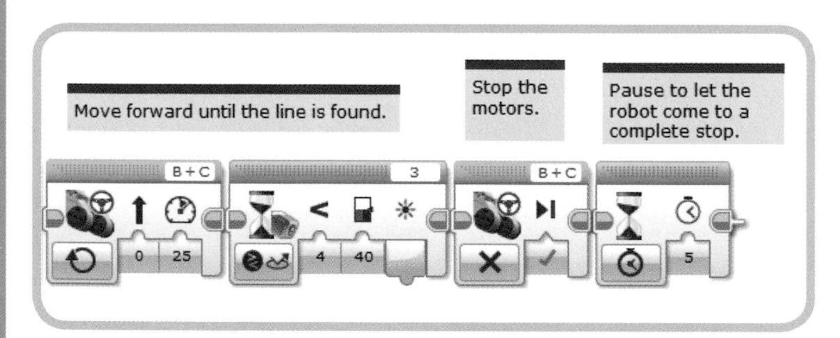

Figure 10-8: The LineFinder *Program from Chapter 5*

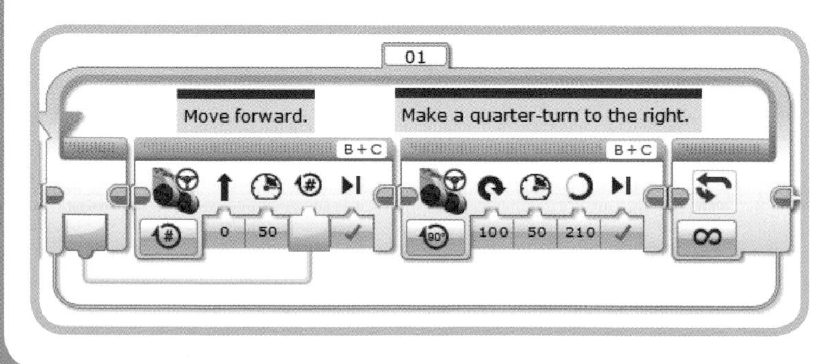

Figure 10-10: Following a rectangular spiral

Figure 10-10 shows one way we could make the robot follow a spiral. The first Move Steering block uses *On for Rotations mode*, in which the number of rotations is set by the Loop Index. Each time through the loop, this block moves the TriBot farther than the last time. The second Move Steering block makes the TriBot spin in place for a quarter-turn.

NOTE If you are using the tires in the Education Edition, set the Duration of the second Move Steering block to 160 instead of 210.

Download and run this program and see what happens. You should see the TriBot move in a rectangular spiral pattern—with one possibly unexpected behavior. The first move that the TriBot makes is a turn, rather than a move forward. Why does this happen?

Remember: the Loop Index starts at 0. This means that the first time through the loop, the first Move Steering block is set to move for 0 rotations, so it doesn't move the robot at all. For this program, it shouldn't matter which direction the robot is pointing in at the start, so having it turn once before moving forward shouldn't cause a problem. If you want to change this behavior, you can use a Math block to add 1 to the Loop Index before passing the value to the Move Steering block.

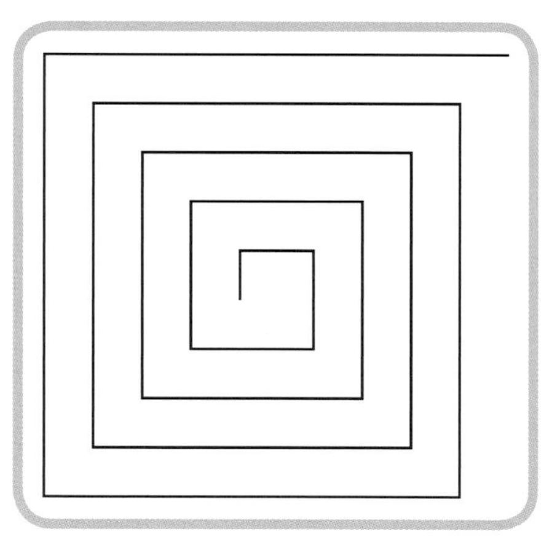

Figure 10-9: A rectangular spiral

For our *SpiralLineFollower* program, we'll do things a bit differently so that the program can constantly check for a line while it's moving forward. We'll still use the Loop Index to control how far the robot moves forward, but we'll pass that value to a Loop block, as you'll see in a moment.

detecting a line while moving in a spiral

The *LineFinder* program (shown earlier in Figure 10-8) uses a Wait block in Color Sensor-Reflected Light Intensity mode to detect the line, while a Move Steering block in On mode keeps the robot moving forward. The *SpiralLineFollower* can't simply use a Move Steering block in On mode, however, because we want the robot to move in a more sophisticated pattern.

We'll use a Switch block inside a Loop block to check the Color Sensor while the robot is moving forward and decide when to stop the motor. Start by adding the blocks in Figure 10-11. The first block resets the Rotation Sensor for Motor B to 0. Then the Move Steering block in On mode starts the robot moving, and the Loop block keeps the motors running until the Rotation Sensor reaches the Loop Index. So if the Loop Index value is 2, the robot moves forward until motor B completes two rotations. After the Loop block, a Move Steering block makes the robot turn.

At this point, the program should make the robot move in a spiral, just like our earlier program did in Figure 10-10. This time, however, the Loop block gives us a place to put a Switch block to check the Color Sensor. Be sure to test whether the program works as expected before adding the rest of the code.

Next, we need to add a Switch block inside the Loop block. The Switch block checks the Color Sensor and stops the motors if the reflected light reading is below a threshold (meaning that the robot detected a line). To stop the program when it detects a line, we'll use a Loop Interrupt block to exit the outer loop.

Figure 10-12 shows the Switch block that needs to be added to the program. The Switch block uses the same Threshold value as the Wait block in the original program. When the condition is true, the motors stop and the loops exit. The name

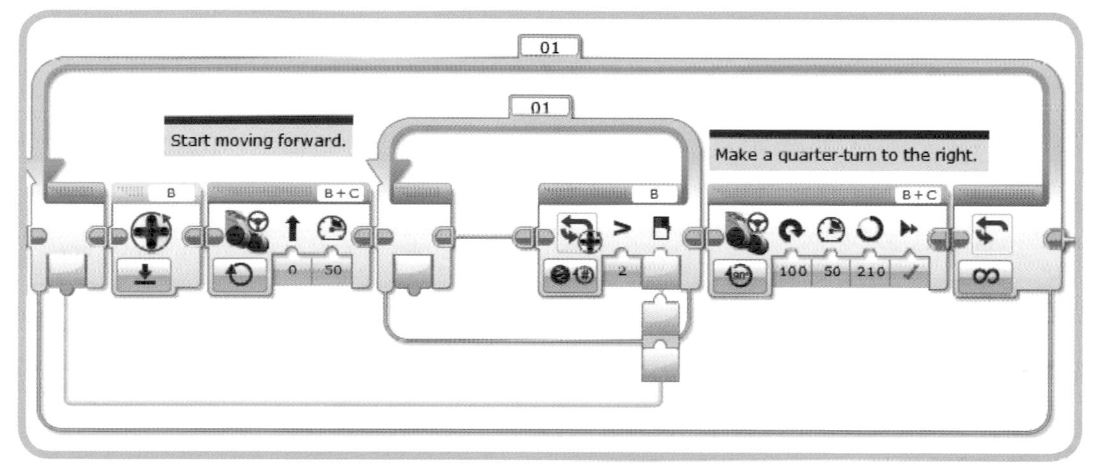

Figure 10-11: Using a loop and a Rotation Sensor to move forward

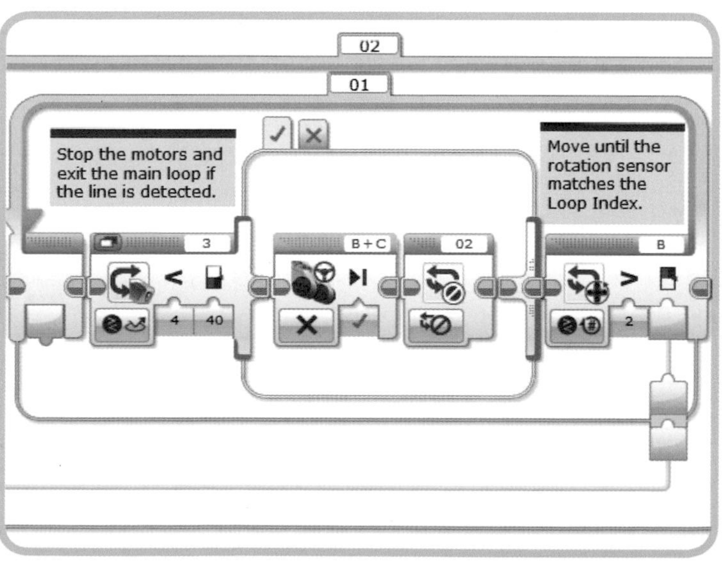

Figure 10-12: Stop the motors and exit the loop when the line is found.

of the outer Loop block is set to *02*, and the Loop Interrupt block is configured to match.

When you run the program, the TriBot should move in a rectangular spiral pattern until it finds a dark line, at which point the TriBot stops and the program ends. Experiment with different values for the speed, turning block duration, and Switch block threshold to find the values that give the best results for your test area.

CHALLENGE 10-2

The distance between each successive path around the spiral depends on how much each forward movement is increased. Increasing each move by one rotation results in a two-rotation distance between each spiral. You can see this by drawing the rectangular spiral on a piece of graph paper and looking at how adjacent lines are separated. You can increase or decrease the distance between each pass by multiplying or dividing the Loop Index (respectively) before passing the value to the Move Steering block. If your target line is small, then a tighter spiral will reduce the chance of missing the target, while a larger spiral will make the TriBot cover a larger area more quickly. Experiment with different values and target sizes to see how changing the Move Steering block's duration affects how well and how quickly the program finds a target.

using the gyro sensor for better turns

As the TriBot searches in the spiral pattern, it makes many turns, but it's impossible to make the TriBot turn exactly a quarter-circle each time. You can (and should) experiment with different values for the duration to get as close as you can to a 90 degree turn, but the movements are not perfectly repeatable, and there are too many other factors that affect how a robot moves (such as battery level, smoothness of the floor, dog hair on the wheels, and so on). Each time the TriBot turns a corner, there is a small amount of error, and as the program runs, these errors add up. After running for a while, you should notice that the rectangle of the spiral starts tilting a little to one side.

If you have a Gyro Sensor (from the Education Edition, or if you purchased one separately), you can use it to eliminate some of the error. The key is to use the Gyro Sensor to tell when the TriBot reaches increments of 90 degrees as it turns, instead of trying to make the robot turn precisely 90 degrees each time. Figure 10-13 shows the Move Steering block that performs the turn in our original program. Figure 10-14 shows the replacement blocks that perform the same operation using the Gyro Sensor.

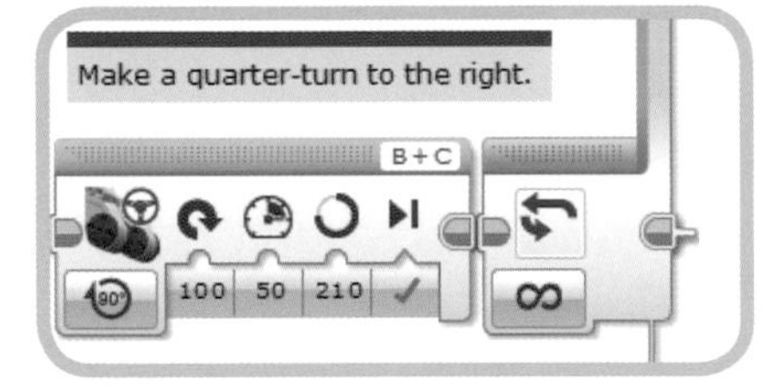

Figure 10-13: Turning using a Move Steering block alone

The first of the new blocks starts the TriBot turning using On mode so that the program can continue running other blocks while the robot is moving. Each time through the loop, the TriBot should turn 90 degrees. The Math block multiplies the Loop Index by 90 to calculate the Threshold value used by the Wait block. The Wait block uses *Gyro Sensor Compare Angle mode* to tell when the robot has moved far enough. So when the Loop Index is 1, the robot turns until the Gyro Sensor measures 90 degrees. When the Loop Index is 2, the robot turns until the Gyro Sensor measures a total of 90 × 2, or 180 degrees. This amounts to another 90-degree turn, as the Gyro Sensor measurement isn't reset anywhere. The final block in the sequence stops the motors.

Download and test the program, and you should find that the path that the TriBot follows doesn't tilt to one side after the program has been running for a while. In this version, each turn the TriBot makes is not necessarily more accurate than the version that uses the single Move Steering block. The robot might still turn a little past 90 degrees on each turn. The difference with this version is that the errors from each turn don't accumulate. The second time the robot turns, the program turns the robot until the Gyro Sensor reads 180, regardless of any errors from the previous turn.

You might also notice that the robot now moves forward first, rather than turning first. This is again because the Loop Index starts at 0. This means that the first time through the loop, the robot only turns until the Gyro Sensor reading is greater than or equal to 0, so the turn stops immediately.

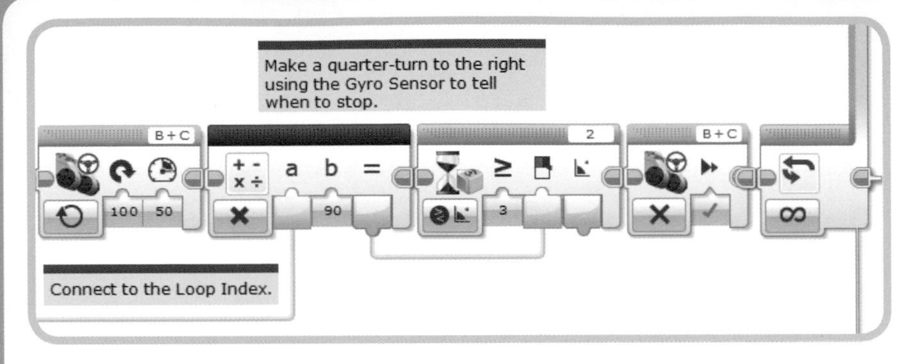

Figure 10-14: Turning with the help of the Gyro Sensor

further exploration

Try these activities for some more practice with Loop blocks and data wires:

1. Try making the TriBot follow a circular spiral rather than a rectangular one. You'll need to adjust the Steering parameter each time through the loop, starting near 100 and then gradually decreasing to make the spiral slowly increase in size.
2. Use the Loop block to count button presses from the Infrared Remote. You only need a Wait block set to wait for either of two buttons—say, button 1 and button 2. If button 1 is pressed, the loop should repeat, and if button 2 is pressed, the loop should exit. When the loop exits, the Loop Index will match the number of times button 1 was pressed. Hint: You'll actually need two Wait blocks; the first waits for a button to be pressed, and the second waits for the button state to change. Without the second Wait block, the loop repeats several times for each button press.
3. With the code from the last challenge, use the number of button presses to set a parameter later in the program. For example, you could use this technique to set the power level of a Move Steering block in order to test a program at different speeds. At the beginning of the program, add code to count the number of button presses and then use the number of presses times 10 as the motor speed.

conclusion

The Loop block can use data wires for both the Loop Index and the loop condition. In this chapter, you modified the *GentleStop* program to use information from the Infrared Sensor block to control when the loop exits. You'll use a data wire to set the loop condition again in Chapter 13, where you'll learn how to combine conditions from multiple sensors.

Once you get used to the Loop Index starting at 0, it's easy to use it to control blocks in the loop body. The *SpiralLineFinder* program uses the Loop Index to control the Duration of the MoveSteering block, which makes the TriBot follow a spiral pattern. The Loop Index can be used anytime you want a block's setting to increase or decrease each time through the loop.

variables

Variables let you store a value and use it later in the program. For example, say you want to take a sensor reading and then compare that reading to other sensor readings in the future. To do this, you would save the first sensor reading to a variable, which you could access later and compare to other readings. In this chapter, I'll show you how to use variables and describe the types of problems they solve. I'll also show you how to use the Constant block, which lets you use a single value to control multiple blocks throughout your program.

the variable block

Think of a *variable* as a place in the Brick's memory where you can store a value. The Variable block, found on the Data Operations palette, can either store or retrieve a value from a variable. You can store any value from a data wire in a variable (this is called *writing to the variable*). Then, later in your program, you can retrieve that value (this is called *reading the value*) and use it as the input to other blocks.

To demonstrate how variables are used, we'll create the *VariableTest* program, which stores the reading from the Color Sensor in a variable, reads the value out of the variable, and displays it. A Color Sensor block in Measure Color mode provides the reading, followed by a Variable block to store the value.

The first thing we need to do with a Variable block is set the mode, which selects the operation (read or write) to perform on the variable and the data type of the variable (see Figure 11-1). To store the reading from the Color Sensor (which is a Number value), select Write Numeric mode. I'll use only simple data types (Text, Numeric, and Logic) in this chapter. I'll discuss Array data types in Chapter 15.

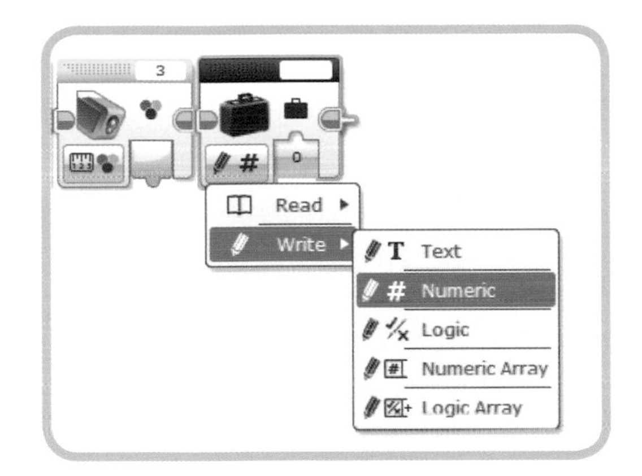

Figure 11-1: Setting the Variable block's mode

The Write operation stores a value in the variable. That value can be supplied by manually entering a value or by using an input data wire. For this program, we'll connect the Color Sensor block's output plug to the Variable block's input plug (see Figure 11-2). In other programs, you could manually set a variable's initial value and then use a data wire to change the value later in the program.

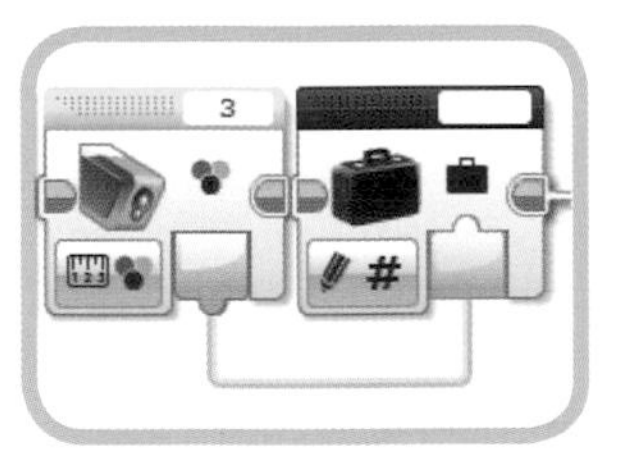

Figure 11-2: Storing the Color Sensor reading in a variable

After setting the mode, set the name of the variable by clicking the box at the top right of the block. A menu appears that shows all the variables of the selected data type that have previously been defined, along with an option to add a new variable (see Figure 11-3). Because we haven't added any variables to this project, the only choice is Add Variable.

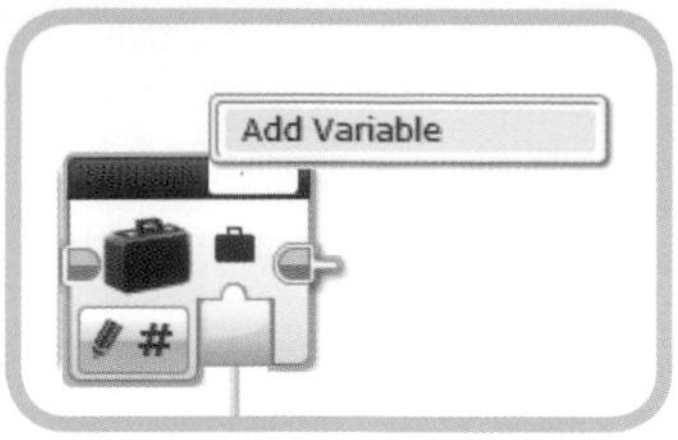

Figure 11-3: Setting the variable name

Selecting **Add Variable** displays the New Variable dialog (see Figure 11-4). Type the name you want to use to create a new variable. Each variable can only hold values of one specific data type (Text, Numeric, Logic, Numeric Array, or Logic Array). When you create a variable, the mode of the Variable block determines the data type that the variable can hold. For this program, we'll name the variable **Color**.

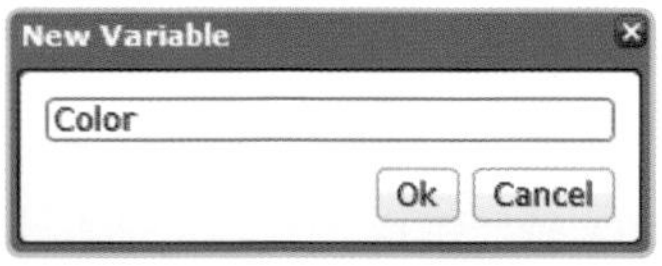

Figure 11-4: The New Variable dialog

Now we need to add another Variable block to read the value, and a Display block to show the value on the EV3 screen. A Wait block at the end of the program prevents the display from clearing before you have a chance to see the result. Figure 11-5 shows the completed program.

After you add the three new blocks to the program, set the new Variable block's mode to **Read Numeric**. The Variable Name is automatically set to Color because that's the only Numeric variable in the project, but in most cases you'll need to click the Variable Name box and select the variable. The final step in building this program is connecting the output of the Variable block to the Display block's Text parameter.

When you run this program, it reads the Color Sensor, stores the reading in the Color variable, reads the variable, and finally displays the value. To see the value passed into and out of the Variable blocks, connect the EV3 to your computer and run the program from the EV3 software. Then you can examine the values on the data wires while the program is running, as shown in Chapter 8 (you might want to set the Wait block to pause longer so you have more time to see the values).

the RedOrBlueCount program

In this section, I'll take you through the steps for creating the *RedOrBlueCount* program based on the *RedOrBlue* program from Chapter 6 (shown in Figure 11-6). The new program uses two variables, named Red Total and Blue Total, to keep track of the number of red and blue objects. As the program runs, it displays the running totals on the EV3 screen. When the program first starts, the count for both colors is zero, so it should display "Red: 0" and "Blue: 0". After that, it should update the display as the count increases. Listing 11-1 shows the pseudocode for the program with the parts you'll add to the original *RedOrBlue* program in bold. Note that the program doesn't count the objects that are neither red nor blue.

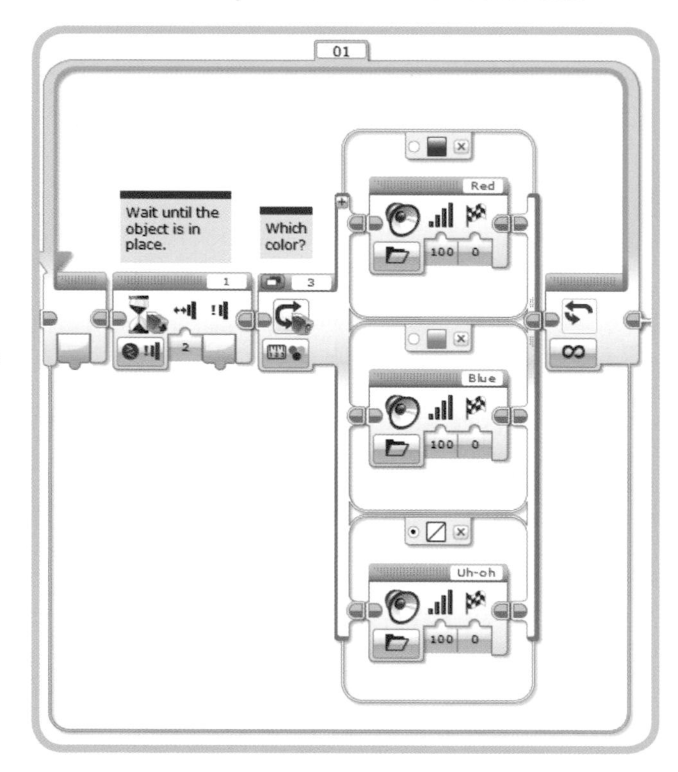

Figure 11-6: The RedOrBlue program

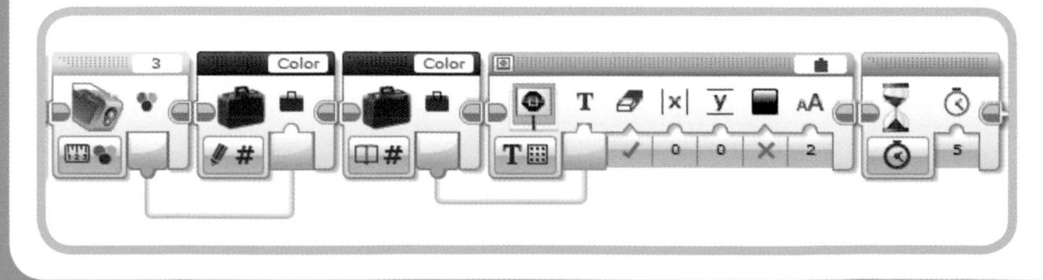

Figure 11-5: The VariableTest program

```
set Red Total to 0
set Blue Total to 0
display "Red: 0"
display "Blue: 0"
begin loop
   wait for the Touch Sensor to be bumped
   if the object is red then
      use a Sound block to say "Red"
      read the Red Total value
      add one to the Red Total value
      write the new value to Red Total
      display "Red: " followed by the Red Total
         value
   else if the object is blue then
      use a Sound block to say "Blue"
      read the Blue Total value
      add one to the Blue Total value
      write the new value to Blue Total
      display "Blue: " followed by the Blue Total
         value
   else
      use a Sound block to say "Uh-oh"
   end if
loop forever
```

Listing 11-1: The RedOrBlueCount *Program*

creating and initializing the variables

The first step is to create the two variables and give each a starting value. This is called *initializing the variables*. To count the number of red objects, the Red Total variable needs to start at zero and increase by one each time a red object is detected. When the program begins, a Numeric variable will have 0 for a value by default. Still, I recommend always setting a variable's initial value just in case you later move the code around or reuse part of the code in other programs.

Start building the *RedOrBlueCount* program by following these steps:

1. Make sure the *Chapter11* project is open.
2. Open the *Chapter6* project, and copy the *RedOrBlue* program to the *Chapter11* project. Rename the new program (in the *Chapter11* project) *RedOrBlueCount*.

CHOOSING VARIABLE NAMES

Choosing an appropriate variable name makes your program easier to understand. For example, in a program that counts red and blue objects, a name like Red Total makes it clear how the variable is being used (both for other people reading your program and for yourself if you look back at the program later). It's helpful to be consistent with your variable names, too. For example, if you use Red Total for the number of red objects, then you should use Blue Total for the number of blue objects, not something different, such as Blue Count. By the same token, avoid names that are difficult to decipher or abbreviations that are too short, such as TtlR or just R. Also keep in mind that the space available for the name on the Variable block is limited, so avoid using really long variable names that start the same way. For example, if you used the names Total Red and Total Blue, then the Variable block looks the same with either name selected (see Figure 11-7), which makes it harder to tell what the program is doing. To see the full name of the variable, hover the cursor over the Name box.

Figure 11-7: Can't tell if this block uses the variable Total Red or Total Blue

3. Add a Variable block to the beginning of the program. Leave it in Write Numeric mode, with the value set to 0 (the default mode and value).

   Click the block's Variable Name box. A menu pops up that contains the **Add Variable** choice and any Numeric variables that you've already created (see Figure 11-8).

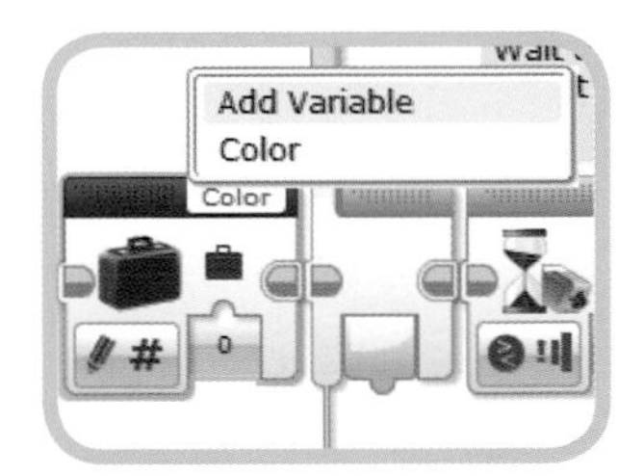

Figure 11-8: Setting the variable name

4. Click **Add Variable**, and the New Variable window appears. Type **Red Total** (see Figure 11-9) and click **Ok**. The Variable block should now show the first few letters of the variable name (see Figure 11-10).

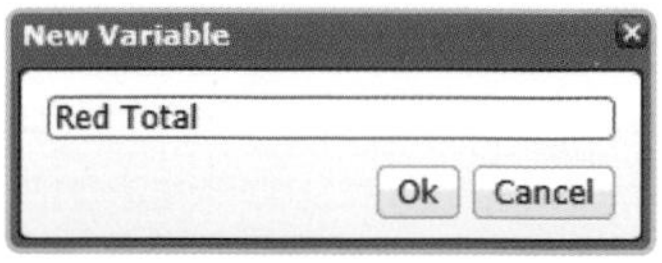

Figure 11-9: Creating the Red Total variable

Figure 11-10: Setting the Red Total variable to 0

5. Add another Variable block, just to the right of the first one.
6. Type the Variable Name **Blue Total**.

At this point, the beginning of the program should look like Figure 11-11. These two blocks initialize the two variables to 0.

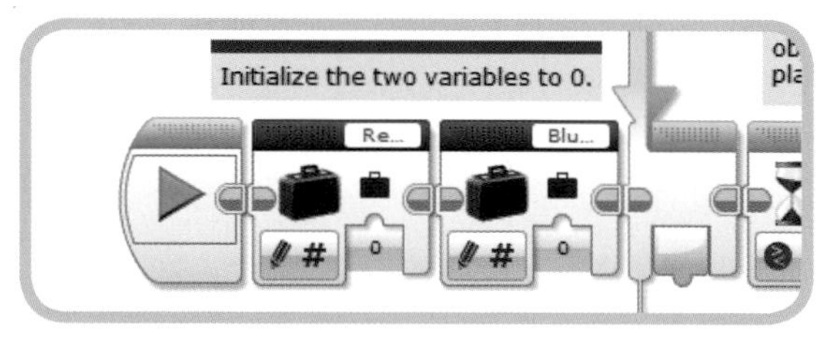

Figure 11-11: Initializing the Red Total and Blue Total variables to 0

displaying the initial values

To display the initial values, use two Display blocks placed before the Loop block:

7. Add a Display block after the second Variable block. Set the mode to **Text Grid** and the row to **2**. Set the text to **Red: 0**.
8. Add another Display block after the first one. Set the mode to **Text Grid** and the row to **4**. Set the text to **Blue: 0**.
9. Set the Clear Screen option to **False**.

The beginning of the program should now look like Figure 11-12.

counting the red objects

When a red object is detected, you want the program to add one to the Red Total variable and display the new value. To make that happen, you'll use three blocks: a Variable block to read the current value and send it out on a data wire, a Math block to add one to the current value, and a second Variable block to store the new value. Here's how to do that:

10. Add a Variable block to the red case of the Switch block, after the Sound block.

    Set the mode to **Read Numeric**, and make sure the Variable Name is set to **Red Total**. The Switch block should now look like Figure 11-13.

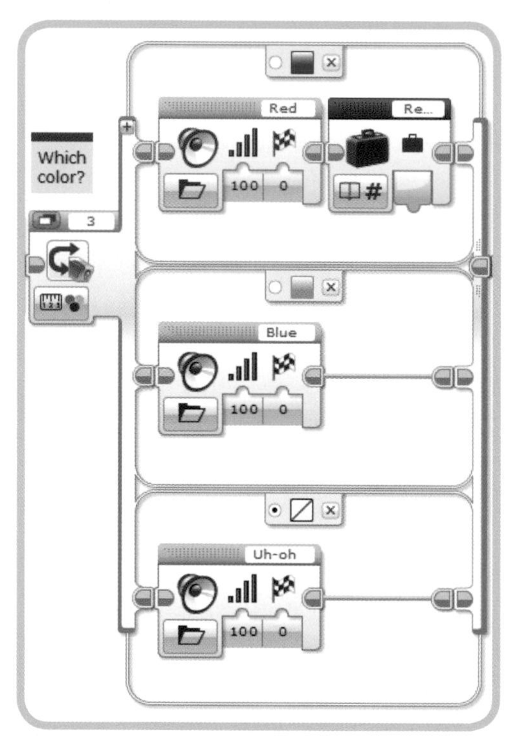

Figure 11-13: Reading the current Red Total

11. Add a Math block after the Variable block. Make sure the *b* value is set to **1.**
12. Add another Variable block after the Math block. Make sure the mode is set to **Write Numeric**. Select **Red Total** for the Variable Name.

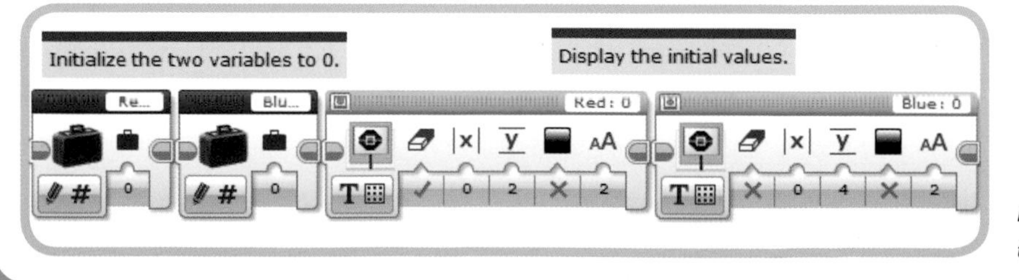

Figure 11-12: Displaying the initial values

13. Connect the blocks by drawing a data wire from the first Variable block to the Math block's *a* input, and another data wire from the Math block's output to the second Variable block's input, as shown in Figure 11-14.

After the total has been updated, the program should display the new value using a Text block and a Display block. (This is the same technique that you used in the *SoundMachine* program in Chapter 8.)

14. Add a Text block after the second Variable block, and set the *a* value to **Red:** . (Be sure to include a space after the colon).
15. Connect the Math block's Result output to the Text block's *b* input.
16. Add a Display block after the Text block, and set the mode to **Text Grid**. Set the row to **2**. Be sure to set the Clear Screen option to **False** so that you don't erase the blue total while writing the red total.
17. Click the text box in the upper-right corner of the Display Box, and select **Wired** from the pop-up menu.
18. Connect the Text block's Result output to the Text input plug that appears on the Display block.

This section of the program should now look like Figure 11-15.

This section of code contains a pattern that you'll often see to update a variable: Read the current value (with a Variable block), modify it (with a Math block), and write the new result to the variable (with another Variable block).

NOTE Before continuing, test your program to make sure it counts and displays the total number of red objects. You'll be copying this code for the blue case, so if there's a bug in the code, it's better to catch it now rather than risk repeating it.

counting the blue objects

The code for counting the blue objects is almost identical to the code for counting the red objects, so instead of writing the code by hand, you can copy the code with a few changes. Here's how to duplicate the code:

19. Select the five new blocks on the Switch block's upper case (from the first Variable block to the Display block) by drawing a selection rectangle around them or by selecting the Variable block and then pressing SHIFT while clicking the other blocks.
20. While pressing CTRL, click one of the blocks and drag it to the lower case. (When you drag with the CTRL key pressed, the blocks are copied instead of moved.)

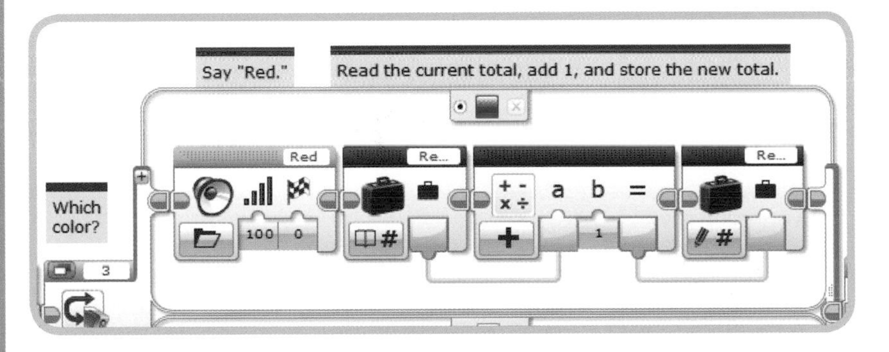

Figure 11-14: Adding one to the Red Total variable

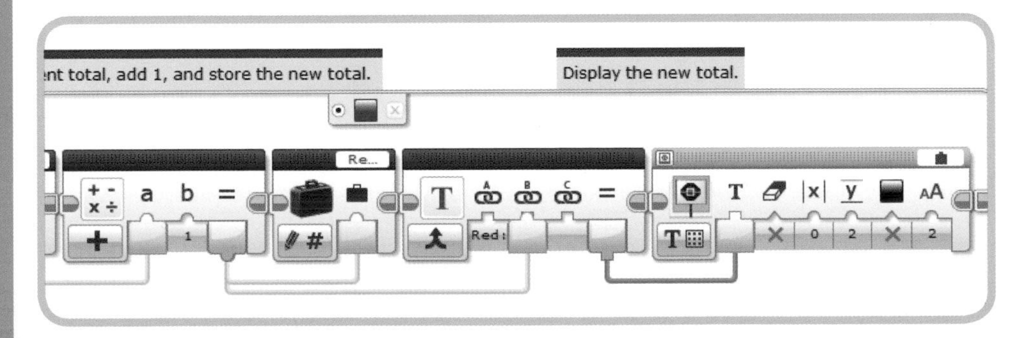

Figure 11-15: Displaying the new Red Total value

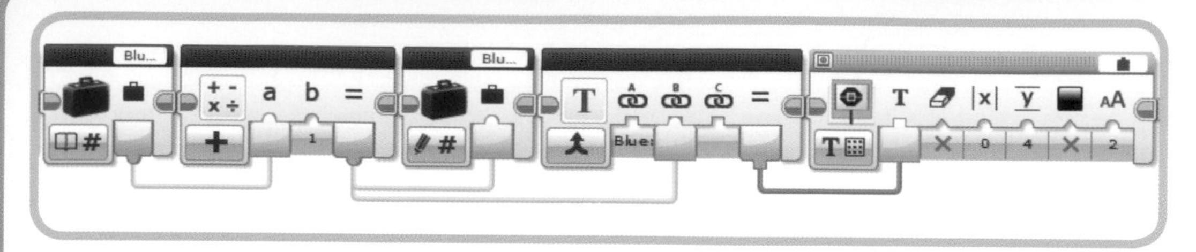

Figure 11-16: Counting the blue objects

After the blocks are in place, make these changes:

21. Set both Variable blocks to the **Blue Total** variable instead of **Red Total**.
22. Set the label in the Text block to **Blue:** (with a space after it).
23. Set the Display block to use row **4**.

Figure 11-16 shows the section of the program on the lower case.

With these blocks copied and the parameters changed, your program should now correctly count and display the totals for both red and blue objects. Download and test the program to be sure it works for both colors.

managing variables using the project properties page

For the *RedOrBlueCount* program, you created the two variables by using Variable blocks. You can also create—and delete—variables using the Variables tab on the Project Properties page, which you open by clicking the wrench icon to the left of the program name tabs (see Figure 11-17).

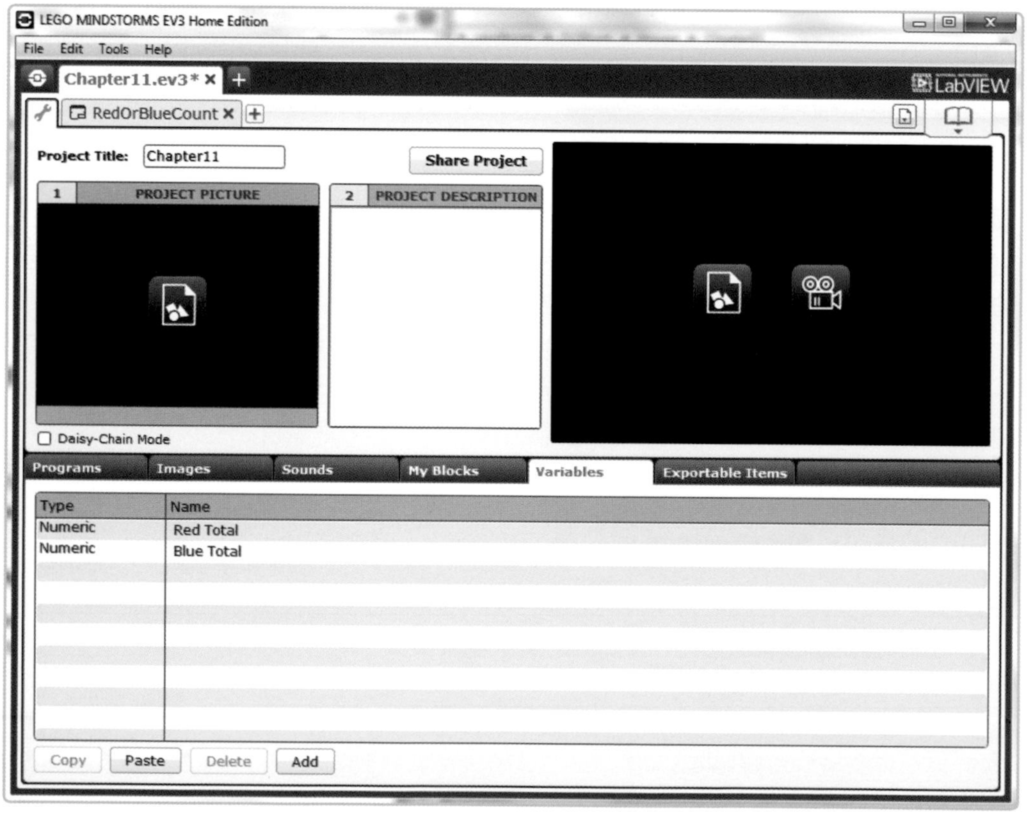

Figure 11-17: The Variables tab on the Project Properties page

The Variables tab shows the name and data type of all the variables used in the current project. To add a variable to the project, just click the **Add** button. The New Variable window pops up (see Figure 11-18) so you can set the name and data type of your new variable. You can also delete a variable if you've decided not to use it in your programs. To delete a variable, select it in the list and click the **Delete** button at the bottom of the window. If you delete a variable that a program still uses, the program continues to work but the variable won't be listed on the Project Properties page or in the list displayed when you click the Variable block's Name box.

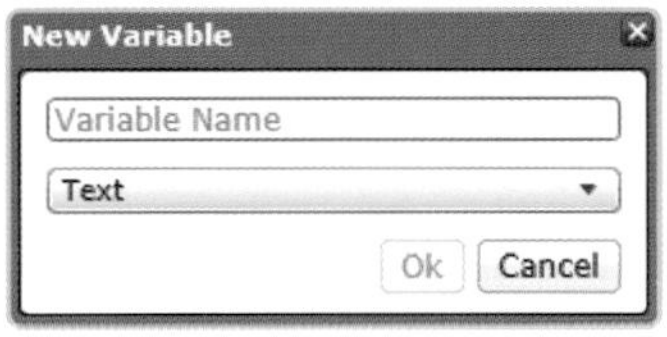

Figure 11-18. Adding a new variable

the compare block

For the next program, you'll use the Compare block to compare two numbers (see Figure 11-19). Let's take a quick look at how this block works. The *Compare block* is on the Data Operations palette. You can supply the two Input values using data wires or by manually setting one or both of the parameters. The block compares these two values according to the mode selected, as shown in Figure 11-20, and the result is made available in the block output. For example, the block shown in Figure 11-19 is in Equal To mode, so it checks to see if the two values are equal and sends the result to the output plug.

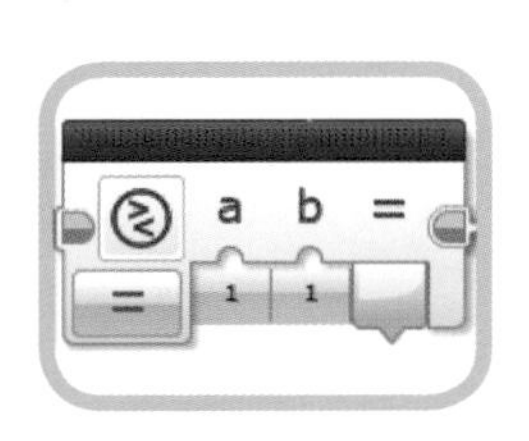

Figure 11-19: The Compare block

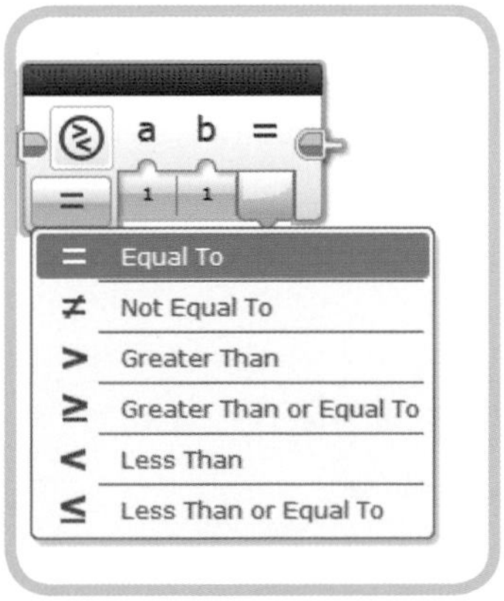

Figure 11-20: The Compare block modes

The result from the Compare block is always a Logic value (true or false). When using the Compare block, it's like you're asking a question such as "Is the value of the Ultrasonic Sensor greater than 20?" The answer will be a Logic value.

The Compare block is useful for making decisions because the resulting Logic value can be used to control Switch and Loop blocks, and it gives you more flexibility than just using a Switch or Loop block alone. For example, you can compare the readings from two Rotation Sensors, or you can use a Math block to modify a sensor value before comparing it with a Target value.

Typically, the Compare block is used as shown in Figure 11-21. Here, data wires give the block two numbers to compare. In this example, the mode for the block is set to **Less Than**.

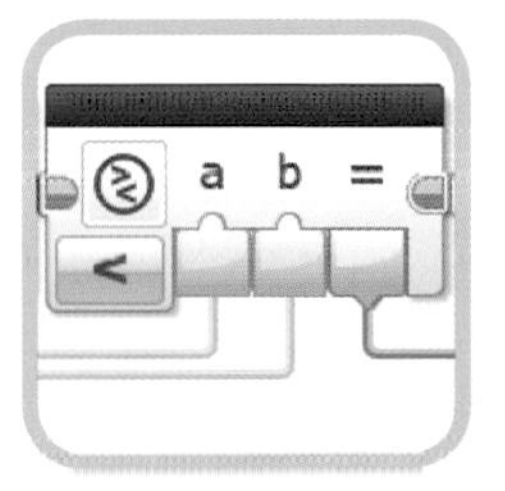

Figure 11-21: Compare block with connections

When the block runs, it compares the two input numbers and puts the resulting Logic value on the output data wire. For example, if the A value is 7 and the B value is 12, the result is true because 7 is less than 12. On the other hand, if A is 25 and B is 8, the result is false because 25 is not less than 8.

the LightPointer program

The *LightPointer* program demonstrates how to use variables to remember values that you want to use later in your program. This program uses the Color Sensor to point the TriBot at a light source by spinning the robot in a circle and remembering where the sensor detected the brightest light. The code and ideas you develop here could be used as part of a larger program, such as a flashlight-following program or a robot that finds the longest clear path through a field of obstacles.

For this program, you can place the Color Sensor at the front of the TriBot (Figure 11-22) or on its side (Figure 11-23). You can, of course, adjust the placement of the sensor depending on the height of the light source.

Figure 11-22: Color Sensor on the front of the TriBot

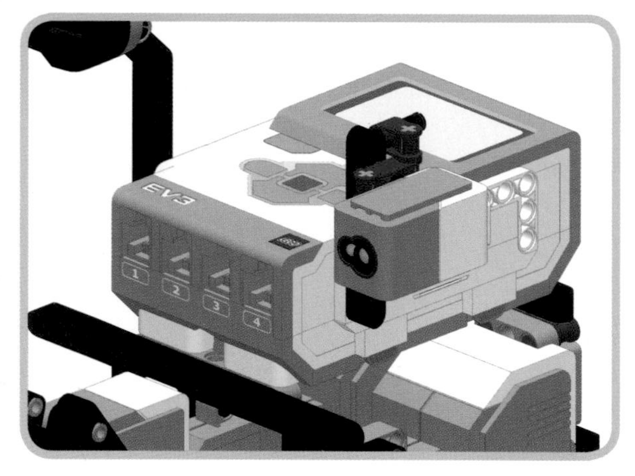

Figure 11-23: Color Sensor on the side of the TriBot

This program has two distinct stages: First it searches for a light source, and then it points the TriBot at that light source. For the first part, the TriBot slowly spins in a circle as the sensor continuously measures the amount of ambient light. Each reading is compared with the largest reading seen so far, and when a larger reading is measured, the position is recorded.

Figure 11-24 shows how the sensor reading changes as the TriBot spins. The robot starts facing away from the light, so the reading is low (10). As the robot spins toward the flashlight, the sensor reading increases to 40, as shown in the second image. When the TriBot is pointing directly at the flashlight, the reading is at its highest level (70 in this example). The sensor reading decreases as the TriBot spins past the flashlight, as shown in the final image.

The second part of the program turns the TriBot back to the position with the largest reading, which should result in the robot pointing at the light source.

defining the variables

This program needs two variables to hold two different Numeric values. The first one, Max Reading, holds the brightest sensor reading so far. The other variable, Position, holds the robot's position where Max Reading was measured. Create these two Numeric variables using the Variables tab on the Project Properties page. Remember to set the type to Numeric in the New Variable window. After creating the two variables, the Variables tab should look like Figure 11-25.

finding the light source

The first step in finding the light source is to spin the robot around. We'll use a Move Steering block with the Steering value set at -100 to do this. We also want to track the robot's position as it spins, and for that we'll use the rotation value from the C motor, which increases as the TriBot spins.

The TriBot should spin in a complete circle so that it can find the light source in any direction. A little experimentation shows that a Duration parameter of 900 degrees moves the robot in a full circle. This value doesn't have to be exact; the program will work just fine if the robot rotates a little past its starting point.

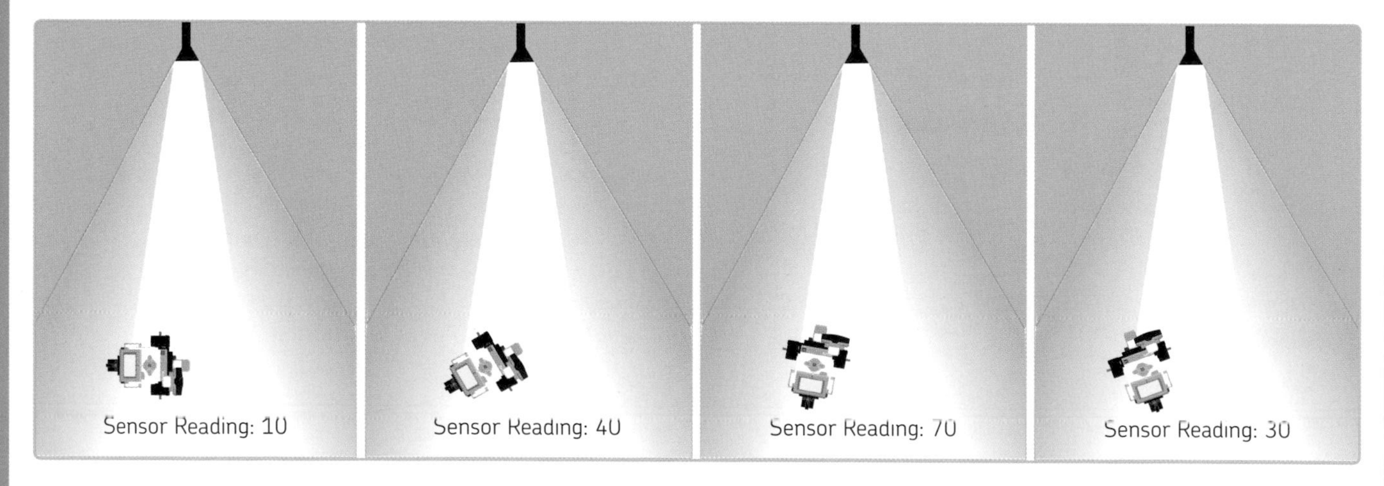

Figure 11-24: Sensor readings at four positions

Programs | Images | Sounds | My Blocks | Variables | Exportable Items

Type	Name
Numeric	Red Total
Numeric	Blue Total
Numeric	Max Reading
Numeric	Position

Copy | Paste | Delete | Add

Figure 11-25: The Variables tab after creating the new variables

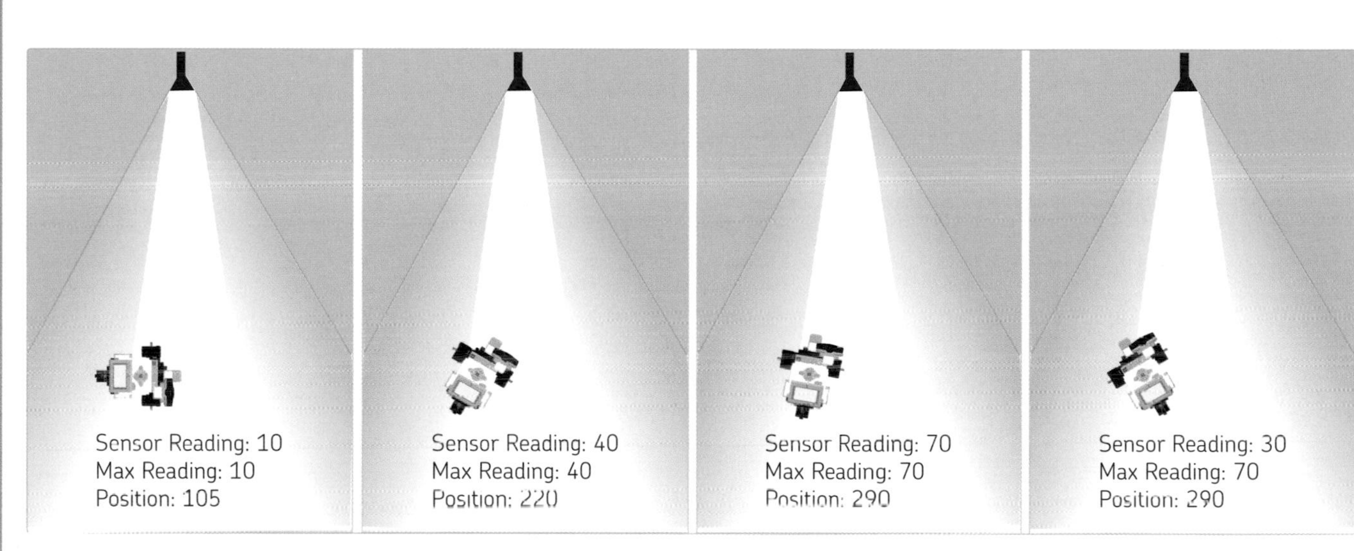

Figure 11-26: Max Reading and Position variable values as the robot spins

NOTE For the Education Edition tires, use 700 degrees for a full circle.

As the TriBot spins, the program compares the reading from the Color Sensor with the highest reading seen so far. If a higher reading is found, Max Reading is set to the new (higher) value and Position is set to the C motor's position. Figure 11-26 shows what the variable values would be for the positions shown in Figure 11-20. Notice that in the fourth panel, the Max Reading and Position aren't updated with the newest position and sensor reading because the sensor reading at that position is less than a reading the robot detected earlier.

Listing 11-2 shows the pseudocode for this section of the program. Now that you have some experience with variables, I've shortened the notation a bit:

```
Max Reading = Color Sensor reading
```

The line above is a more concise way of saying "take the Color Sensor reading and store it in the Max Reading variable."

This notation matches how you would update a variable in many other programming languages, where the equal sign means "store the value on the right in the variable on the left."

```
start the robot spinning slowly
begin loop
   if Color Sensor reading > Max Reading then
      Max Reading = Color Sensor reading
      Position = B motor Rotation Sensor reading
   end if
loop until B motor Rotation Sensor > 900
```

Listing 11-2: Finding the light source

creating the LightPointer program

The first three blocks in the *LightPointer* program initialize the Max Reading and Position variables and reset the C motor's Rotation Sensor, as shown in Figure 11-27.

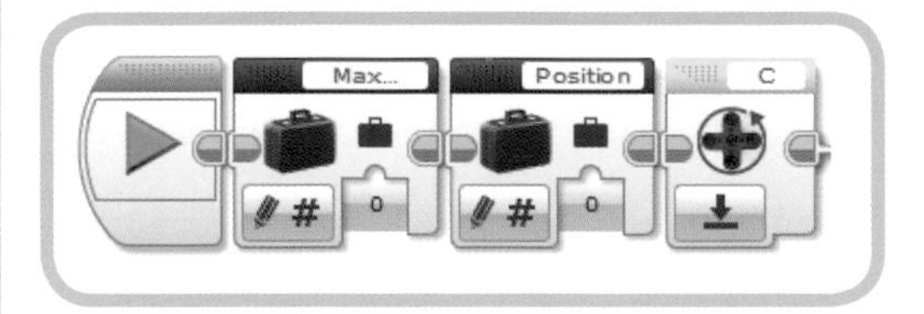

Figure 11-27: Initializing the variables and resetting the Rotation Sensor

With initialization complete, you can begin writing the code for locating the light source using the pseudocode developed in the previous section as a guide. First, start the TriBot spinning by using a Move Steering block, as shown in Figure 11-28. Set the Power to 20 to make the robot spin slowly so that it doesn't miss the light source.

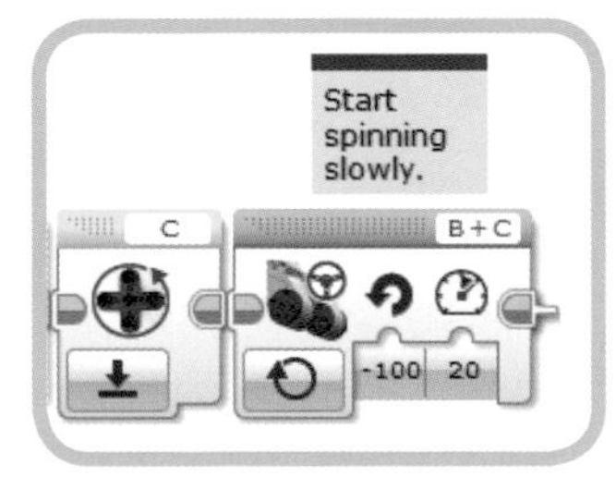

Figure 11-28: Starting the robot spinning

In the next part of the program, shown in Figure 11-29, a Loop block keeps the robot spinning until the Rotation Sensor reads greater than 900 degrees by using the **Motor Rotation - Compare Degrees** mode. Each time through the loop, the Color Sensor reading is compared with the Max Reading value. If the sensor reading is greater than the current Max Reading value, the code in the Switch block is executed. This code updates the Max Reading variable and sets the Position variable to the current motor position using the Rotation Sensor. There are no blocks on the other tab of the Switch block.

Let's go through the blocks in this section one at a time.

1. The Loop block is configured to run until the reading from the Rotation Sensor for motor C is greater than 900. This keeps the loop body repeating until the TriBot has spun in a complete circle.
2. The Color Sensor block uses **Measure - Ambient Light Intensity** mode to measure the brightness of the light in front of the sensor. It sends this value to other blocks using data wires.
3. The Variable block reads the current value of the Max Reading variable and sends it to the Compare block with a data wire.
4. The Compare block compares the reading from the Color Sensor block with the Max Reading value. If the Max Reading value is less than the new sensor reading, the result passed to the Switch block is true; otherwise, the result is false.

INITIALIZING THE VARIABLES

Before writing the code for the first part of this program, think about the values that the variables should have at the start of the program. Even though the code used to initialize the values usually comes at the beginning of the program, you'll often need to design the program (or at least its major parts) first in order to determine how the values should be initialized.

The Max Reading variable is meant to hold the highest reading from the Color Sensor, which will be between 0 and 100. By setting Max Reading to 0 (the lowest possible reading) at the beginning of the program, we ensure that the sensor reading and the robot's position will be recorded the first time the sensor reads a value greater than 0.

Even though the Position variable will be set when the sensor first detects light, it's still a good idea to give it an initial value. (Initializing all your variables is a good programming practice that can help avoid some tricky bugs that are difficult to find later.) In the code shown in the following section, the Position variable is set to 0 at the start of the program.

In addition to the two variables, the code in Listing 11-2 uses the Rotation Sensor for motor C. To make sure that the loop works correctly, reset the Rotation Sensor at the beginning of the program. The Rotation Sensor is automatically set to 0 when the program starts, but if you want to reuse the code for this program in another program, you'll need to reset the Rotation Sensor before the Move Steering block starts spinning the robot. Your programs will be much easier to reuse if you explicitly initialize everything instead of relying on automatic initialization when the program starts.

5. If the result from the Compare block is true, the Switch block runs the three blocks on the true case shown in Figure 11-29. There are no blocks on the false case.
6. The first Variable block within the Switch block stores the latest Color Sensor reading in the Max Reading variable. The next time through the loop, the Compare block uses this value.
7. The Motor Rotation block reads the current position of the C motor.
8. The second Variable block stores the position of the C motor in the Position variable.

The Loop block runs until the TriBot has spun around in a complete circle. After the loop completes, the Position variable should contain the position of motor C where the brightest light was detected. The second part of the program uses this value to make the TriBot spin in the opposite direction and return to that position.

A Move Steering block with the Steering value set to 100 will spin the TriBot the other way around, causing the Rotation Sensor value for the C motor to decrease. The TriBot needs to keep spinning as long as the Rotation Sensor reading is greater than the Position value so that it ends up pointing in the direction where the brightest light was detected. Listing 11-3 shows the pseudocode for this section of the program.

```
start the robot spinning slowly in the opposite
   direction from the first move
read the Position variable
wait until the Rotation Sensor reading reaches the
   Position value
stop the motors
```

Listing 11-3: Moving the TriBot back to point at the light source

Figure 11-30 shows this section of the program. Notice that the Wait block reads the Rotation Sensor value and compares it with the value stored in the Position variable. The comparison is set to Less Than, making the block wait until the Rotation Sensor reading is less than the stored position. When the Wait block ends, the Move Steering block stops the motors.

Now try out the full program. The TriBot should slowly spin all the way around and then reverse direction and stop so that the robot faces the brightest light source. Try testing this behavior in a dark room with a flashlight shining at the robot.

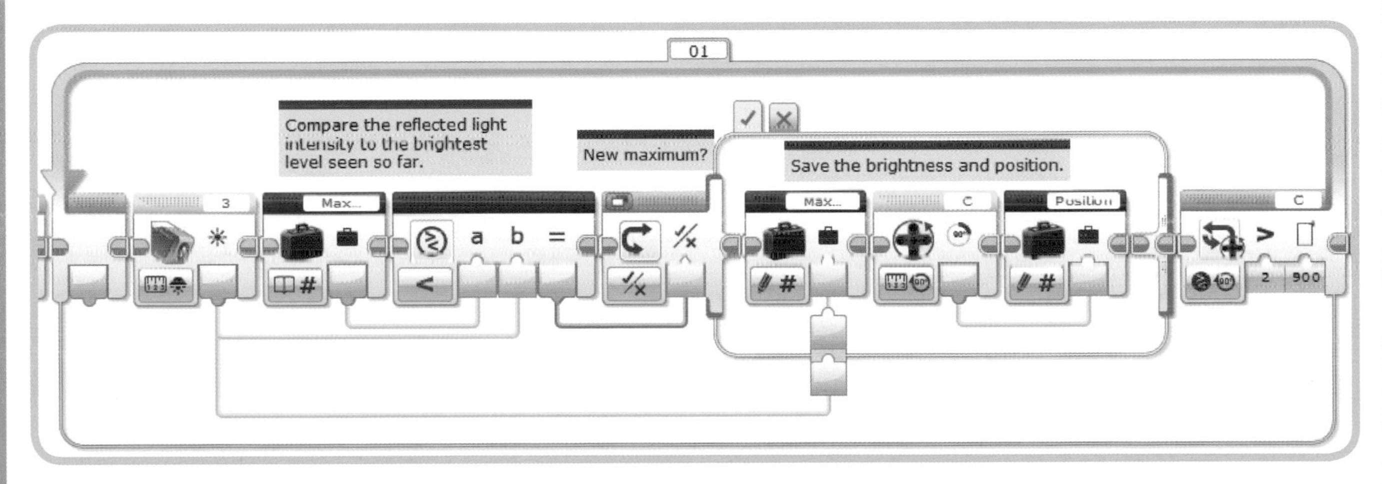

Figure 11-29: Finding the brightest light source

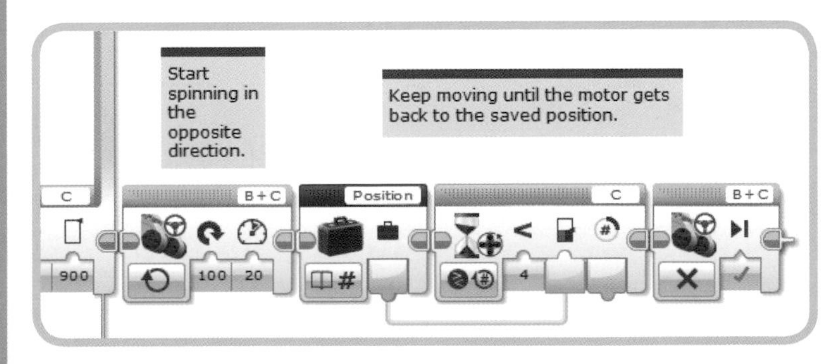

Figure 11-30: Spinning back to the saved position

the constant block

Often, a program uses several blocks with the same settings. For example, the *WallFollower* program contains seven Move Steering blocks that all use the same Power parameter. If you decide to change this parameter from 35 to 45, you need to make sure you change that parameter on all seven blocks. That can be a hassle if you want to test several different Power parameters—plus, you might forget to change the parameter on one of the blocks by mistake.

The *Constant block* lets you save a value that you can then use to set parameters in blocks throughout your program. Figure 11-31 shows the Constant block with the Mode menu open. This block resembles the Variable block except that it only has read modes. That's because you can't write a new value to a Constant block during a program; you can only set the value ahead of time using the box in the upper-right corner of the block.

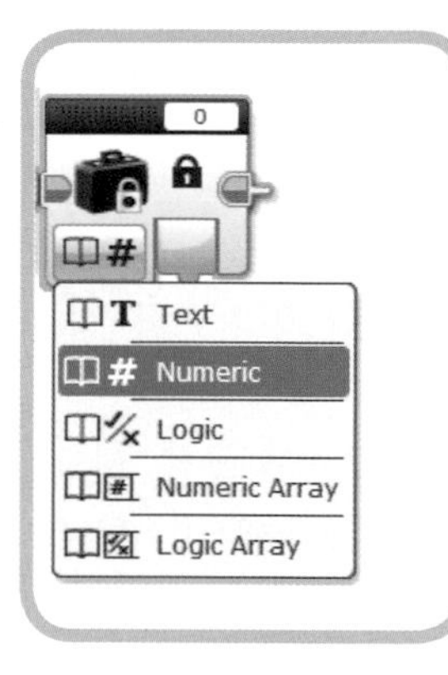

Figure 11-31: The Constant block

For example, Figure 11-32 shows a modified version of the *AroundTheBlock* program from Chapter 4. Here, a Constant block is used to control both Move Steering blocks, rather than setting the Power value for each Move Steering block separately. This way, if you want to test different Power parameters, you only need to change the Constant block instead of changing each Move Steering block. This also prevents you from accidentally changing a setting on one block and not the other. This can be particularly useful when you're working with larger programs, such as *WallFollower*, in which you can use one Constant block to control the Power parameter of seven blocks at once.

further exploration

Here are a few activities involving variables and constants for you to try.

1. Use a Constant block to set the Power item on all the Move Steering blocks in the *WallFollower* program. Then use a variable instead of the Constant block. You'll need one Variable block at the beginning of the program, and then one or more Variable blocks to read the value and pass it to one or more Move Steering blocks. If you use a Variable block, you can avoid the confusion of having long data wires. Do you prefer the solution that uses one Constant block with longer data wires, or the one that uses multiple Variable blocks and shorter data wires? (There's no right or wrong answer; this is more a question of style and preference.)
2. Create a new program called *ObstaclePointer* based on the *ObstacleAvoider* program. Change the program so that the TriBot points in the direction of the nearest obstacle. You'll need the program to find the direction where the sensor

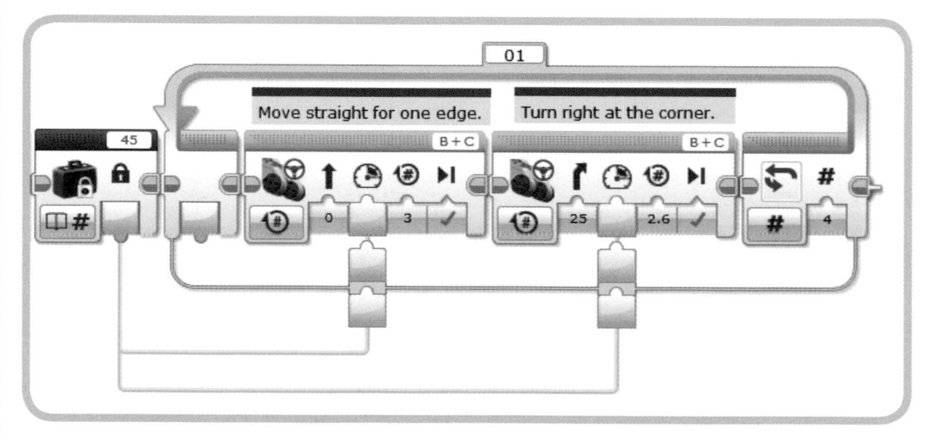

Figure 11-32: Using a Constant block to set the power level

reads the smallest value instead of the largest. Test objects with different shapes, colors, and textures. You should find that the reading from the sensor depends on more than just the distance between the robot and the object. Hint: Instead of a Max Reading variable, you'll need a Min Reading one, initialized to 100 instead of 0.

3. Add a loop around the *ObstacleAvoider* program, and make the robot move forward a little at the end of each loop. In other words, the robot should check for the clearest path, move forward a little (say, five rotations), check again for the clearest path, move forward again, and so on. Repeating this over and over should allow the robot to navigate through an area filled with obstacles.

CHALLENGE 11-1

Create a new program called *ObstacleAvoider* based on the *LightPointer* program. Use either the Infrared or Ultrasonic Sensor to make the TriBot face the direction where the sensor reads the largest value (this should make it face the direction with the clearest path).

CHALLENGE 11-2

If you have a Gyro Sensor, use it instead of the motor C Rotation Sensor to note the position of the highest sensor reading and make the TriBot turn back to face the light. Does this work better than using the Rotation Sensor? Try increasing the speed of the Move Steering blocks and see if that has an effect on which approach works best.

conclusion

Variables let you store and update data to use in your programs. The Variable block is used to create and access the variables within your program. Variables add a lot of flexibility to programs and are essential for solving many types of problems. The programs presented in this chapter have shown you a few ways to use variables. You'll see them used in other ways in the coming chapters.

Constants, created with the Constant block, are useful when you want to use a value that doesn't change to control multiple blocks. This gives you an easy way to consolidate the parameters for several blocks in one place, so you only need to change one value to affect lots of blocks.

This chapter also introduced the Compare block, which lets your program make more complex decisions than it could with the Switch block alone. In Chapter 13, you'll learn more about the Math and Logic blocks, which allow you to make even more sophisticated decisions.

Before delving into more math-related blocks, it's useful to learn about My Blocks, the subject of the next chapter. My Blocks are special blocks that you create from parts of your program and that help make your programs smaller, easier to reuse, and less error prone.

my blocks

Creating a My Block is an easy way to group together and reuse a collection of blocks. A *My Block* is a block that you create from other blocks, which you can then use in your programs just like other EV3 programming blocks. This makes programs shorter and easier to read by grouping a sequence of related blocks into a single block.

In this chapter, you'll learn how to create My Blocks and use them in your programs. I'll walk you through the process of building some My Blocks, starting with a simple one that plays a chime and moving on to a more complex one that displays a number with a label. Along the way, you'll learn all you need to know to create your own blocks.

creating a my block

I'll begin by creating a simple My Block from part of the *DoorChime* program from Chapter 5 (see Figure 12-1). The Sound blocks in this program play a chime with two tones. You'll turn those Sound blocks into a My Block, which makes the program shorter and creates a Chime block that you can use in other programs.

1. Copy the *DoorChime* program from the *Chapter5* project to a new *Chapter12* project.
2. Select the two Sound blocks by drawing a rectangle around them (Figure 12-2) or by holding down SHIFT while clicking both blocks.
3. Select the **Tools ▸ My Block Builder** menu item to create a My Block from the selected blocks. The My Block Builder window appears, as shown in Figure 12-3.

In the My Block Builder window, enter a name and description for the new block. At the bottom of the window is a section where you can pick the icon for the block.

4. Type **Chime** in the Name box.
5. Type **Play a chime using Sound blocks** in the Description box.
6. Choose the Sound block icon (🔊) in the bottom section of the window.
7. Click **Finish**.

Clicking Finish creates the Chime block and replaces the Sound blocks in the *DoorChime* program with the new My Block (as shown in Figure 12-4). After moving the blocks closer together (click the **sequence plug exit** on the right side of each block to collapse the sequence wires), the program looks like Figure 12-5.

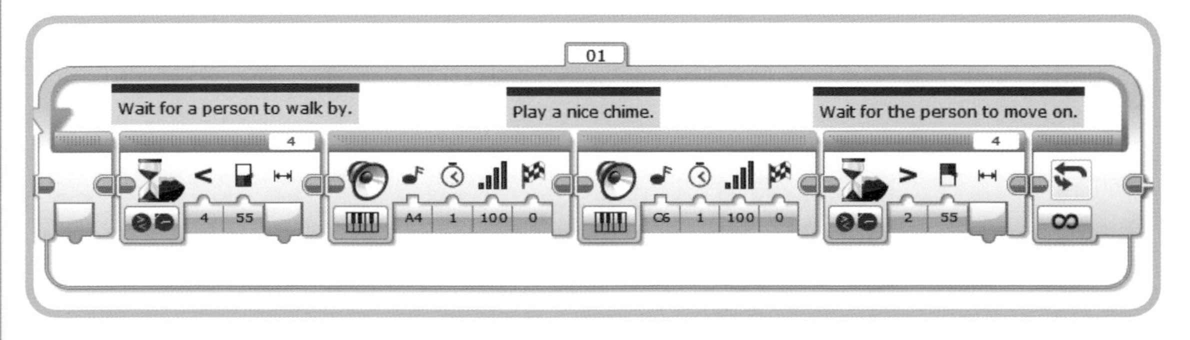

Figure 12-1: The DoorChime *program*

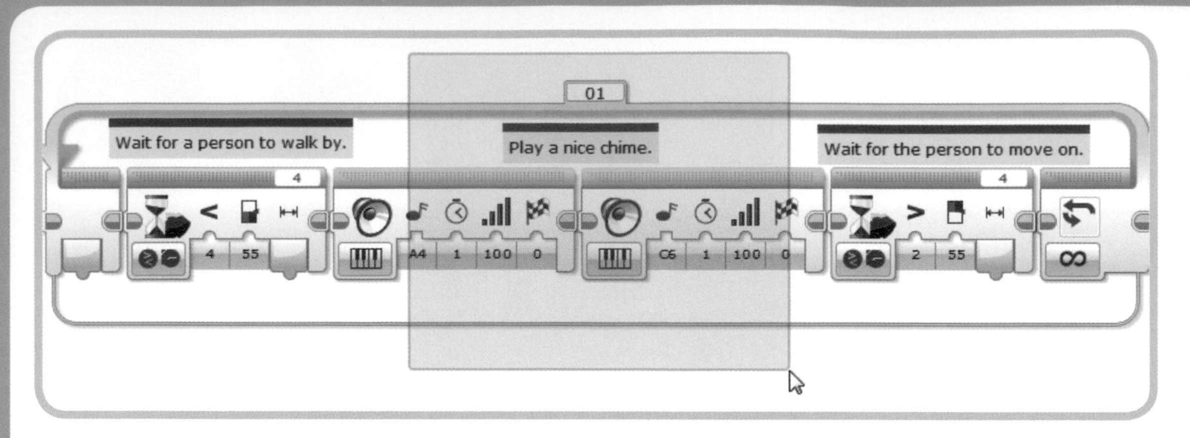

Figure 12-2: Selecting both Sound blocks

Figure 12-3: The My Block Builder window

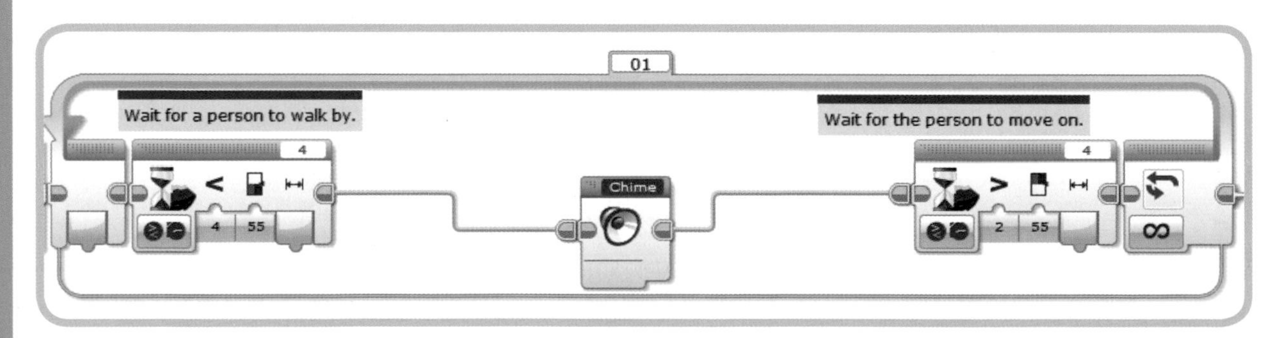

Figure 12-4: The DoorChime *program after creating the Chime block*

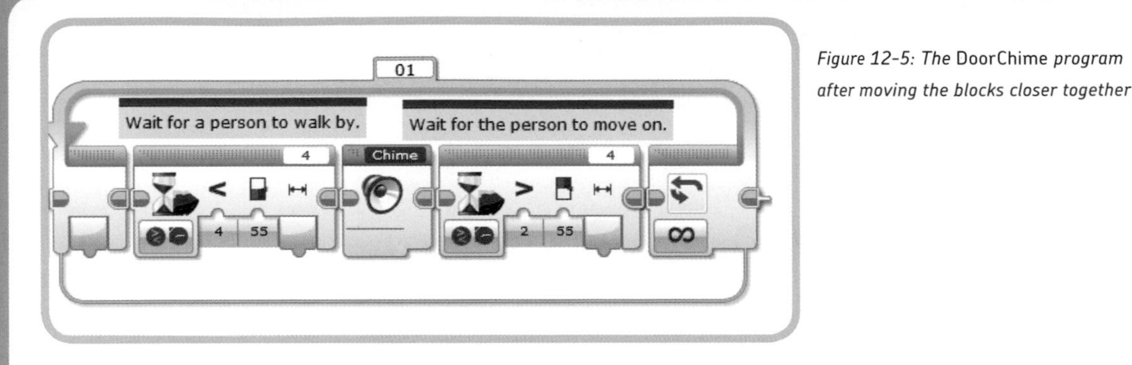

Figure 12-5: The DoorChime *program after moving the blocks closer together*

The program is now smaller and simpler. It's also easier to understand the purpose of a single block than it is two Sound blocks. Although you've changed the arrangement of the blocks, you haven't changed the way the program works; when you run this program, it should behave exactly as it did before you created the Chime block.

the my blocks palette

After you've created a My Block, you can use it in any program in your project, just like any other block. All the My Blocks in your project appear on the My Blocks palette, in the cyan-colored tab on the far right. Each My Block displays the icon you selected when creating it. The name of the My Block appears if you place the mouse cursor over the block, as shown in Figure 12-6.

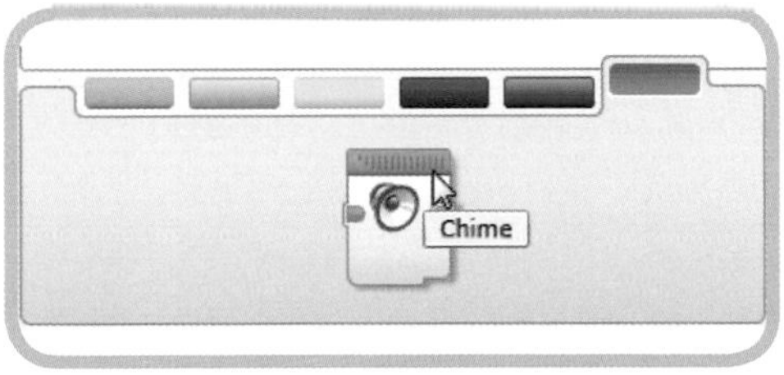

Figure 12-6: The Chime block on the My Blocks palette

Your My Blocks also appear in the My Blocks tab on the Project Properties page (see Figure 12-7). You can use this tab to delete a My Block you're not using, copy and paste My Blocks from one project to another, or export or import the My Block to or from a file on your computer.

editing a my block

To edit a My Block, you can double-click either an instance of the block in a program or on the block name in the My Blocks tab on the Project Properties page. For example, to edit the Chime block, follow these steps:

1. Open the *DoorChime* program if it's not already open.
2. Double-click the Chime block. The two Sound Blocks appear so you can edit them.
3. Add two more Sound blocks. Set the mode to **Play Note**, and select the notes to play (Figure 12-8).

Now when you download and run the *DoorChime* program, it should play all four Sound blocks. Other programs that use the Chime block will also use these new Sound blocks. When you edit a My Block, it affects every program that uses it. This can be helpful if you fix a bug in your My Block because that fix will automatically apply to all programs that use it. On the other hand, you need to make sure that the changes you make to improve one program don't adversely affect other programs.

CHALLENGE 12-1

The *WallFollower* program has three distinct sections: one that keeps the TriBot near the wall, one that moves the TriBot into an opening, and one that turns the TriBot left if it gets to a corner. Turn each of the three sections into My Blocks to make the program shorter and easier to interpret.

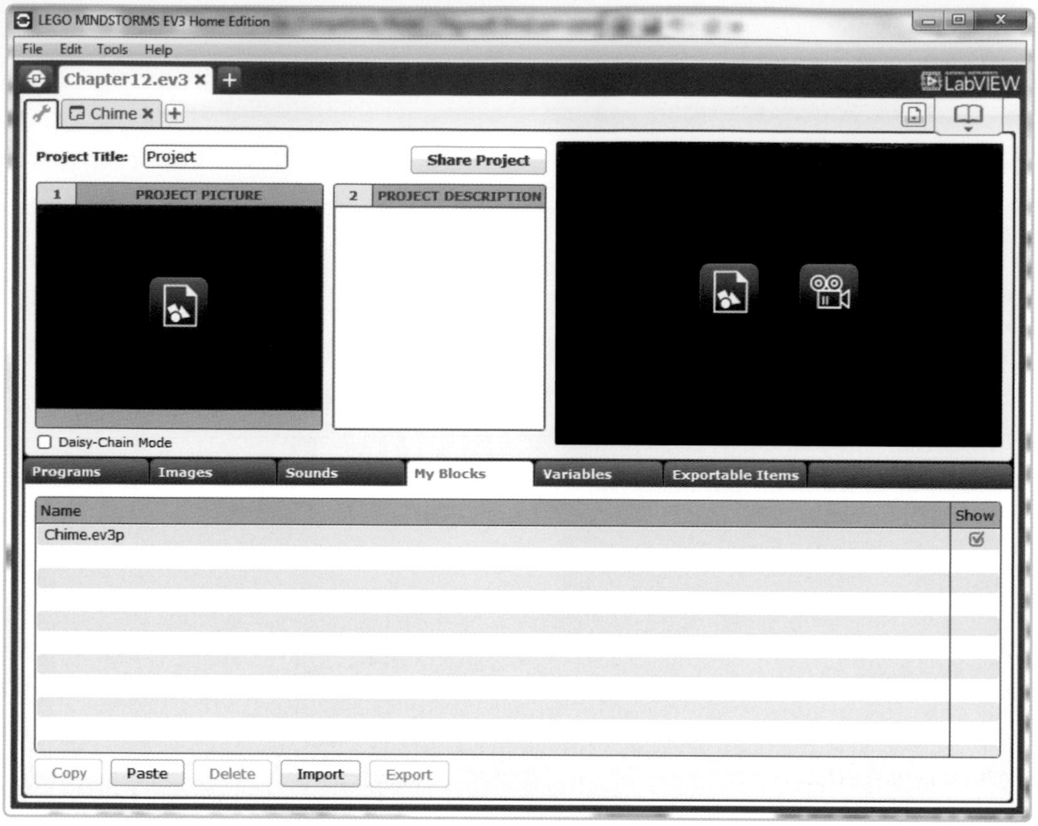

Figure 12-7: The My Blocks tab on the Project Properties page

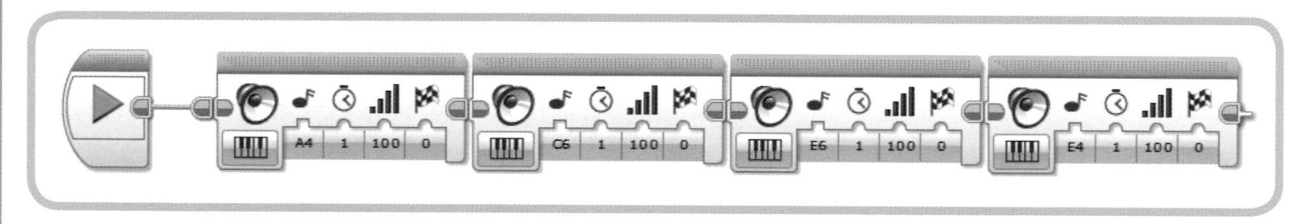

Figure 12-8: Editing the Chime My Block

the LogicToText my block

The Chime block doesn't have any parameters, which means it always does the exact same thing. Most EV3 blocks need parameters that tell the block how to behave. You've also seen how blocks can send output to other blocks using data wires. In this section, you'll create a LogicToText My Block that has an Input parameter and an Output parameter. Let's create a My Block from the Switch block in the *LogicToText* program from Chapter 9 (see Figure 12-9). This block takes a Logic value as an input and creates a Text value as an output. You can then use this block in any program to easily convert a Logic value to text so you can show it on the EV3 screen.

Follow these steps to create the LogicToText My Block:

1. Open the *Chapter9* project, and copy the *LogicToText* program to the *Chapter12* project.

   I want to name the new My Block *LogicToText*, but you can't give a My Block the same name as a program, so let's change the name of the program.

2. Change the name of the program to *LogicToTextBuilder* (double-click the tab with the program name and type the new name).
3. Select the Switch block.
4. Select the **Tools ▸ My Block Builder** menu item.

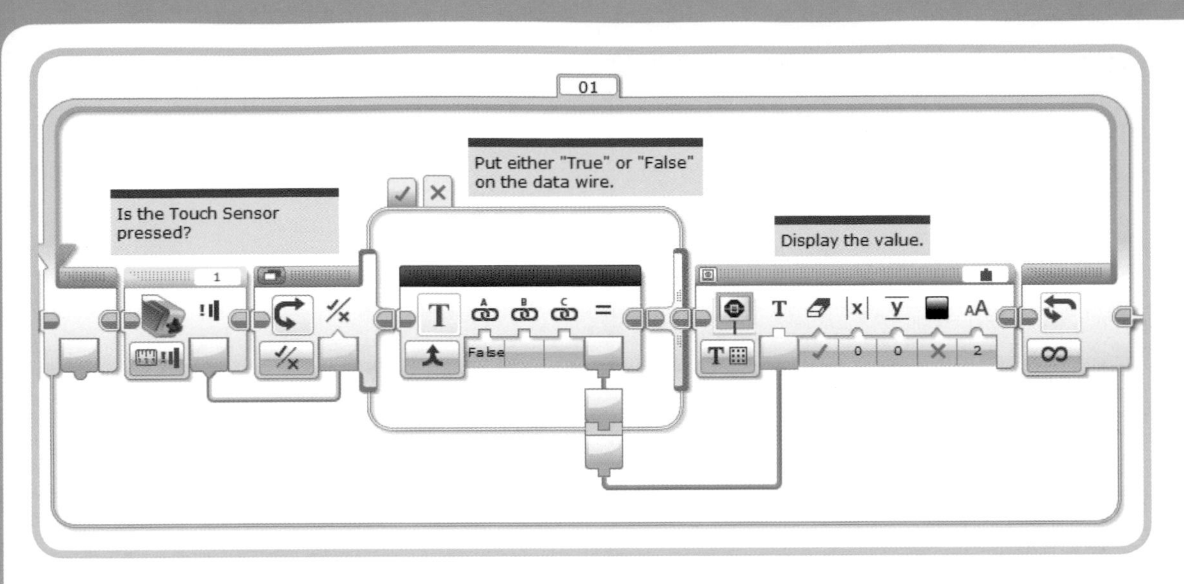

Figure 12-9: The LogicToText *program*

The My Block Builder window should show the My Block with two parameters, highlighted in Figure 12-10. Notice that the Switch block used two data wires: one to accept the Logic value as an input and one to pass the resulting Text value to the blocks that follow. When you create a My Block, the data wires that connect to the blocks you've selected become Input and Output parameters of your new block. In this case, the Logic data wire coming from the Touch Sensor block becomes an Input parameter and the Text data wires going to the Display block becomes an Output parameter.

In the lower section of the window, you should see two new tabs, Parameters Setup and Parameter Icons, which you can use to change how the block's parameters appear. We'll use each of these tabs in a moment.

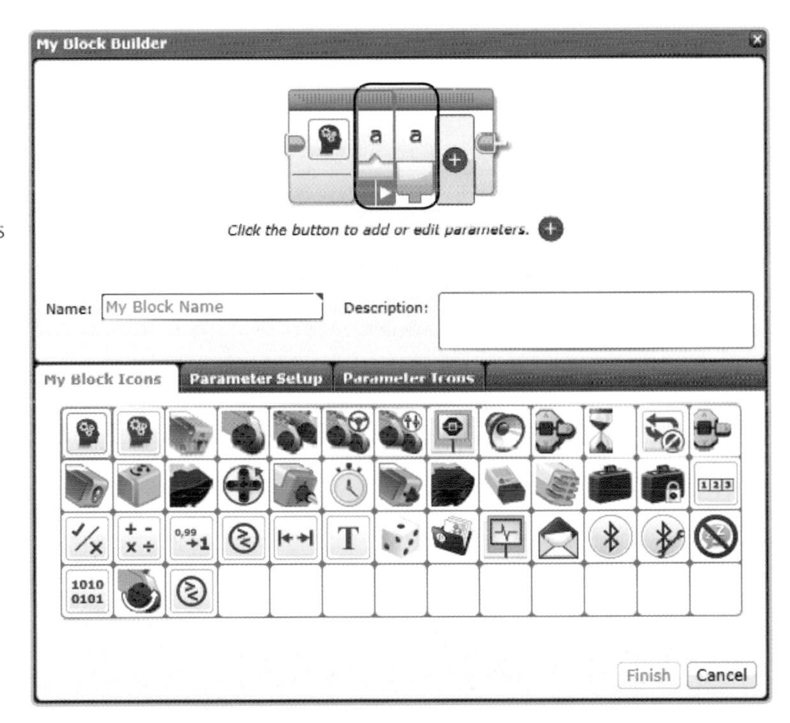

Figure 12-10: The My Block Builder window with the block input and output highlighted

5. In the My Block Builder window, type **LogicToText** in the Name box.
6. Type **Convert a Logic value to a Text value using either "True" or "False"** in the Description box.
7. Because this My Block deals with Logic values, choose the Logic values icon (✓⁄✗).
8. Make sure the first parameter is selected, and then click the Parameter Setup tab. The window should look like Figure 12-11.

This tab allows you to select the parameter's name, type (input or output), data type, and default value. The name is displayed when you hover the mouse cursor over the parameter, just like with other EV3 blocks. The default value you select will be used for the parameter if you don't attach a data wire to it.

Notice that for this parameter, the Parameter Type and Data Type are already set and can't be modified. That's because this parameter was created automatically from a data wire connection. You can also add parameters using the My Block Builder window; when you add parameters that way, you can modify these settings.

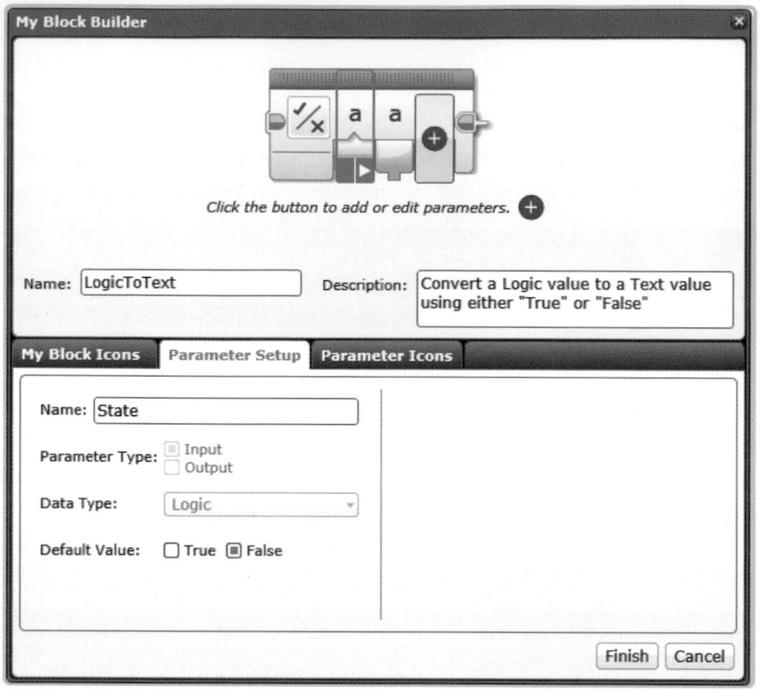

Figure 12-11: The Parameter Setup tab for the first parameter

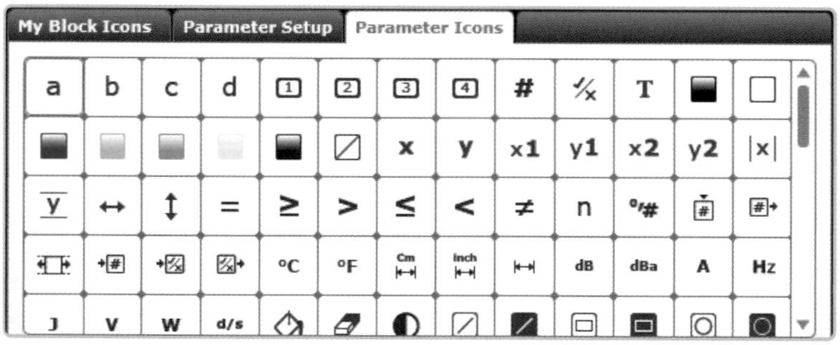

Figure 12-12: The Parameter Icons tab

In this tab, the EV3 software has also filled in the parameter name as State and the Default Value as False. You can leave the Default Value (it doesn't matter too much for this block), but let's give the parameter a better name:

9. Type **Value** in the Name box.

The next step is to select the icon for the Input parameter. There are many icons to choose from so you can almost always find something that is more meaningful than the default.

10. Select the Parameter Icons tab (Figure 12-12).

11. Select the icon for Logic values (✓/×). It should now show up in the block at the top of the window (see Figure 12-13).

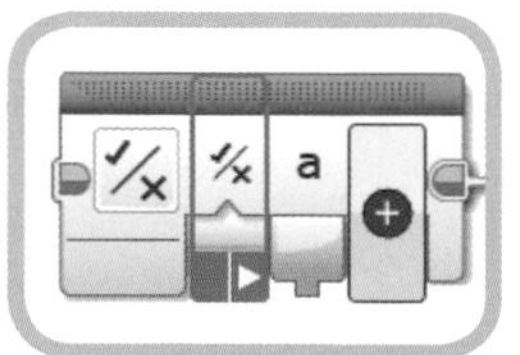

Figure 12-13: The block with the icon you've selected

That's it for the first parameter. Now select the second parameter and set its name and icon.

12. Select the second parameter.

13. Select the Parameter Setup tab.

14. Type **Result** in the Name box.

15. Select the Parameter Icons tab.

16. Select the icon for Text values (T).

The block should now look like Figure 12-14.

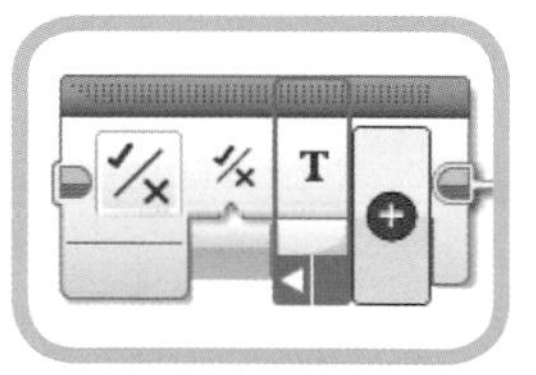

Figure 12-14: Both parameter icons selected

WARNING **Before you click Finish, make sure that you have all the parameters you need and the settings and icons are set the way you want. After you click Finish and the My Block is created, you can't go back to add, remove, or change the parameters or the My Block icon.**

17. Click **Finish** to create the LogicToText My Block.

After moving the blocks closer together, the original program should look like Figure 12-15.

Double-click the LogicToText block to open it in a Programming Canvas, as shown in Figure 12-16. If the Output parameter shows up in an odd location, it's easy enough to drag it over to the right, as shown in Figure 12-17. I like to put the inputs on the left and the outputs on the right. It's also a good idea to add some comments. That way you or another programmer can easily understand what the My Block does and how it does it.

Try running the *LogicToTextBuilder* program. It should display "True" when the Touch Sensor is pressed and "False" when it isn't.

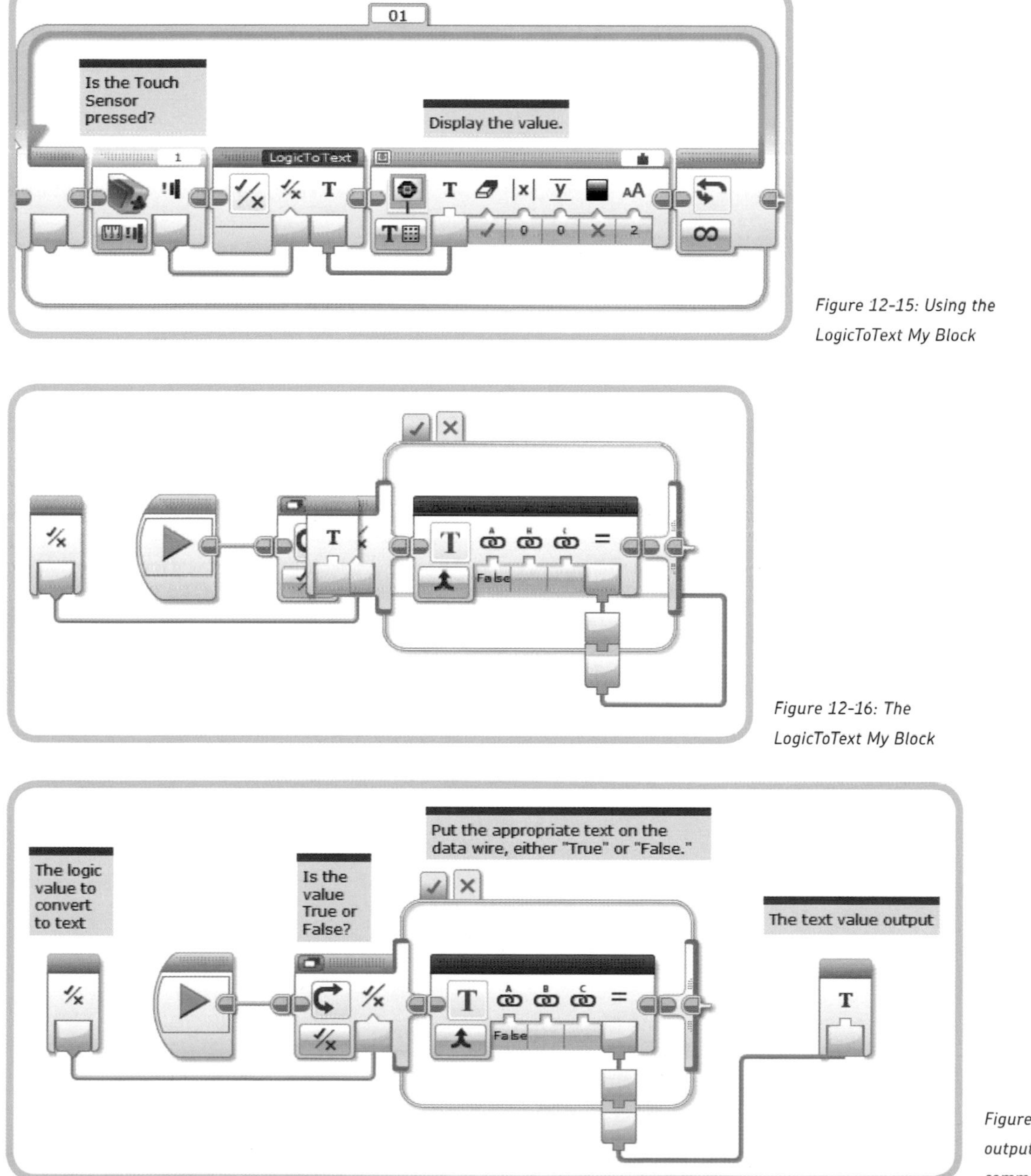

Figure 12-15: Using the LogicToText My Block

Figure 12-16: The LogicToText My Block

Figure 12-17: After moving the output to the right and adding comments

adding, removing, and moving parameters

Let's take a closer look at how to add, remove, and modify parameters in the My Block Builder window. Select a block, open the My Block Builder tool, and add a few parameters. Figure 12-18 shows the controls for a My Block with four parameters.

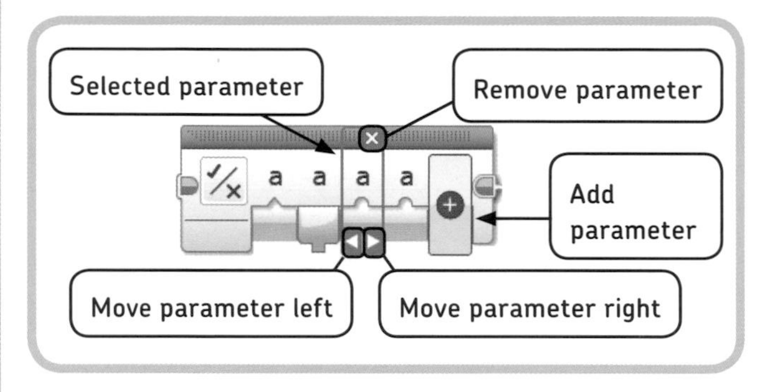

Figure 12-18: Controls for working with block inputs and outputs

- The selected parameter is outlined in blue. You can modify the selected parameter using the options in the Parameter Settings and Parameter Icon tabs.
- The Add Parameter button adds a new parameter to the block.
- The Remove Parameter button removes the selected parameter. You can only remove parameters that you've added with the Add Parameter button. (Parameters that were automatically created from data wire connections cannot be removed.)
- The Move Parameter Left button moves the selected parameter to the left.
- The Move Parameter Right button moves the selected parameter to the right.

the parameter setup tab

The Parameter Setup tab defines how the parameter looks and behaves. We already saw this tab briefly in Figure 12-11 for the LogicToText My Block. Figure 12-19 shows the Parameter Setup tab for one of the parameters I added with the My Block Builder. For parameters added this way, you can choose whether it should be a block input or output and select the data type. If the Parameter Type is Input and the Data Type is Number, the right side of the tab will allow you to select a slider for entering the value if you prefer that over the normal input plug. If you select one of the sliders, you can still type in the value or use a data wire (as with the Move Steering block's Power parameter).

the DisplayNumber my block

In this section, you'll create the DisplayNumber My Block. I use this block often when displaying numbers, either to give the user feedback while the program runs, debug information, or show the result of an experiment.

Figure 12-20 shows the blocks used by the *SoundMachine* program for displaying the volume level, which is controlled by motor B. The Motor Rotation block reads the value, the Text block adds the label and unit, and the Display block shows the resulting text on the EV3 screen. We want our My Block to perform a similar function, except we want to be able to send it any number as input. There are other parameters that would be helpful to be able to configure in this My Block on a case-by-case basis, such as the text for the label and unit.

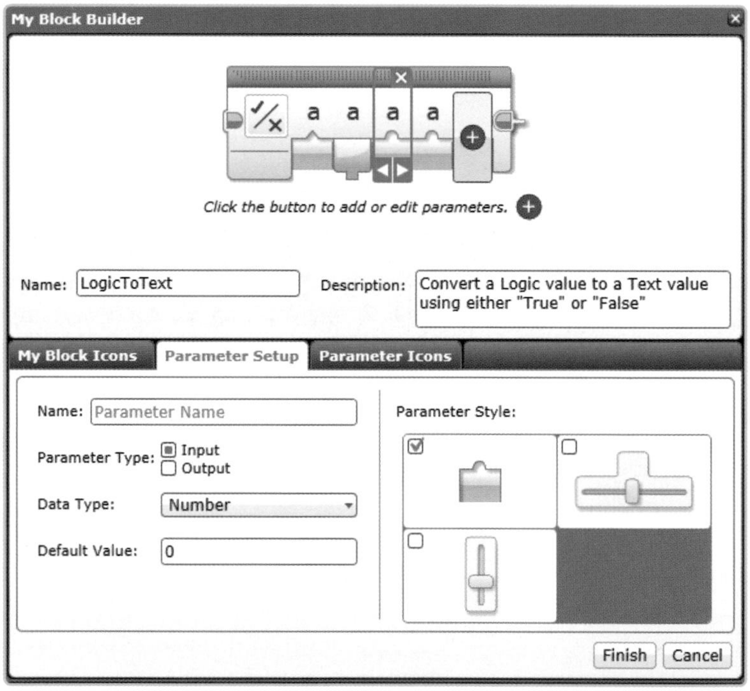

Figure 12-19: The Parameter Setup tab for a new parameter

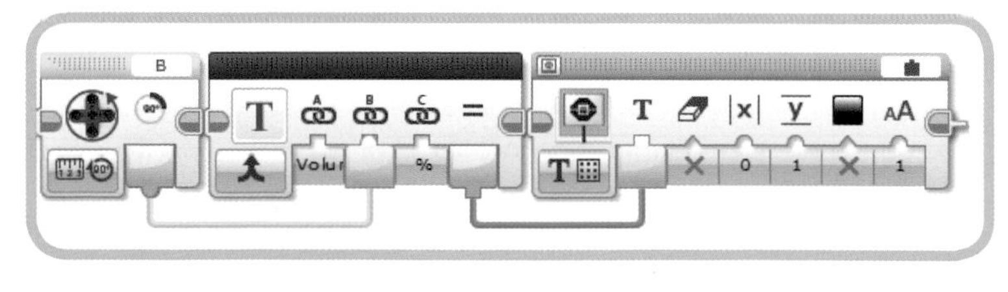

Figure 12-20: Displaying the volume level

Take a look at each of the other inputs to the Text block and Display block in Figure 12-20, and think about which ones should be configurable and what their default values should be. Following is the list of parameters that I would include.

Text block A The label displayed before the number. The default is an empty string.

Text block C The unit displayed after the number. The default is an empty string.

Display block Clear Screen Depending on where the block is used, you may want to clear the screen before displaying the value. I would set the default to True to match the way the Display block works.

Display block Row If you want to display more than one value, the row needs to be changed, so this needs to be configurable. A default of 0 is reasonable, which places the text at the top of the display.

Once you've decided which parameters you want to be configurable in the new My Block, you need to connect data wires to each of the parameters. For block inputs, you can add a Constant block to the beginning of the program. You'll need to set the mode of each Constant block to match the data type of the corresponding parameter (Text for the label and unit, Logic for the Clear Screen parameter, and Numeric for the Row parameter). For block outputs, I typically use a Variable block in Write mode at the end of the program. (The DisplayNumber block has no outputs, so we'll only need Constant blocks here.)

Here's how to create the DisplayNumber block:

1. Create a new program named *DisplayNumberBuilder*.
2. Copy the three blocks shown in Figure 12-20 to the new program.
3. Add a Constant block for each configurable parameter. Set the mode to the correct data type, and draw a data wire from the Constant block to the parameter.

At this point, the program should look like Figure 12-21, and we're ready to create the My Block.

4. Select the Text and Display blocks.
5. Select the **Tools ▸ My Block Builder** menu item.
6. Set the name of the My Block to **DisplayNumber** and add a description.
7. Select an icon for the new block.
8. Select each parameter and set the name, default value, and icon. You can see the names and icons I selected in Figure 12-22.
9. Click the **Finish** button.

When you click the Finish button, two things happen. The DisplayNumber My Block is created and the Text and Display blocks in the *DisplayNumberBuilder* program are replaced with the new block. If you want to restore the builder program to

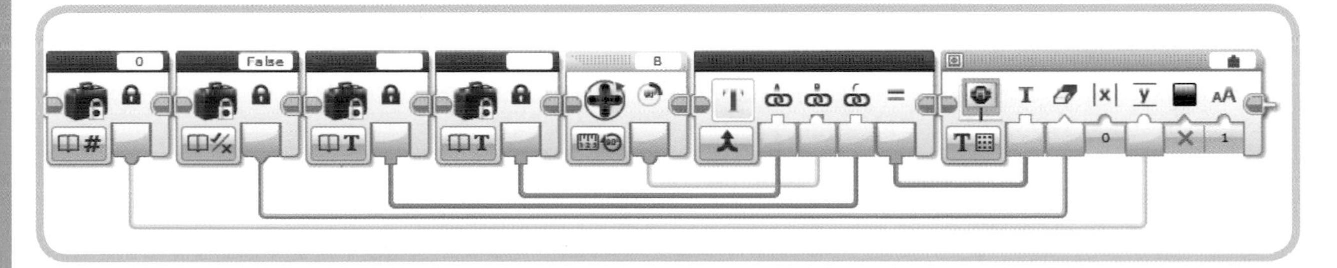

Figure 12-21: The DisplayNumberBuilder *program*

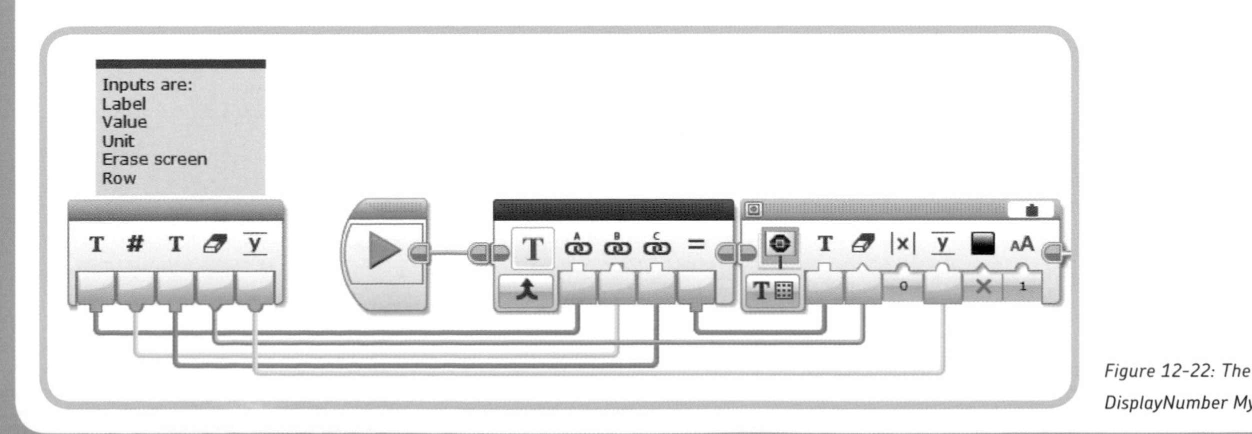

Figure 12-22: The DisplayNumber My Block

its original state, you can select **Edit ▸ Undo** (or press CTRL-Z). This can be helpful if you decide to make changes to any of the parameters in the DisplayNumber block because it returns you to the same program that you used to create the My Block.

After creating a My Block, it's a good idea to open it and make sure it looks how you expect. You might also want to add some comments describing the parameters and how the block should be used. Figure 12-22 shows how the DisplayNumber block should look with some comments added.

After you've created the My Block, you can use it in any of your programs (though to use it in a different project, you'll have to copy the My Block first). For example, Figure 12-23 shows the *SoundMachine* program with the Text and Display blocks replaced by DisplayNumber My Blocks. The program is now shorter, which makes it easier to see the logic of the program, and the purpose of the DisplayNumber blocks is more explicit.

changing my block parameters

As shown in "Editing a My Block" on page 139, it's very easy to edit the contents of a My Block. You simply open the My Block and edit the blocks just as you would a regular program. Unfortunately, there is no easy way to change a My Block's parameters. You'll need to re-create the My Block if you decide you want to add a new parameter or change an existing parameter's name or default value.

Re-creating a My Block is much easier if you already have a program like *DisplayNumberBuilder* to use as a starting point. If you don't have a builder program, you can create one by copying the blocks from the My Block to a new program and adding Constant or Variable blocks for any data wires you need. Once you have a builder program, you can re-create the My Block with these four steps:

1. Open the existing My Block and rename it by adding "Old" to the name (double-click the name on the tab at the top of program, and enter a new name).
2. Create the My Block from the builder program.
3. Go through each program that uses the My Block and replace the old My Block with the new one.
4. Delete the old My Block.

When you rename a My Block, the EV3 software uses that new name in all the project's programs that use the block. So changing the name from DisplayNumber to DisplayNumberOld will update all the programs to use DisplayNumberOld, and you'll have to update the programs manually to use the new DisplayNumber block.

variables and my blocks

You can use variables in your My Blocks just as you do in the main program. The important thing to know is that there is one list of variables that is shared by the program and all the My Blocks in the program. EV3 variables are called *global variables* because they can be accessed from anywhere in a program, including My Blocks.

That means you can use variables to share information between the main program and the My Blocks it uses or between two (or more) My Blocks. For example, if you split the *WallFollower* program into three My Blocks, you could use one variable to control the Power parameter of all the Move Steering blocks even though the Move blocks will be divided among the three My Blocks.

Variables are also useful if you need a My Block to remember a value. For example, take a look at the ScrollDisplay My Block in Figure 12-24. This My Block is designed to display scrolling text on the EV3 screen. The text to display is passed to the block on a data wire. The first time the block is used, it

MY BLOCKS AND DEBUGGING

While your program is running, you can use the EV3 software to see which block is running and the values on data wires. I find these two features incredibly useful when debugging a program. Unfortunately, these features only work for the main program, so you can't use them to tell what's going on inside a My Block. This means that if you have a bug inside a My Block, it can be more difficult to find and fix.

On the other hand, if you thoroughly test your My Blocks and are certain they work correctly, finding bugs in your main program is much easier. It's a good idea to test the code in a Builder program before creating a My Block, and small test programs can help make sure the My Block works as expected. Using My Blocks reduces the size of your program and therefore the number of blocks you need to look at to find a problem. Using well-tested My Blocks for common tasks can greatly reduce the amount of time spent debugging.

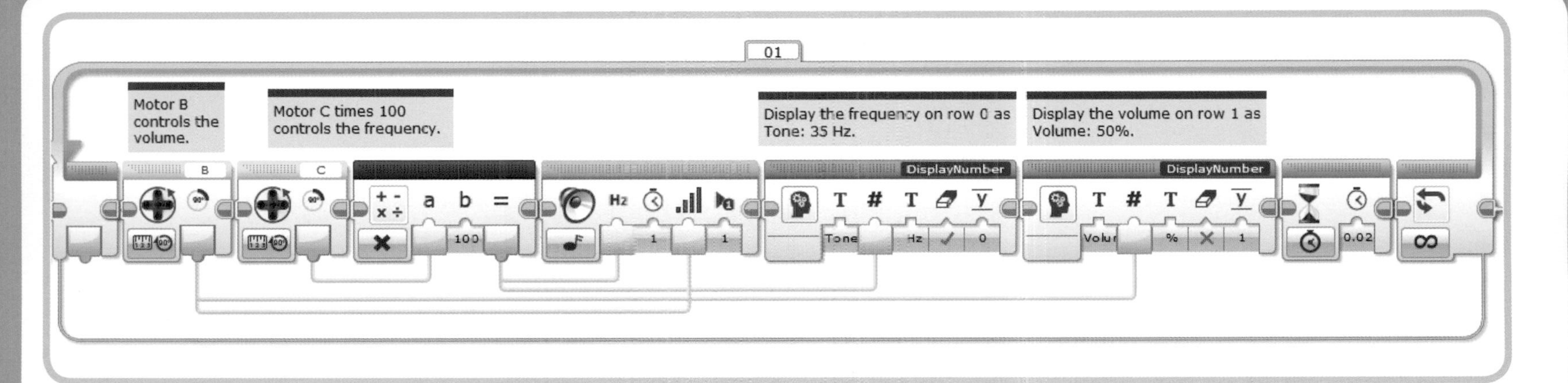

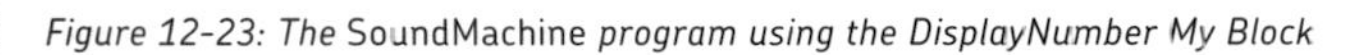
Figure 12-23: The SoundMachine program using the DisplayNumber My Block

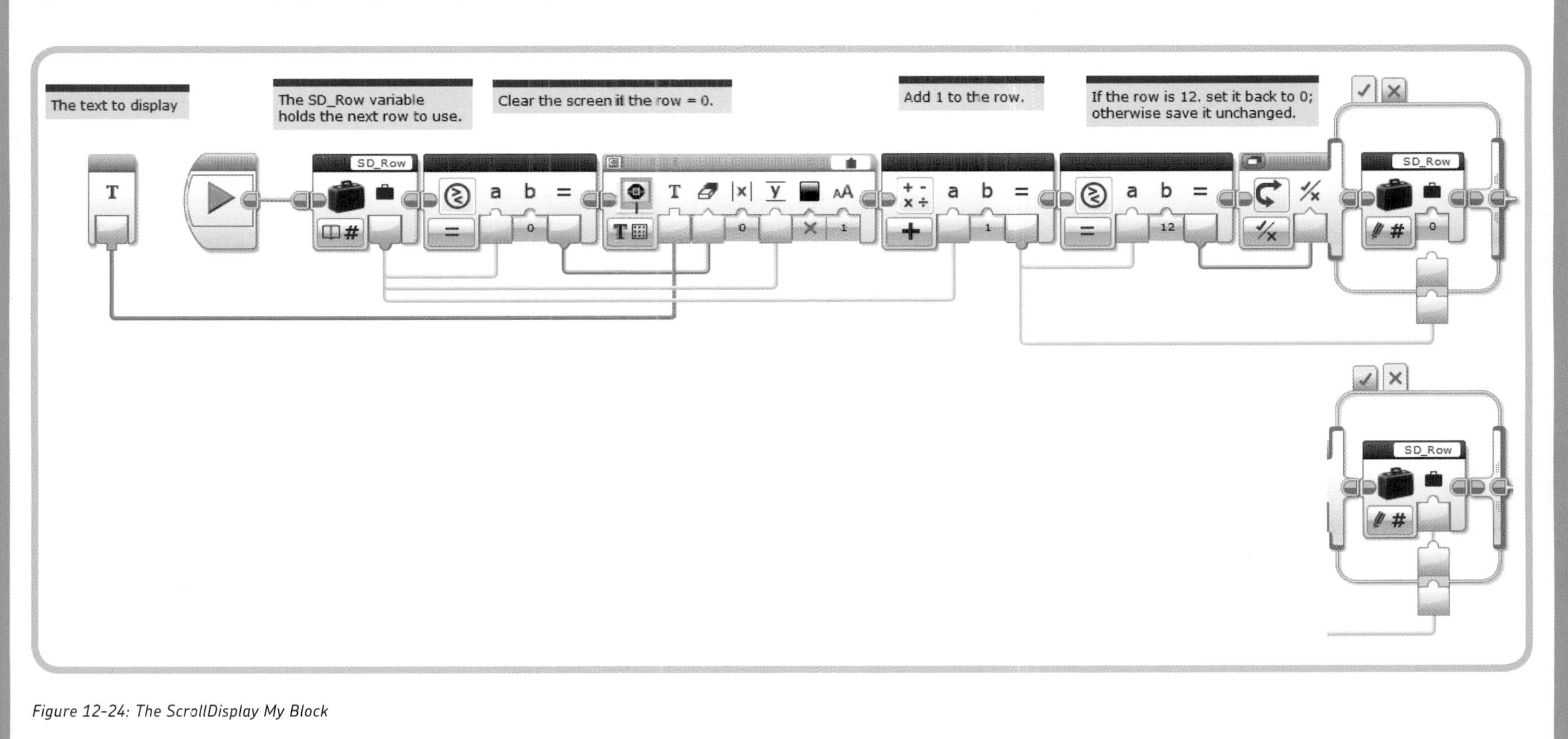

Figure 12-24: The ScrollDisplay My Block

clears the screen and displays the text on the first line. The next time the block is used, it displays the text on the second line; the next time on the third line; and so on, until it displays the text on the last line. Then the next time the block is used, it clears the screen and displays the text on the first line again.

The block uses a variable named SD_Row to keep track of which row to use. I used the prefix SD_ because the variable is used in the ScrollDisplay block. Using a prefix based on the block name helps avoid accidently using and changing a variable that already exists in the main program or another My Block.

This block uses a lot of data wires, but the basic mechanism is actually fairly simple. The EV3 display can show 12 lines of text, with the rows numbered 0 to 11. Each time the block is used, we want text to appear on the next line. When the text reaches the last row (11), we want it to wrap around to the top line (row 0) the next time and clear the screen.

The SD_Row value is passed to the Display block as the Row parameter. A Compare block checks to see if the Row value is 0 to determine if the screen should be cleared. After displaying the text, the Row value is incremented by 1. Then the Compare and Switch blocks check the value: If it's 12, 0 is stored in the SD_Row variable; otherwise, the new incremented value is stored. This updated value will be used to set the Row value the next time the My Block is used.

further exploration

Here are two more activities for you to practice building My Blocks:

1. Create the DisplayLogic block, which works like the DisplayNumber block but for Logic values. Use the LogicToText block within your new block to turn the Logic value into a Text value.
2. The *LightPointer* program from Chapter 11 has two parts: one to find the direction to the light source and another to point the TriBot in that direction. Create two My Blocks for the two sections of the program. Use data wires to pass the position between the two sections rather than using the Position variable.

CHALLENGE 12-2

Add volume and duration parameters to the Chime My Block, and apply these to each of the Sound blocks. Each parameter should have reasonable default values (for example, 0 is not a good default for volume). Check that the *DoorChime* program uses your updated block.

conclusion

Creating your own blocks is a simple yet powerful way to reuse code and makes your programs easier to understand and less prone to error. In this chapter, you learned how to create My Blocks with varying degrees of complexity. Simple blocks like the Chime block allow you to easily reuse code and help keep programs to a manageable size. More complex blocks with many configuration options, such as the DisplayNumber block, can save you from rewriting the same complicated code over and over.

13

math and logic

In this chapter, you'll learn about the Math block's Advanced mode, which lets you perform more sophisticated calculations in your programs. You'll also learn about the closely related Logic block, which lets you combine logical values so your programs can make complex decisions. Other blocks that work with numbers are the Range block, Random block, and Round block, which are covered in this chapter as well.

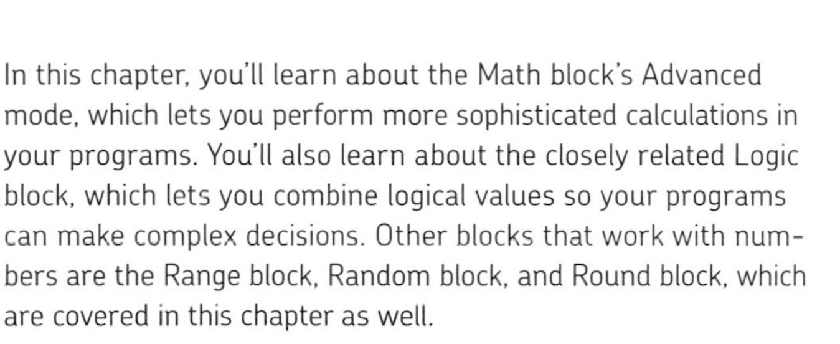

the math block's advanced mode

The Math block's Advanced mode lets you perform complex calculations. You've used the *Math block* in several programs by selecting a single operation (addition, multiplication, and so on) and setting two parameters. The *Advanced mode* (Figure 13-1) lets you combine multiple operations and up to four Input parameters (named *a* through *d*) to calculate the result of more complex mathematical expressions.

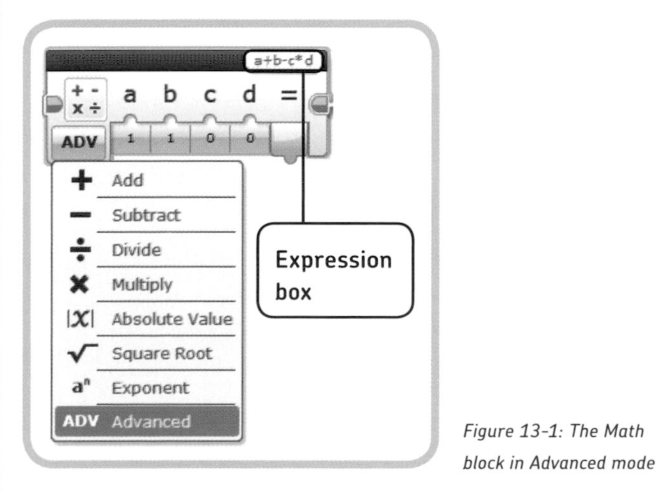

Figure 13-1: The Math block in Advanced mode

To enter your expression, click the Expression box, and a small window appears that contains an entry box on top and a list of operators and functions below (Figure 13-2). You can simply type your expression using common math operator symbols, or you can click one of the operators or functions in the list to add it to the current expression.

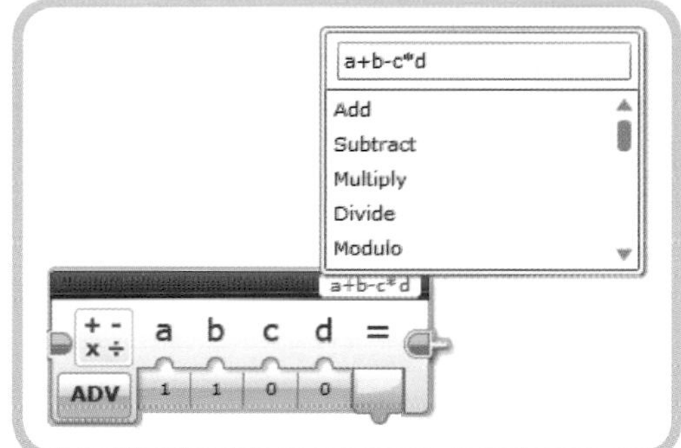

Figure 13-2: Entering the expression

supported operators and functions

Table 13-1 shows the list of supported operators, most of which you probably recognize. To use these operators, you can simply type the appropriate symbol, or you can select the operator from the list in the Expression box window.

table 13-1: supported operators

Operator	Description
+	Add
-	Subtract or negate. 5–3 is subtraction. –*a* is negation.
*	Multiply
/	Divide
^	Exponent. 2^3 means two raised to the third power, or 2^3.
%	Modulo. Gives the remainder when the first number is divided by the second number. 5 % 2 is 1 because 2 goes into 5 twice with a remainder of 1.

Table 13-2 shows the list of supported functions, along with a brief description. Most of these deal with fairly advanced math, and we'll use only a few of these functions in this book.

table 13-2: supported functions

Function	Description
floor()	Rounds a value down to the nearest integer. floor(4.7) is 4, floor(4.1) is 4, and floor(-4.4) is -5.
ceil()	Rounds a value up to the nearest integer. ceil(4.7) is 5, ceil(4.1) is 5, and ceil(-4.4) is -4.
round()	Rounds a value to the nearest integer. round(4.7) is 5, round(4.1) is 4, round (-4.4) is -4, and round(-4.7) is -5.
abs()	Gives the absolute value. abs(5) is 5, and abs(-5) is 5.
log()	Gives the base 10 logarithm of a value.
ln()	Gives the natural logarithm of a value.
sin()	Gives the sine of an angle. All the trigonometric functions work in degrees.
cos()	Gives the cosine of an angle.
tan()	Gives the tangent of an angle.
asin()	Gives the arcsine of an angle.
acos()	Gives the arccosine of an angle.
atan()	Gives the arctangent of an angle.
sqrt()	Gives the positive square root of a value. sqrt(16) is 4.

Clicking a function in the list adds the function name and an opening parenthesis; then you need to enter the value that you want to perform the function on, followed by a closing parenthesis. For example, if you want to round the *a* parameter to the nearest integer, click the Round item so that *round(* appears. Then type ***a)*** (see Figure 13-3).

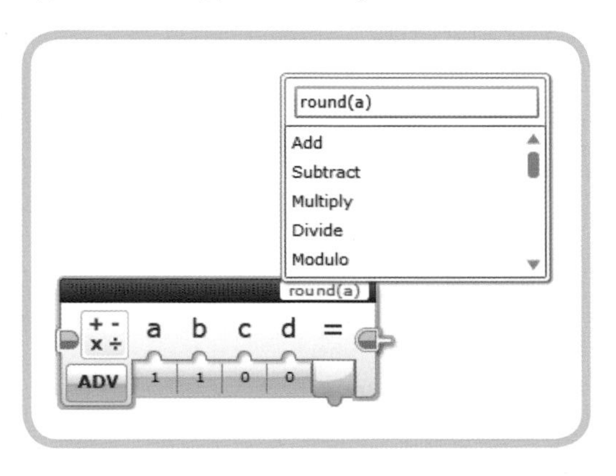

Figure 13-3: Entering an expression

the modulo operator

The *modulo operator* (%) gives the remainder when one number is divided by another. For example, 7 % 4 is 3, because 4 goes into 7 once with a remainder of 3. (The expression 7 % 3 is read "seven mod three"). This operator has some properties that make it very useful in computer programming.

Table 13-3 shows the result of the expression *a* % 3 for increasing values of *a*. Notice that the result starts at 0, increases to 2, and then starts at 0 again. The modulo operator is useful when you want a value to increase but then reset to a starting value each time it reaches a certain point. We saw this situation is Chapter 12 with the ScrollDisplay My Block, where the row starts at 0, increases to 11, and then goes back to 0. Figure 13-4 shows the part of the ScrollDisplay block that increases the row number, checks to see if it has reached 12, and then stores the new value. The value coming into the Math block from the left is the previous row value.

table 13-3: the behavior of the modulo operator

a	*a* % 3
0	0
1	1
2	2
3	0
4	1
5	2
6	0
7	1

By using a Math block in Advanced mode and the modulo operator, we can accomplish the same thing with just two blocks (Figure 13-5). The expression (*a* + 1) % 12 adds 1 to *a* (which is the previous row number) and then gives the remainder when that value is divided by 12. So the row number starts at 0, increases to 11, and then returns to 0, just like in the original code.

math block errors

If you enter an expression that the Math block can't calculate properly, it produces an error. The Math block responds to different errors in different ways, and these errors can cause problems when you pass the Math block result to other programming blocks.

You can check for Math block errors by looking at the value on the data wire coming out of the block, or by showing the value on the EV3 screen using a Display block. For example, if you try to divide a number by 0, then the data wire coming

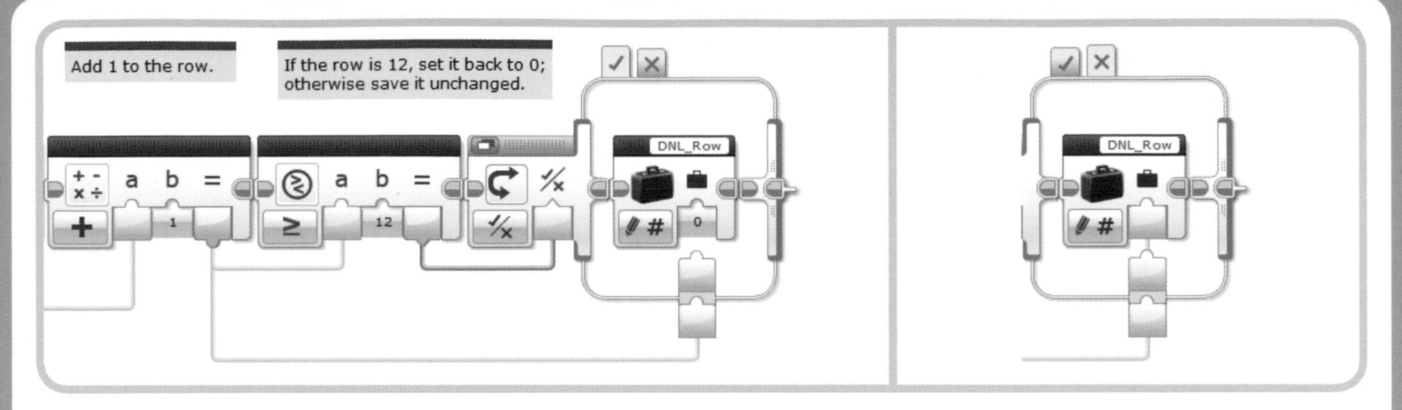

Figure 13-4: ScrollDisplay blocks for computing and saving the next row number

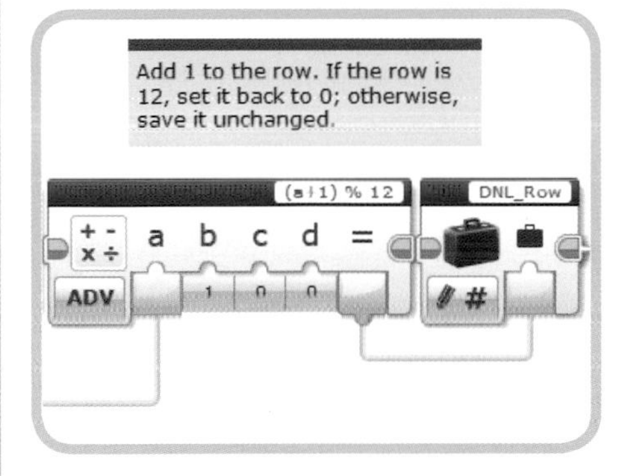

Figure 13-5: Computing and saving the next row number with two blocks

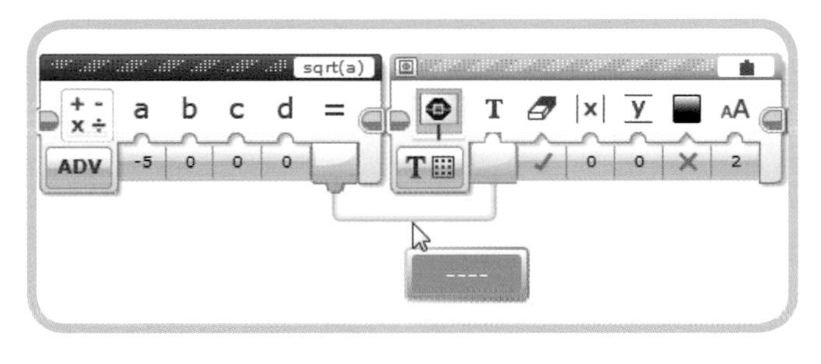

Figure 13-7: Math block error on the data wire when taking the square root of a negative number

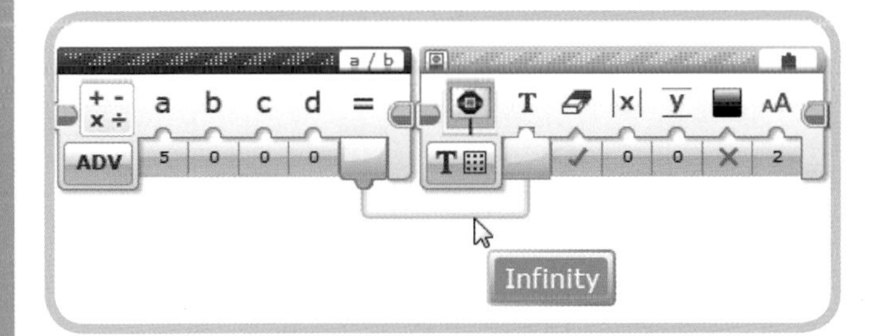

Figure 13-6: Math block error on the data wire when dividing by 0

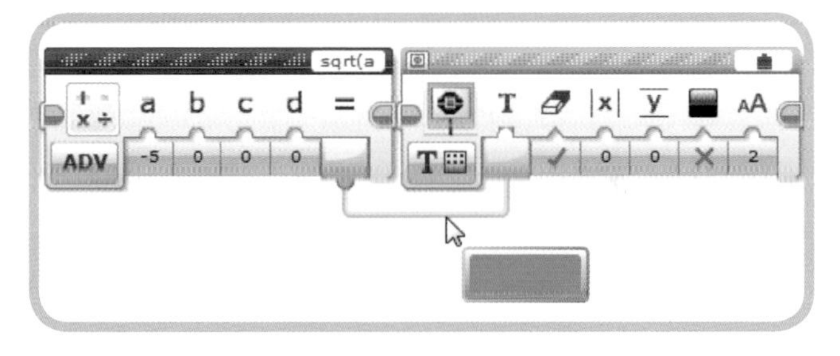

Figure 13-8: Math block error on the data wire when entering an invalid expression

from the Math block will show Infinity, as in Figure 13-6, and will be printed on the EV3 display as "Inf".

Another common error occurs when you try to take the square root of a negative number. This produces a special Error value, which is displayed as ---- on either a data wire or the EV3 screen (see Figure 13-7). Similarly, if you enter an expression that's missing a value or a parenthesis and can't be evaluated, such as sqrt(*a*, or *a* + *b**), the value on the data wire might appear blank, as it does in Figure 13-8, and will appear on the display as "----".

Using the value as an Input parameter to a block often gives strange results. For example, if you use the square root of a negative number for the Move Steering block's Power parameter, the block acts as if the Power is set to 100. The same value used for the Duration in degrees makes the block behave as if you set the parameter to 0. Using the value for the Steering parameter causes the two motors to oscillate back and forth, a behavior that doesn't correspond to any Steering value. Keep an eye out for these Error values because they can cause bugs.

a proportional LineFollower

In this section, we'll turn back to our *LineFollower* program and use an advanced Math block to improve how the program adjusts the robot's steering. The part of a line-follower program that adjusts the steering based on sensor readings is called a *control algorithm*. Improving the control algorithm makes the TriBot move more smoothly and follow lines with tighter turns.

The control algorithm developed in Chapter 6 and reworked in Chapter 9 (see Figure 13-9) is called a *three-state controller* because the program does one of three things based on the sensor reading: go straight, turn left, or turn right. The main problem with this method is that when the robot needs to turn, it always turns the same amount; whether the robot encounters a sharp or gentle turn, it uses the same fixed Steering value.

It would be better if the Steering value depended on the sharpness of the line, with gradual steering for straighter curves and sharp steering for sharper corners. This makes the program respond more quickly to changes in the direction of the line while still moving smoothly when the line is straight. This approach is called a *proportional controller* because the change made to the steering is proportional to, or directly related to, the robot's distance from the edge of the line.

A proportional controller changes a control variable (in this case, the steering direction) based on a *Target value* and an *Input value*. In our case, the Input value is the reading from the Color Sensor. The Target value is the Color Sensor reading when the sensor is directly over the edge of the line. We determined this value back in Chapter 6 by taking the average of the Color Sensor readings with the TriBot off the line and centered over the line, which in my case was 52.

The difference between the Target value and the Input value is called the *Error value*. You can think of the Error value as the difference between where we want the robot to be and where it actually is. We multiply the Error value by the Gain value to get the Steering value. The *Gain value* determines how quickly the robot reacts to changes in the Error value. A smaller gain makes the robot move slowly, which means that it might not react quickly enough for tight turns, but results in less

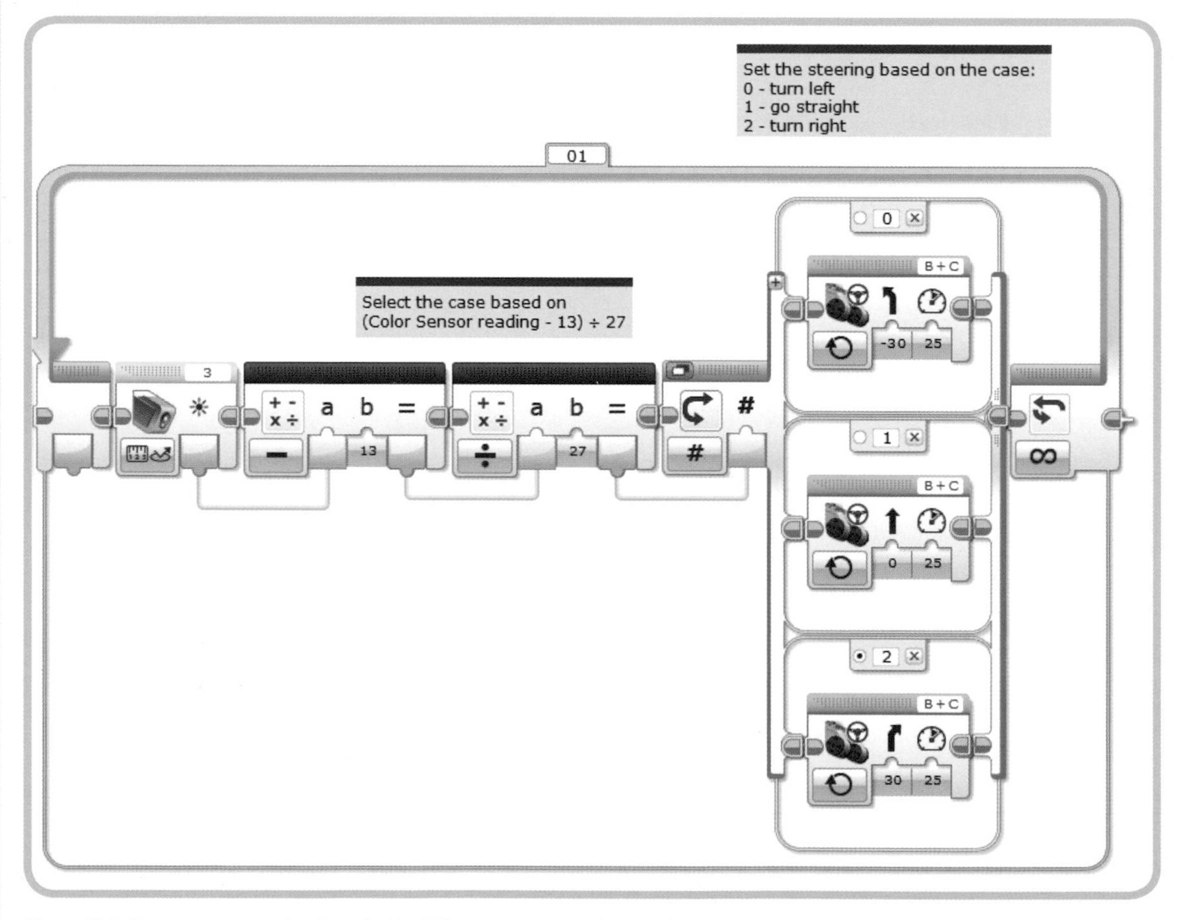

Figure 13-9: Three-state controller from the LineFollower *program in Chapter 9*

side-to-side motion when the line is fairly straight. A larger Gain value means a quicker reaction but can cause jerkier motion. Selecting the Gain value is called *tuning* the controller and usually involves some trial and error.

These are the equations for the Error value and Steering value:

Error value = Target value – Sensor reading
Steering value = Error value × Gain value

We can combine these two expressions into (Target – Sensor Reading) × Gain and use a single Math block in Advanced mode to calculate the Steering value. Two of these values (the Target value and the Gain value) are constants, and in the program, we'll use Constant blocks to supply the values to the Math block—this makes them easier to adjust. Changing the value shown by a Constant block requires fewer mouse clicks and is less error prone than changing the expression in the Math block.

The complete program is shown in Figure 13-10. I set the Gain value to 0.7, which works well for my test line. Try different values to see what works best for your setup. These are the Input parameters to the Math block:

a The Color Sensor reading
b The Target value
c The Gain value

Using these parameters, the expression for the Steering value as described in the previous paragraph would be $(b - a) \times c$. However, that expression yields values that steer in the opposite direction from the previous *LineFollower* program. (This would keep the robot on the right side of the line, instead of the left side like our earlier program did.) To make this more consistent with previous versions, I changed the sign of the result by making the expression $-(b - a) \times c$.

Test this program with your own Target value and Gain value. You should be able to get a result that works better than the three-state *LineFollower* program.

EV3 timers

In the next program, you'll use EV3 timers. The EV3 has eight built-in timers that act like stopwatches. You can use an *EV3 timer* to tell you how long your program has been running or to measure how long it takes the robot to perform a particular task. Typically, you reset the timer to 0 before beginning a task, and then read the timer when the task is complete. Notice that this is similar to how we've used the Rotation Sensor and Gyro Sensor in some of our programs—you can think of a timer as simply a time sensor.

Because the EV3 has eight timers, you can perform several timing tasks within the same program, for a variety of purposes. Here are a few ideas:

* Time how long it takes your entire program to run, and use that information to compare different approaches. For example, you might measure how long it takes your robot to solve a maze using different programs and then use that information to choose the faster solution.
* Time parts of your program to see whether you can speed up certain sections.
* Use timers to make your program perform a periodic action. For example, as part of an experiment, you could use a timer to read a sensor every 10 seconds over a period of 5 minutes.
* Use a timer to limit how long you wait for a sensor to reach an expected Target value. This technique can help you avoid situations where your program stops working completely if something unexpected happens.

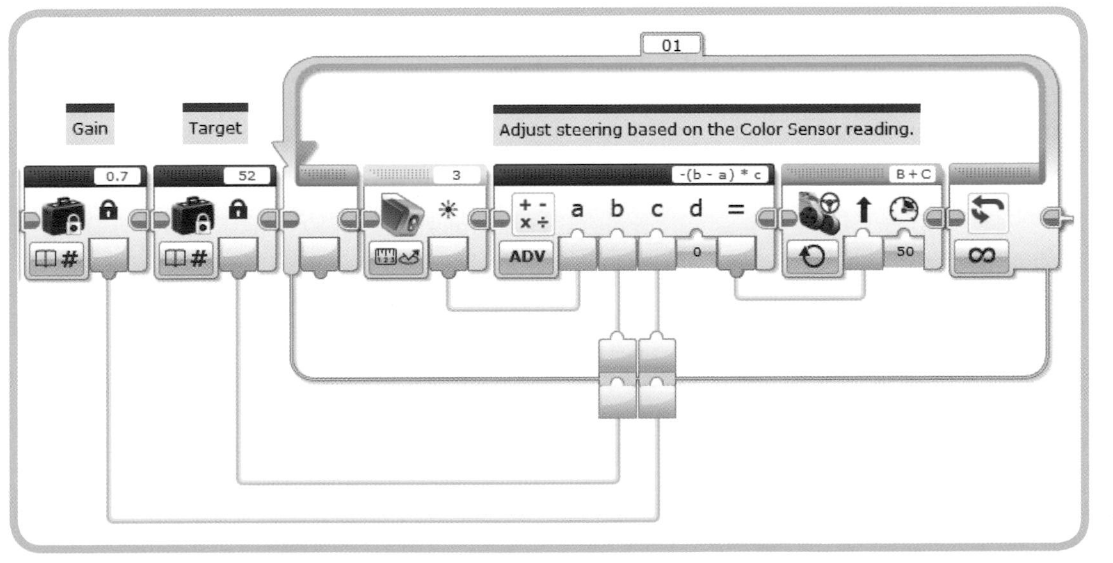

Figure 13-10: A proportional LineFollower *program*

CHALLENGE 13-1

The *GyroTurn* program introduced in Chapter 5 has the TriBot make a quarter-turn by moving until the Gyro Sensor reads 90. This works well if the robot moves slowly and less well when it moves quickly. Improve on this by using the Gyro Sensor reading to control the speed, making the robot move more quickly when far from the end of the turn and slowing it down as it gets close to the target. This lets the TriBot make an accurate quarter-turn quickly, completing it in a second or two rather than the several seconds it takes at a constant, slow speed. Keep the TriBot from spinning too slowly by making sure that the Math block always gives some minimum value.

HINT 10 + (Target – Sensor reading) × Gain will always be at least 10 if the target is greater than or equal to the reading (and the gain is positive).

Timers can be selected from the Sensor list of the Wait, Switch, and Loop blocks. You can also control a timer using the Timer block, which appears on the Sensor palette. The *Timer block* (Figure 13-11) has three modes: *Measure mode* reads the current value of the timer (in seconds), *Compare mode* compares the value with a threshold and gives a Logic value result, and *Reset mode* resets the timer to 0. The block also has a parameter to select which of the eight timers to use.

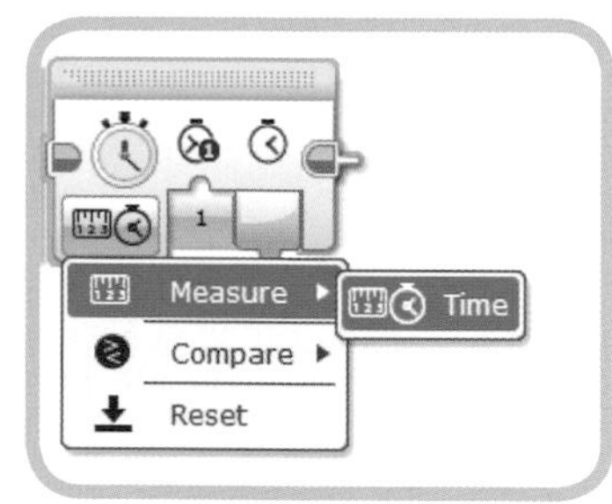

Figure 13-11: The Timer block

the DisplayTimer program

The *DisplayTimer* program combines the power of the Math and Timer blocks to show a running timer on the EV3 screen. The program takes a reading from the Timer block and displays it in the typical minutes:seconds format; for example, 0:03 for three seconds, and 2:15 for two minutes and fifteen seconds. The program uses a Loop block to keep running until you stop it, and each time through the loop, a Timer block reads the timer and the new value is displayed.

Note that the Timer block reads out fractional values, such as 7.46 or 11.038, and the program displays those fractions as well (2:15.947). But to keep the discussion of this program simple, I'll use whole numbers in any examples.

splitting the timer reading into minutes and seconds

In this program, we take the reading from the Timer block and divide it into minutes and seconds. For example, if the Timer block reading is 127 seconds, we want to display this as 2 minutes and 7 seconds. We can do this with two simple formulas:

Seconds = Timer reading % 60
Minutes = (Timer reading – seconds) / 60

To calculate how many seconds to display (between 0 and 59), we take the remainder of the Timer block reading divided by 60. So when the Timer block reading is 127, we get 7, because 127 % 60 is 7.

Subtracting the number of seconds from the Timer block reading gives us a value that is a multiple of 60. Using the same example, 127 – 7 is 120. Dividing this value by 60 gives us the number of minutes, because there are 60 seconds in a minute.

Figure 13-12 shows the first part of the program. The Timer block reads the elapsed time and passes the value to the two Math blocks. The Math blocks each use Advanced mode to calculate the number of seconds and minutes to display. The block to calculate the number of seconds needs to come first because its result is used by the other Math block.

building the text to display

The next step is to take the minutes and seconds values, which are numbers, and combine them with a colon (:) to create a Text value in the form minutes:seconds. We could send the minutes and seconds values straight to a Text block to combine them, but there is one problem: If the number of seconds is less than 10, then the result won't be correct. For example, if the number of minutes is 2 and the number of seconds is 7, then the Text block would generate 2:7 instead of 2:07. Before we send values to the Text block, we have to add a leading 0 to the Seconds value when it's less than 10.

Figure 13-13 shows the code that constructs the correctly formatted Text value when the number of seconds is less than 10, along with the two Math blocks that supply the number of minutes and seconds. The program uses a Compare block to check if the seconds value is less than 10. The Switch block's

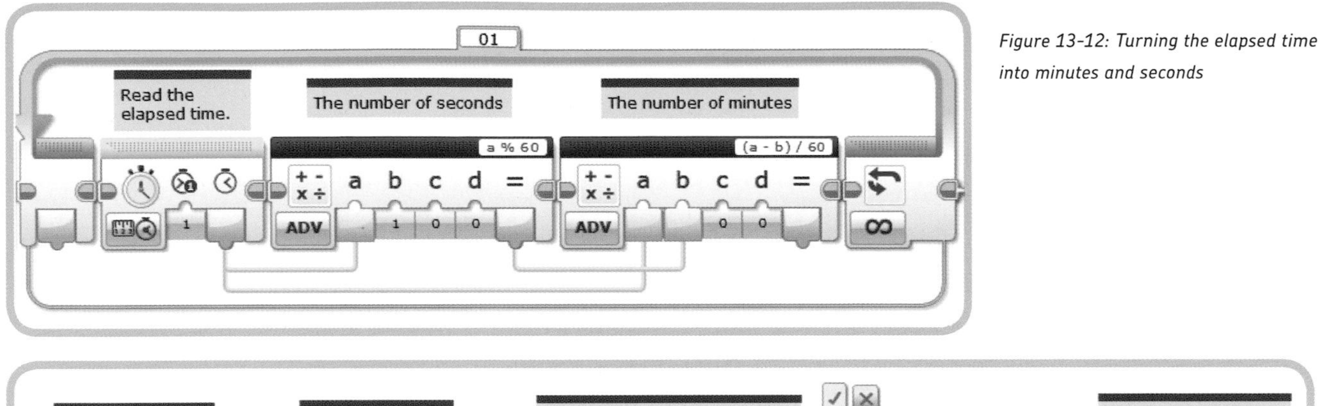

Figure 13-12: Turning the elapsed time into minutes and seconds

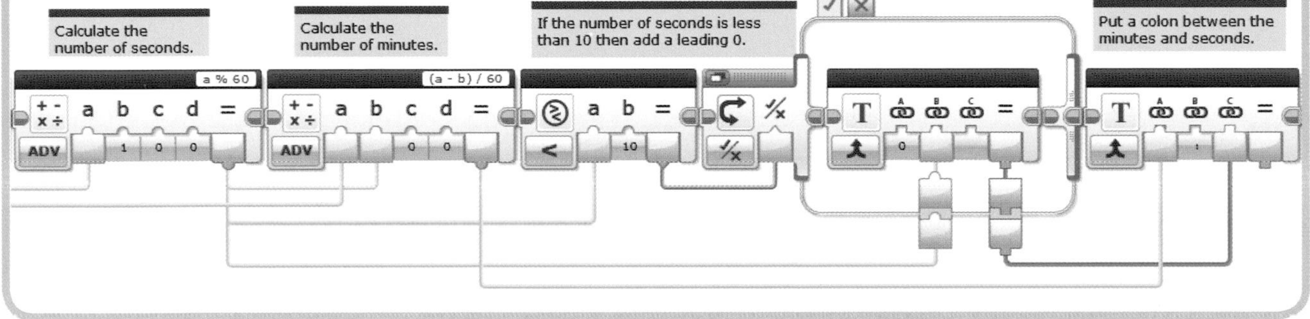

Figure 13-13: Adding a 0 if the number of seconds is less than 10

true case uses a Text block to add a 0 to the number of seconds and passes the result out of the Switch block to another Text block, which combines it with the number of minutes and a colon to generate the formatted value.

The final Text block now gives the correct value, 2:07, when the number of minutes is 2 and the number of seconds is 7.

What happens when the number of seconds isn't less than 10? The false case of the Switch block is used, and we want to pass the value to the final Text block unchanged, so we can just draw a data wire from the block input plug directly to the block output plug, as shown in Figure 13-14.

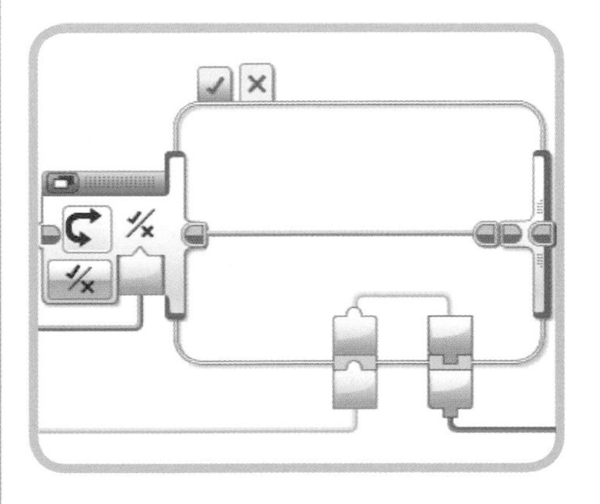

Figure 13-14: Passing the value out of the Switch block as a Text value

Notice that the data wire going into the Switch block contains a Numeric value from the Math block, which explains why it's yellow. The data wire coming out holds a Text value (this plug was created by the Text block inside the true case), which explains why it's orange. When you build this program, be sure to fill in the true case first so that you get an orange Text output plug going out of the Switch block. If you fill in the false case first (passing the Numeric data wire through to the next block), this would create a yellow Numeric output plug, and then you wouldn't be able to connect the data wire from the Text block in the true case to that plug. Remember, you can connect Numeric values to a Text plug (and the data type will convert automatically), but you can't connect a Text value to a Numeric plug.

The final part of the program is a Display block to show the result of the final Text block on the EV3 screen (Figure 13-15). I've set the Row parameter to 4 because the text is a little easier to see when it's closer to the middle of the screen.

Run the program and you should see the Time value counting up. Let the timer run for at least a minute so you can make sure it does the right thing when the number of seconds is less than 10, when the number of seconds is greater than 10, and when the minute should change from 0 to 1. The display will include three digits after the decimal point, and these digits will change very quickly. You'll see how to hide or truncate those fractional values in the next section using the Round block.

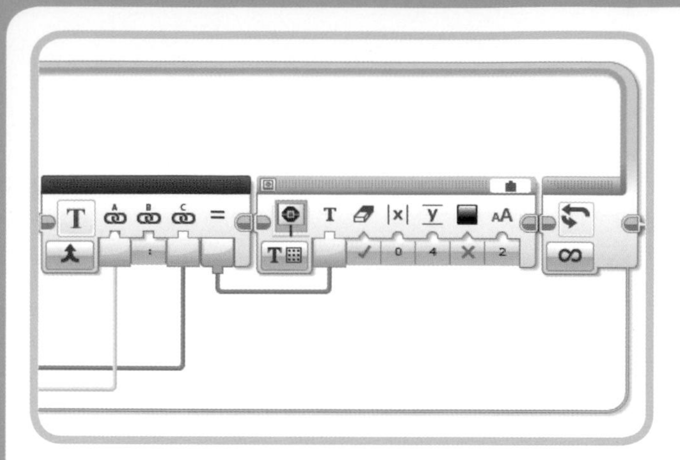

Figure 13-15: Displaying the Time value

the round block

The *Round block* (Figure 13-16) gives you an easy way to round numbers. The four modes correspond to different ways you may want the number rounded. *To Nearest* rounds the value to the nearest integer, *Round Up* rounds the value up to the next largest integer, and *Round Down* rounds the value down to the next-smallest integer. These three modes do the same thing as the Math block's round, ceil, and floor functions (available in Advanced mode).

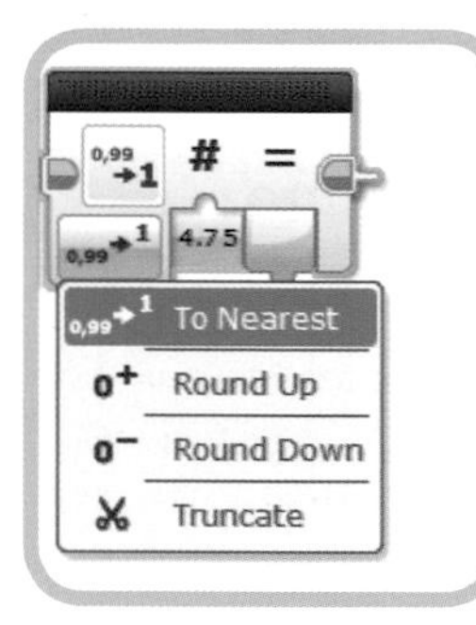

Figure 13-16: The Round block

Table 13-4 compares the results given by the three modes for some sample Input values. Notice that if the Input value is an integer, then all three modes pass the value along unchanged. The way the Round Up and Round Down modes work with negative numbers is a little counterintuitive but makes sense once you walk yourself through it; for example, since -4 is larger than -5, -4.2 rounds up to -4 and rounds down to -5.

table 13-4: round block mode comparison

Input	To Nearest	Round Up	Round Down
4.0	4	4	4
4.2	4	5	4
4.5	5	5	4
4.7	5	5	4
-4.2	-4	-4	-5
-4.5	-5	-4	-5
-4.7	-5	-4	-5

The Round block's *Truncate mode* has an additional parameter, *Number of Decimals*, which allows you to set how many digits to keep after the decimal point (see Figure 13-17). Any digits after the number you specify are eliminated (no rounding takes place).

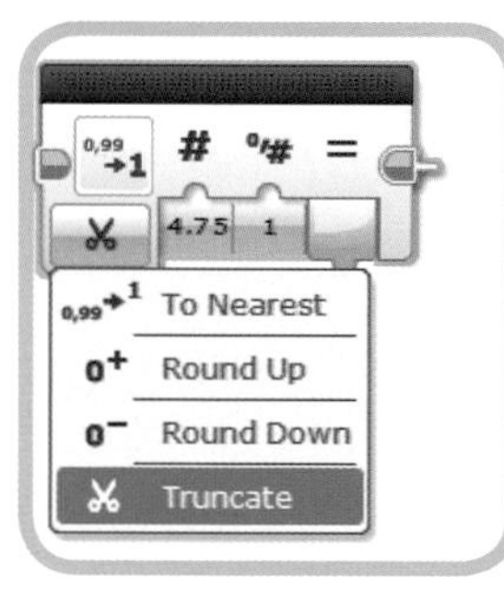

Figure 13-17: The Round block's Truncate mode

To make the *DisplayTimer* program show only one digit after the decimal point, add a Round block in Truncate mode after the Timer block. Be sure to connect the result of the Round block to the *a* Input parameter of both Math blocks (Figure 13-18). If you prefer to only show a whole number of seconds, set the Round block's Number of Decimals parameter to 0.

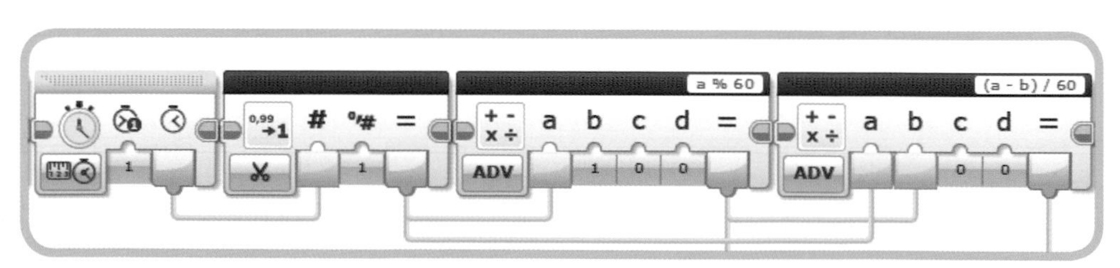

Figure 13-18: Truncating the elapsed Time value

CHALLENGE 13-2

Create a TimeToText My Block from the blocks shown in Figure 13-14. The My Block should take the elapsed Time value as an Input parameter and give the formatted Text value as an Output parameter.

the random block

Another math-related block on the Data Operations palette is the *Random block*, shown in Figure 13-19. A die is used for the picture on the Random block, and like a die, the Random block generates random numbers, which you can use to create robotic games or to add some randomness to your robot's behavior. Often a robot that is a little unpredictable can be more interesting and have more personality.

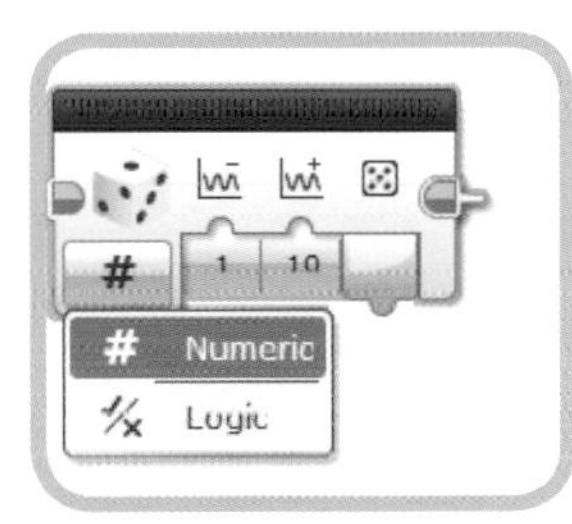

Figure 13-19: The Random block

With the Random block in *Numeric mode*, you can set the Lower Bound and Upper Bound parameters, and the block will generate an integer value between the two bounds. The default bounds are 1 and 10, giving an Output value that can be as small as 1 or as large as 10. You can change the range to suit your program; for example, to create a virtual die, you would set the Lower Bound to 1 and the Upper Bound to 6.

In *Logic mode* (Figure 13-20), the block generates a random Logic value (true or false) based on the probability you specify. The Probability of True parameter, between 0 and 100, determines how often the result will be true. For example, 80 means that there is an 80 percent chance that the result will be true and a 20 percent chance that the value will be false.

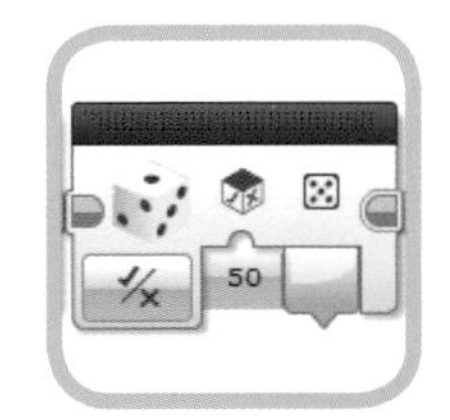

Figure 13-20: The Random block in Logic mode

CHALLENGE 13-3

The DisplayNumber My Block (from Chapter 12) shows numbers using up to three decimal places. Add a parameter to specify the number of decimal places to round the value to before displaying it. The Round block's Truncate mode doesn't help here because we want the number rounded, not truncated. To Nearest mode alone also won't do the trick because it only rounds to the nearest integer.

HINT Say the number is 34.567 and we want it rounded to two decimal places so that the EV3 screen shows 34.57. Here's one way to accomplish this:

1. **Multiply 34.567 by 10^2 to get 3456.7. (Notice that the exponent is the same as the number of decimal places we're interested in.)**
2. **Round the value, which gives us 3457.**
3. **Divide 3457 by 10^2 to get 34.57.**

adding a random turn to BumperBot

In this section, you'll make a small change to the *BumperBot* program to make it a little more interesting. Recall that when the TriBot bumps into something, it backs up and turns in a different direction. The distance the robot turns doesn't need to be any particular value; you simply want to have the robot point in a new direction. You can use a Random block to control the distance the robot turns, which will make the program less predictable.

You can start by copying the latest version of the *BumperBot* program from your Chapter 6 project to your Chapter 13 project. Figure 13-21 shows the part of the *BumperBot* program that you need to change. This is the code that runs after the Touch Sensor is pressed. The first group of blocks makes the robot back up, and the final Move Steering block turns the robot.

Right now, the Move block is set to move 225 degrees. To make the turn less predictable, add a Random block before the Move Steering block to control how much the robot turns. Then set the range of values that the Random block can generate. The original program used 225 degrees, which turns the robot a little more than a quarter-turn. I'll use 200 degrees for

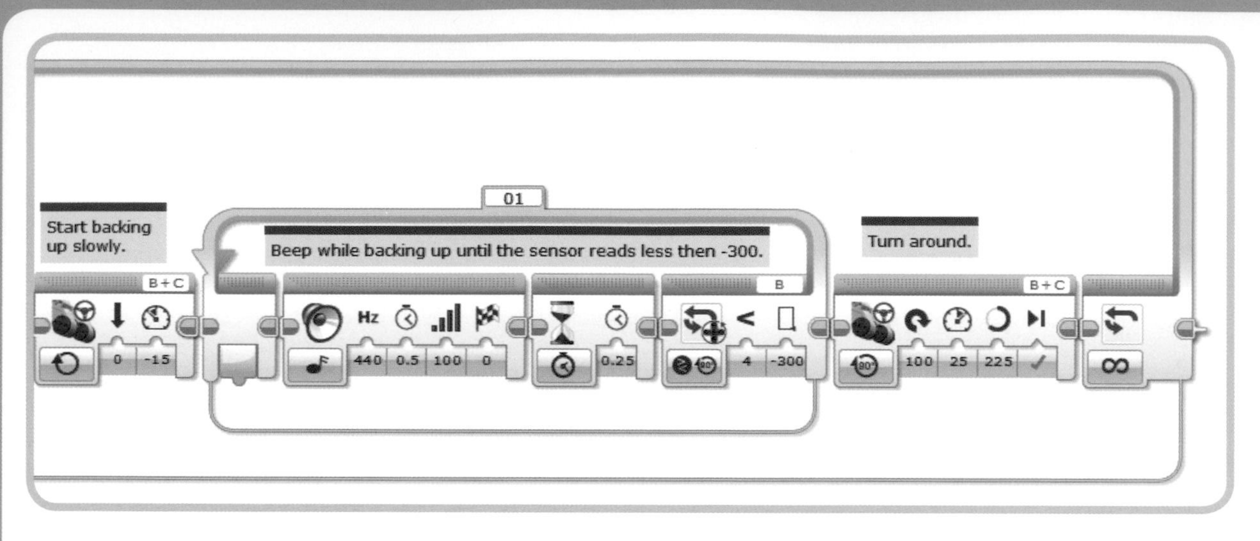

Figure 13-21: Backing up and turning around

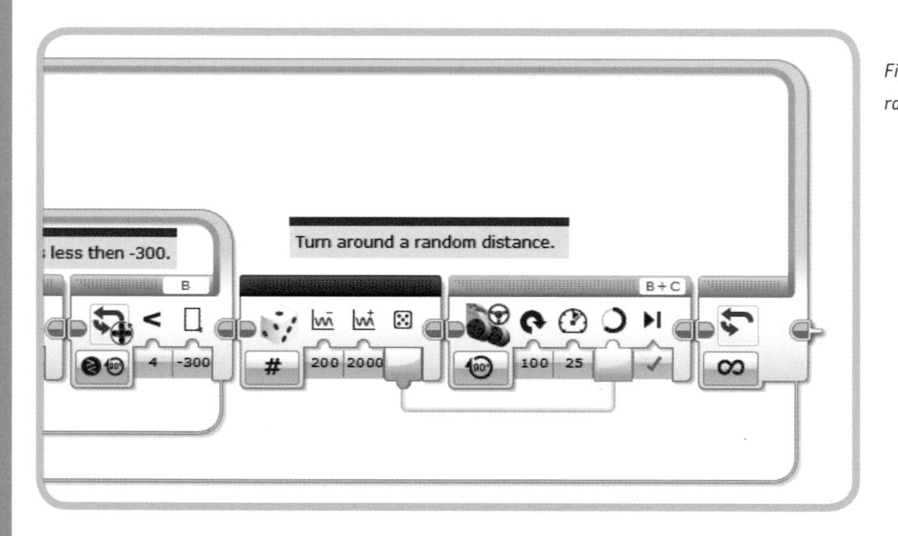

Figure 13-22: Turning a random distance

the Lower Bound and 2000 for the Upper Bound. With these values, the robot sometimes turns quickly and starts off again, and sometimes spins in place for a while before resuming its journey around the room. Figure 13-22 shows this part of the program with these changes made.

Run the program with these changes, and the TriBot should vary the amount it turns after bumping into something.

the logic block

Many of the programs presented so far make decisions that involve a single condition, usually comparing the value from a Sensor to a Target value, with the result (either true or false) used in a Switch or Loop block. To put it another way, the programs are asking simple questions like "Is the Touch Sensor pressed?" or "Is the reading from the Color Sensor less than 50?"

The *Logic block* lets you combine multiple conditions, allowing your program to make more complex decisions. This lets your program ask questions like "Is the Touch Sensor pressed *and* the Color Sensor reading greater than 50?" You can find the Logic block, shown in Figure 13-23, with the other math-related blocks on the Data Operations palette.

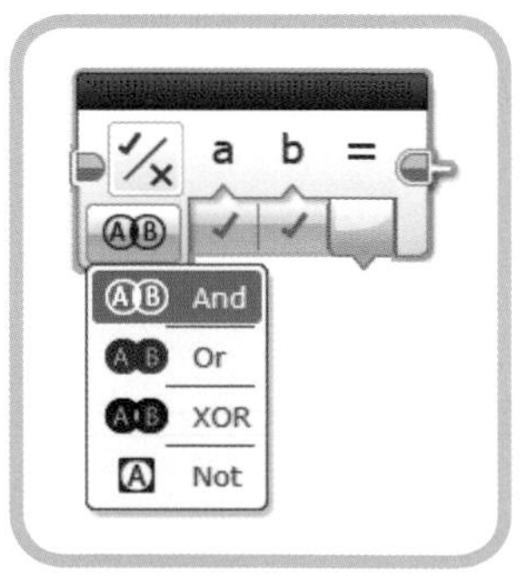

Figure 13-23: The Logic block

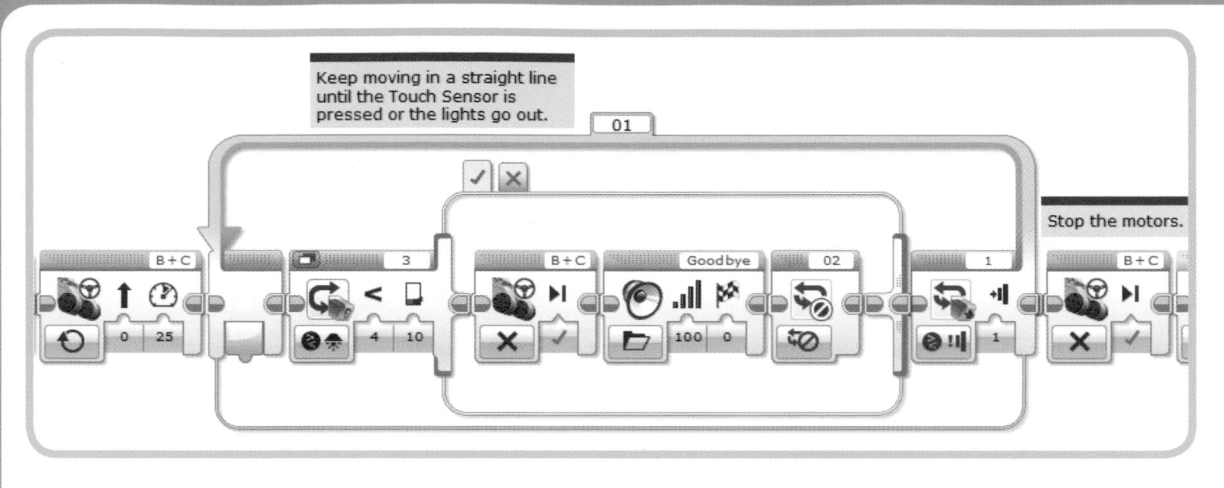

Figure 13-24: Moving forward until the Touch Sensor is pressed

The Logic block supports four operations: *And*, *Or*, *XOR*, and *Not*. Here's how each operation works:

And The result of the And operation will be true only if both Input values are true. If either Input value is false, then the result will be false.

Or The result of the Or operation will be true if either Input value is true or if both Input values are true. The result will be false only if both Input values are false.

XOR XOR is an abbreviation for *Exclusive Or*. This is similar to the Or operation except that the result is false if both Input values are true. This is the way the word *or* is often used in English: If your mother tells you that you can have ice cream or candy, she probably doesn't mean you can have both; she expects you to pick one or the other.

Not This operation only takes one Input value and generates the opposite value. If the Input value is true, then the Output value will be false, and if the Input value is false, then the Output value will be true.

Table 13-5 shows a table that lists all the possible Input values and the result for each operation. (This kind of table is called a *truth table*.) Note that the result of the Not operation depends only on the Input *a* value.

table 13-5: truth table for the logic block

Input *a*	Input *b*	Or	And	XOR	Not
False	False	False	False	False	True
False	True	True	False	True	True
True	False	True	False	True	False
True	True	True	True	False	False

adding some logic to BumperBot

In this section, you'll make a change to the *BumperBot* program using the Logic block. Recall that the program keeps the TriBot moving forward until it runs into something. What if you want to limit how long the robot moves forward, perhaps to keep it from wandering too far? You'll change the program so that the TriBot stops and turns around if it bumps into something *or* if it moves forward for more than 20 seconds.

Figure 13-24 shows the code that moves the TriBot forward. The Move Steering block starts the TriBot moving, and it keeps going until the Loop block ends when the Touch Sensor is pressed.

How can you tell whether the robot has traveled for more than 20 seconds? You can use a Timer block to reset the timer before starting to move and then use another Timer block within the loop to tell when 20 seconds have passed.

The Loop block can be configured to check the Touch Sensor or the Timer, but it can't use both. To make the loop stop when either condition is true, you need to check both conditions, combine the results using a Logic block, and use the result of the Logic block to repeat or exit the loop. I'll take you through these changes step by step.

1. Add a Timer block to the left of the Move Steering block, and set the mode to **Reset** (Figure 13-25).

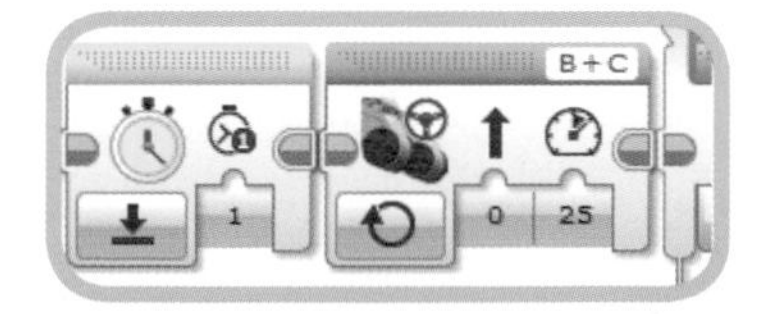

Figure 13-25: The placement of the Timer block

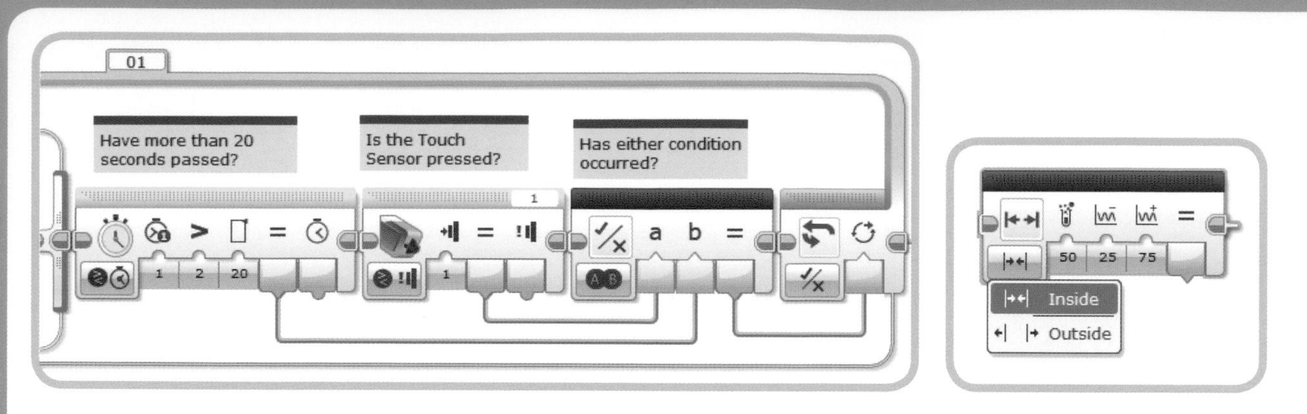

Figure 13-26: Exiting the loop if 20 seconds have passed or the Touch Sensor is pressed

Figure 13-27: The Range block

2. Add a Timer block after the Switch block. Set the mode to **Compare Time** and the Threshold parameter to **20**.
3. Add a Touch Sensor block after the Timer block, and set the mode to **Compare State**.
4. Add a Logic block to the right of the Touch Sensor block, and set the mode to **Or**.
5. Draw a data wire from the Touch Sensor block's Compare Result output to the Logic block's *a* input.
6. Draw a data wire from the Timer block's Compare Result output to the Logic block's *b* input.
7. Select the Loop block, and change the mode to **Logic**.
8. Draw a data wire from Logic block's Result output to the Loop block's Until True input.

Figure 13-26 shows the program with these changes.

Now when you run the program, the TriBot should go forward for a maximum of 20 seconds. If it doesn't bump into something within that time, it should turn and go off in a different direction.

the range block

The final math-related block is the *Range block*, which determines whether a number is inside or outside a range of numbers. The block has three parameters: the Test value (which you'll usually supply with a data wire) and the Lower and Upper Bounds of the range you're interested in.

The Range block has two modes: Inside and Outside (Figure 13-27). In *Inside mode*, the block asks "Is the test value inside the range (between the lower and upper limit)?" In *Outside mode*, the block asks "Is the test value outside the range (less than the lower limit or greater than the upper limit)?" If the Test value equals the Upper or Lower Bound, then it's considered to be inside the range.

the TagAlong program

The *TagAlong* program uses the Range block to make the TriBot follow you, staying a small distance behind, as you move across a room. This is a simple program that only moves the TriBot forward or backward. (It won't follow you if you move to the side.) The program uses the Infrared Sensor and the Range block to see if the robot is within the desired range. If the robot is out of range, then a Move Steering block moves it forward or backward by using the Infrared Sensor reading to set the Power parameter. If the robot is within the desired range, the motors are stopped.

The program is shown in Figure 13-28. Let's go through the blocks one at a time to explain exactly how this program works.

Loop block Keeps the program running until you stop it.

Infrared Sensor block Uses Measure - Proximity mode to read the distance to an object in front of the robot (which is you!).

Range block Checks whether the Infrared Sensor reading is outside the range. I set the range to 40–60, which keeps the robot reasonably close. The result is true if the reading is less than 40 or greater than 60, and it's false otherwise.

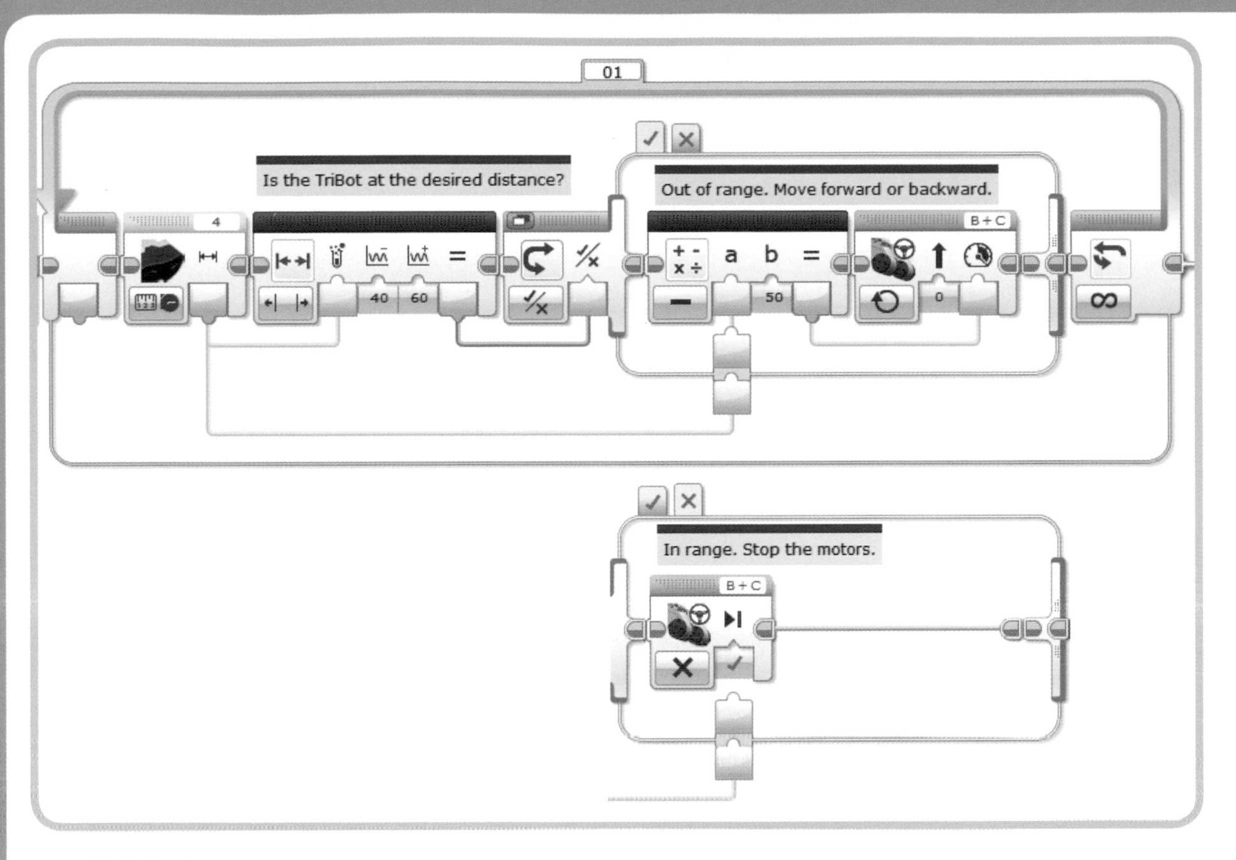

Figure 13-28: The TagAlong *program*

Switch block The true case is used when the robot is out of range and needs to move. The Power parameter of the Move Steering block is calculated by subtracting 50 from the Infrared Sensor reading. 50 is used because it's halfway between 40 and 60. So if the reading is over 60, the Power parameter ends up being at least 10, and the robot moves forward. If the reading is less than 40, the Power parameter ends up being -10 or less, and the robot moves backward.

The false case of the Switch block is used when the robot is in the correct range, with the Sensor reading between 40 and 60. In this case, a Move Steering block is used to stop the motors.

NOTE If you're using the Ultrasonic Sensor, use an Ultrasonic Sensor block in Measure - Distance Inches mode. Set the Range block bounds to 12 and 24 and the *b* parameter of the Math block to 18 (the average of 12 and 24).

Run this program and the TriBot should move forward and backward as you move away from or toward it.

the GyroPointer program

The *GyroPointer* program is a variation of the *TagAlong* program that keeps the TriBot pointing in the same direction while you spin the robot on a rotating platform. If you are using the EV3 Home Edition and don't have a Gyro Sensor, read this section and try applying the concepts in Challenge 13-4.

To test this program, place the TriBot on a turntable (or a lazy Susan, rotating stool, or some other rotating surface) and slowly rotate the turntable while the program runs. The TriBot spins to keep pointing in the general direction in which it started.

Figure 13-29 shows the program. It has the same basic structure as the *TagAlong* program with a few notable changes:

* A Gyro Sensor block is used instead of the Infrared Sensor block.

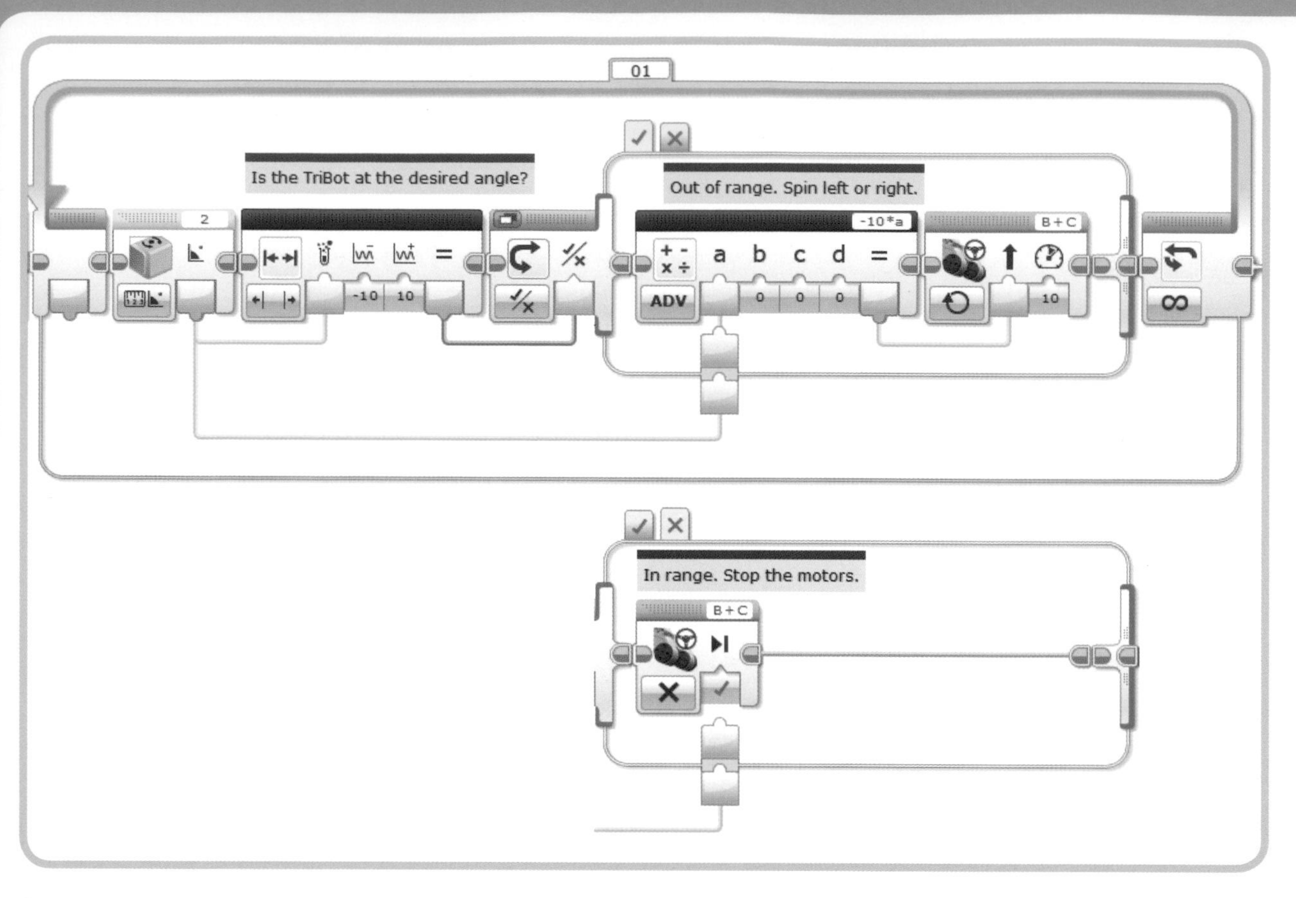

Figure 13-29: The GyroPointer *program*

* The parameters of the Range block check that the sensor reading is between -10 and 10. The reading at the start of the program will be 0, so this range keeps the robot pointing close to where it starts.
* The Math block's result is connected to the Move Steering block's Steering parameter rather than the Power parameter. To make the TriBot spin, we want this setting to be either -100 or 100; otherwise, the robot will move forward as well as turn.
* The Math block multiplies the sensor reading by -10. If the sensor reading is less than -10, the result will be more than 100, and the Move Steering block will treat the value as 100. If the sensor reading is more than 10, the result will be less than -100, and the Move Steering block will treat the value as -100.

When you run this program, the TriBot sits motionless until you move the turntable at least 10 degrees (in either direction). Then the Tribot spins to keep facing the direction in which it started. Move the turntable in one direction and then the other, and the TriBot should adjust to keep pointing in the same general direction.

CHALLENGE 13-4

If you have the EV3 Home Edition, use the Infrared Sensor and Infrared Remote to create a *RemotePointer* program that works the same way as the *GyroPointer* program. The program should spin the TriBot so that it keeps pointing in the general direction of the Infrared Remote by using the beacon heading to control the Move Steering block.

further exploration

Try these activities for more practice with math-related blocks:

1. Write a *CountDown* program that shows the time counting down from two minutes on the EV3 screen. The Loop block should exit when the time gets to zero.
2. Combine the *TagAlong* and *RemotePointer* programs to make a program that will maintain both its distance and direction. The resulting program should make the robot follow the Infrared Remote around a room.
3. Use the Random block to create a *MagicEightBall* program. When you ask a question and trigger the robot (such as by pressing the Touch Sensor), the robot should select an answer randomly from several possible answers. Use the Display block to show the answer on the screen. You can also use the Sound Editor tool to record your own answers and the Sound block to play them.
4. This activity uses some knowledge of trigonometry; feel free to skip over this one if you haven't reached that level of mathematics yet. The sine function starts at 0 and oscillates between 1 and -1, so if you graph this function, it creates a snake-like curve. You can make the robot follow a winding, serpentine path by using the sine function to control how the robot steers. Create a *Slither* program using a timer and a Math block that uses the sine function to control the Steering parameter on a Move Steering block. Hint: Using sin(Elapsed time) directly won't be very interesting because it takes 6 minutes (360 seconds) for the value to go from one extreme to the other, and the value only goes between -1 and 1. However, if you multiply the elapsed time by 10, the entire range will be covered in only 36 seconds, and if you multiply the result by 50, the Steering value will oscillate between 50 and -50.

conclusion

In this chapter, you learned how use blocks that work with numbers and logic. The Math block's Advanced mode gives you all the power you need to calculate complex equations, which allowed you to improve the *LineFollower* program by using a proportional controller. You also learned about the modulo operator and saw it in action in the DisplayNumberNextLine My Block and the *DisplayTimer* program.

The Logic block lets you write programs that make complex decisions, such as combining the input from multiple sensors. The Range block gives you a convenient way to perform the common operation of testing a value to see whether it's in a certain range. The other block introduced in this chapter was the Random block, which you can use to add a little unpredictability to your programs and personality to your robots.

14

the EV3 lights, buttons, and display

The EV3 has five buttons and a display screen that you can use to interact with your programs, much like you use a keyboard to interact with a computer. The buttons are illuminated with colored lights, and in this chapter, you'll learn how to control them using the Brick Status Light block. You'll also learn about some new features of the Display block that give you more control over the EV3 screen.

the EV3 buttons

You can use the five large buttons on the front of the EV3 (shown in Figure 14-1) to control your program. For example, you can make the program wait for you to press a button or choose an action based on which button you press. Like with the Touch Sensor, your programs can detect when a button is pressed, released, or bumped (pressed and released). The Back button can't be used by a program; pressing it while your program is running will end the program.

The Wait, Switch, and Loop blocks each have Brick Buttons modes (see Figure 14-2), or you can work with the buttons using the *Brick Button block*, which is similar to the other Sensor blocks. Each button has three modes: Compare, Measure, and Change.

Compare mode lets you test whether one or more buttons is in a particular state (pressed, released, or bumped). Figure 14-3 shows the button selection menu, which has a list of numbers to identify each button. These numbers are known as *Button IDs*. If you select more than one button, the test succeeds when either button is in the desired state. When the Brick Button block is in Compare mode (Figure 14-3), it generates two Output values: a Logic value to tell you whether one of the selected buttons is in the selected state and a Number value to tell you the Button ID of the matching button.

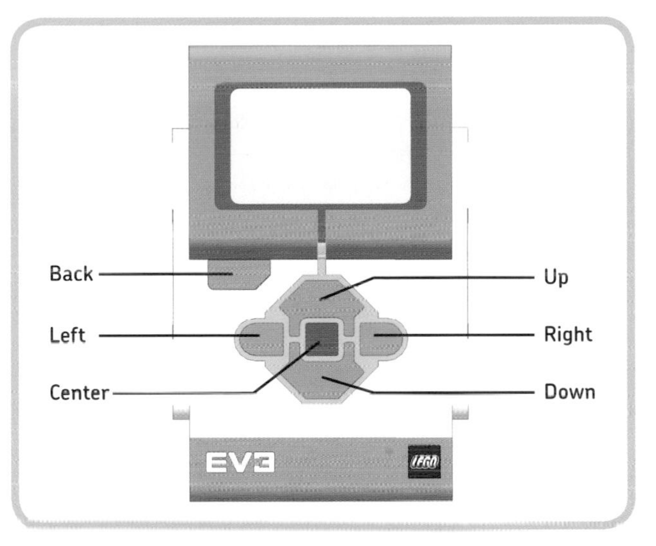

Figure 14-1: The EV3 buttons

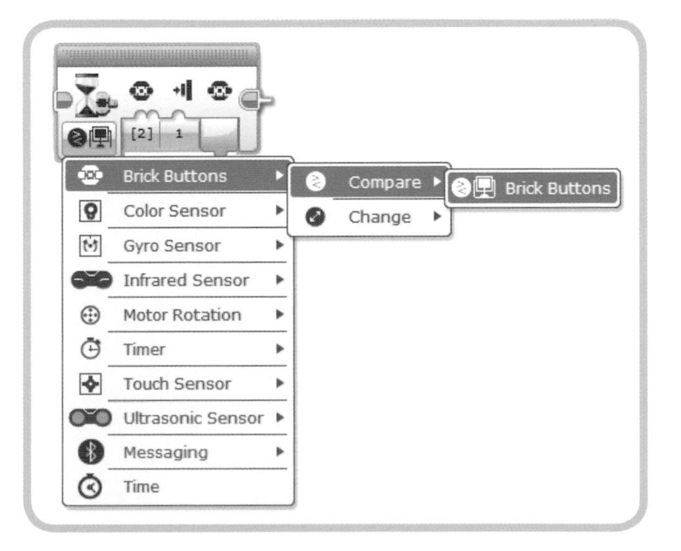

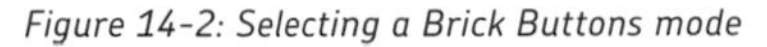

Figure 14-2: Selecting a Brick Buttons mode

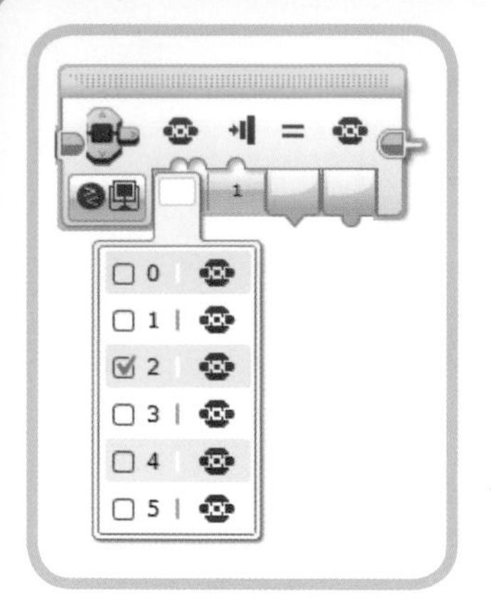

Figure 14-3: Selecting the buttons for Compare mode

Measure mode lets you determine which button is currently pressed. *Change mode* is only supported by the Wait block, which waits for the state of any of the buttons to change.

NOTE If more than one button is pressed at the same time, one button might override the other so that only one button press is sensed.

the PowerSetting program

In Chapter 11, I mentioned that you could improve the *Wall-Follower* program by using a variable to control the Power parameter used by all seven Move Steering blocks. This makes it easier to change the value because you only need to change it in one place. You can control the parameter with Brick buttons to make it even more convenient to test out different values. In this section, I'll describe the *PowerSetting* program in order to show you how to do this. With just a few simple changes, you can use this code anytime you want to set a value with Brick buttons at the beginning of a program.

The *PowerSetting* program uses a variable called *Power* to store the current value and display it on the EV3 screen. Pressing the Right button increases the value by one, and pressing the Left button decreases the value by one. Pressing the Center button accepts the current value. Listing 14-1 shows the pseudocode for the program.

```
set Power to 50
begin loop
   display the current value
   if the Right button is bumped then
      Power = Power + 1
   end if
   if the Left button is bumped then
      Power = Power - 1
   end if
loop until the Center button is bumped
```

Listing 14-1: The PowerSetting *program*

Some of these lines of pseudocode require a few programming blocks to accomplish. For example, the line `Power = Power + 1` is a short way of saying, "Take the current value of the Power variable, add one to it, and store the result in the Power variable," which requires three programming blocks.

the initial value and the loop

The first thing the program does is use a Variable block to set the Power variable's initial value, followed by a Loop block that holds the rest of the program (as shown in Figure 14-4). I'll set the initial Power value to 50 because that's in the middle of the Power range of 0 to 100. The Loop block uses Brick Buttons - Compare mode to keep repeating until the Center button is bumped.

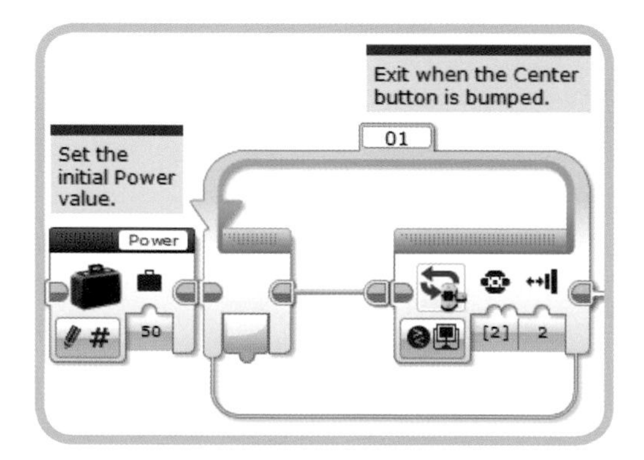

Figure 14-4: Initializing the Power variable and configuring the loop

Making the loop wait for the button to be bumped, rather than just pressed, is often a better choice because it waits for the button to be pressed *and* released before moving to the next part of the program. This way, the buttons will be back to their normal state (released) and a problem is less likely to occur if the next part of your program also uses the buttons.

displaying the current value

Each time the loop repeats, the program reads and displays the current value using a Variable block and a DisplayNumber My Block (created in Chapter 12), as shown in Figure 14-5. I've set the Row parameter of the DisplayNumber block to 6 so that the value is displayed closer to the center of the EV3 screen. The Label parameter is set to "Power: ".

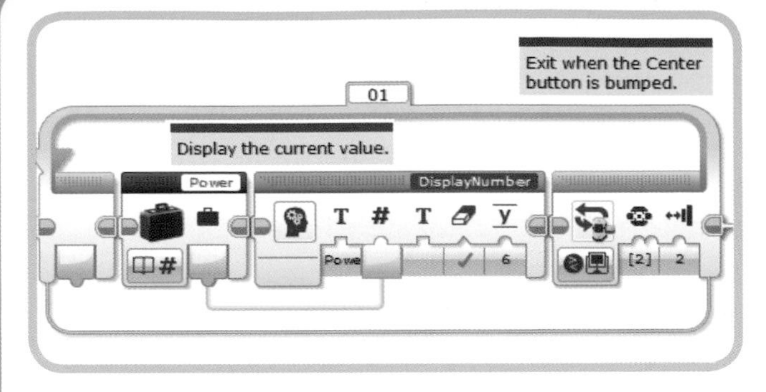

Figure 14-5: Displaying the labeled value

adjusting the power value

When the current Power value is displayed, you can use the EV3's Left and Right buttons to adjust it. The code shown in Figure 14-6 deals with the Right button. When the Right button is bumped, it triggers the true case on the Switch block, which adds one to the Power variable using the Math and Variable blocks. There are no blocks on the false case because you don't need the program to do anything when the button isn't bumped.

The code for dealing with the Left button is almost identical, except that it subtracts one from the Power value when the button is pushed. Figure 14-7 shows the entire program with the new blocks added.

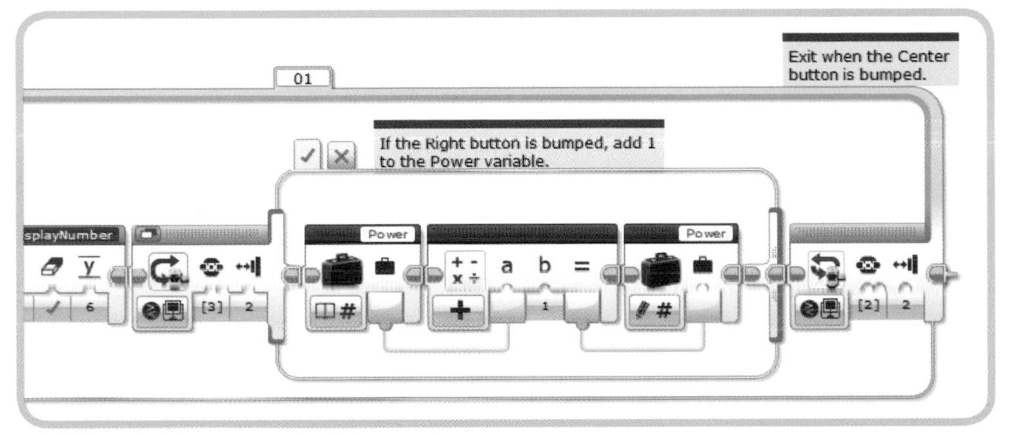

Figure 14-6: Adding one to Power if the Right button has been bumped

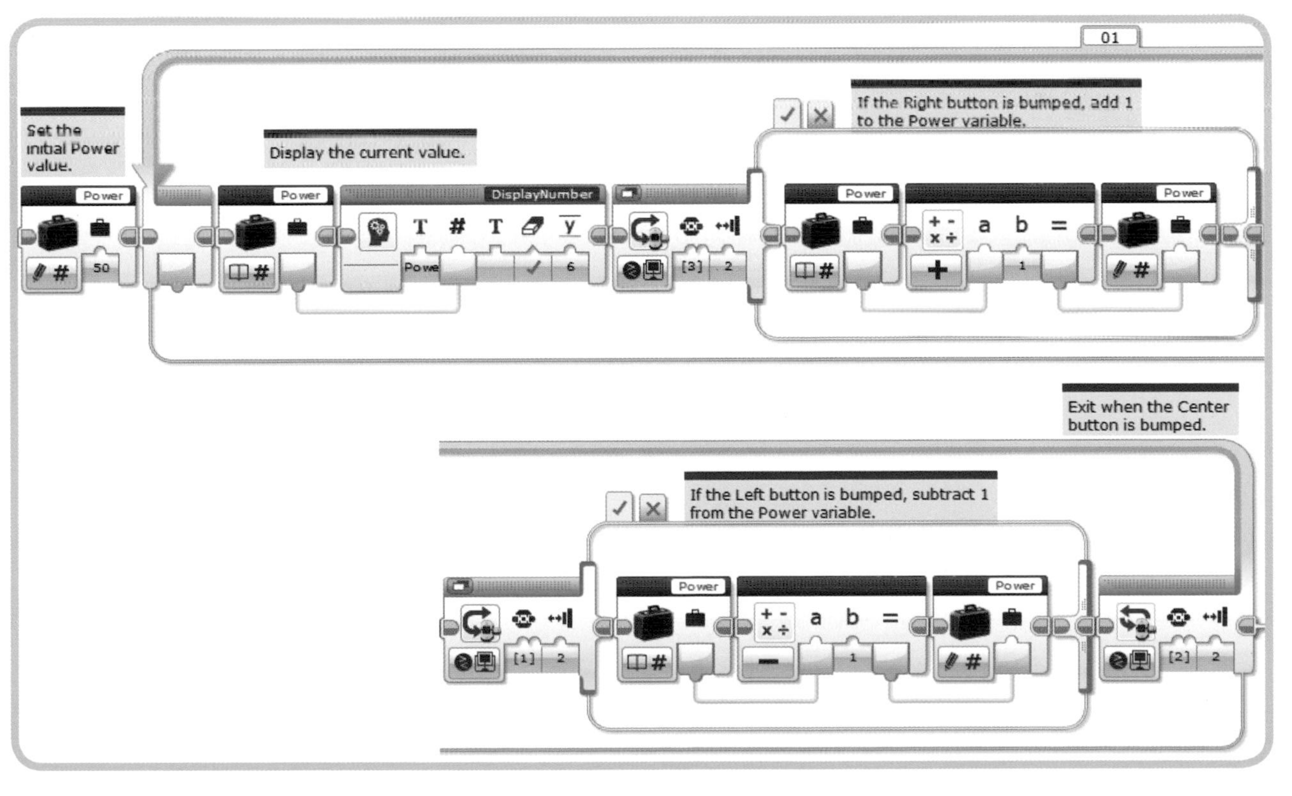

Figure 14-7: The full PowerSetting program

testing the program

When you run this program, it should first display "Power: 50". Press the Right and Left buttons to change this value. Press the Center button to end the program.

When the program ends, the Power variable is set to the value you selected. You can use the blocks in this program at the beginning of a larger program to set a variable. Once you place more blocks after the Loop block, pressing the Center button starts the rest of the program (instead of exiting).

changing the value faster

This *PowerSetting* program only changes a setting by one each time you press and release the button, so it can take awhile if you want to change a value by a lot. How can you speed this up?

Right now, the two Switch blocks in the loop are set to Bumped, which means you need to press and release the button to change the value. The program responds more quickly if you just check for the button being pressed instead of waiting for it to be pressed and released. What happens if you change the State parameter of both Switch blocks to Pressed, as shown in Figure 14-8?

The value certainly changes faster, but now it changes too fast because the loop repeats several times even if you just press the button for a moment. To make the program usable, you need to slow the loop down a little by adding a Wait Time block (Figure 14-9) at the end of the loop body. I've found that a setting of 0.2 seconds provides a good balance between changing the value quickly and being able to stop at the value I want. Experiment with different values to see what works best for you.

When you have the *PowerSetting* program working to your satisfaction, you can put the code to work in the *WallFollower* program. To make the code easier to reuse, you could start by creating a My Block from the entire *PowerSetting* program. Then add the new block to the beginning of the *WallFollower* program and use the Power variable to control the Move Steering blocks.

the brick status light

The *Brick Status Light* illuminates the area around the EV3 buttons. When your EV3 is on, the light is solid green, and when a program is running, it flashes a blinking green pattern unless your program changes it. The *Brick Status Light block*,

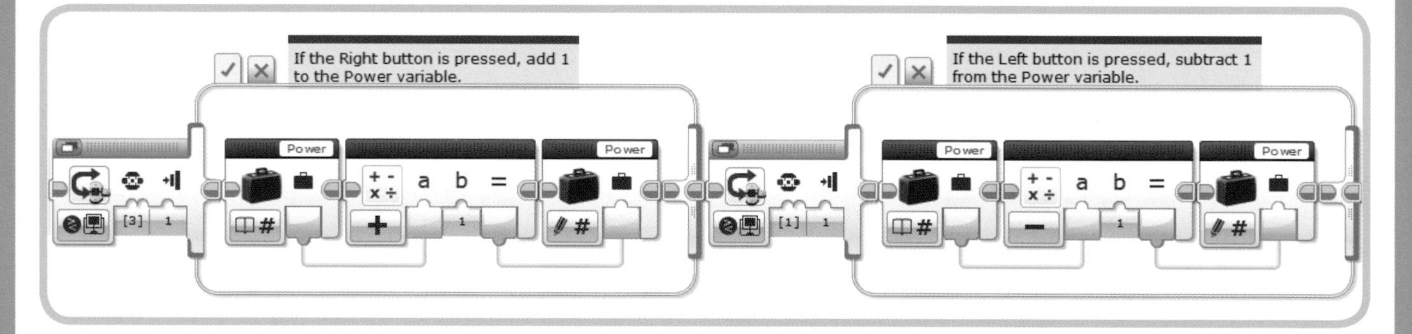

Figure 14-8: Change the State from bumped to pressed

CHALLENGE 14-1

To make large changes even easier, enhance the *PowerSetting* program by using the Up button to add 10 to the value and the Down button to subtract 10 from the value. Does it make more sense to use Bumped or Pressed when checking these two buttons?

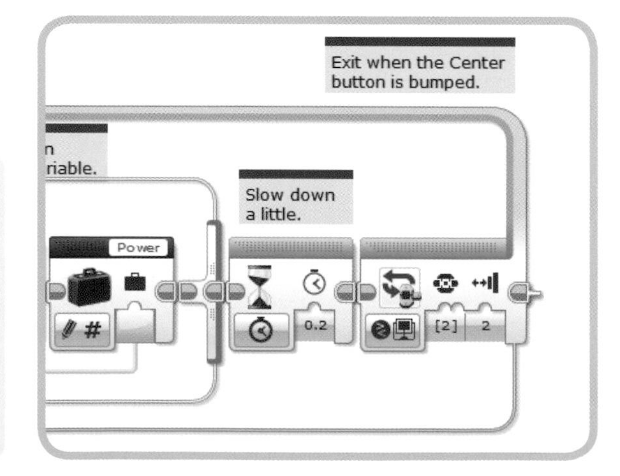

Figure 14-9: Adding the Wait Time block to the end of the loop body

found on the Action palette, lets you control the light in your programs (Figure 14-10).

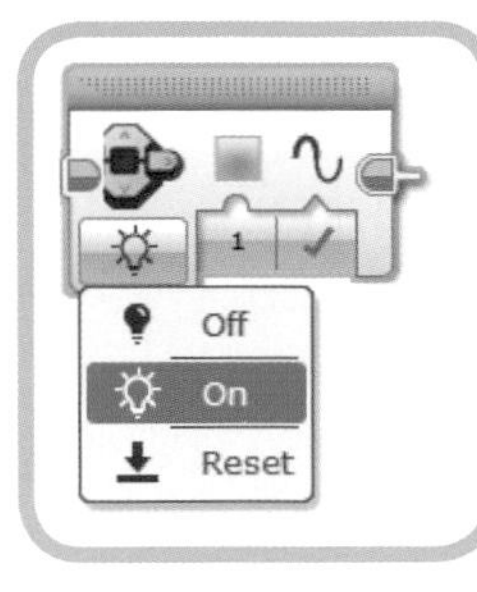

Figure 14-10: The Brick Status Light block

The Brick Status Light block has three modes: Off, On, and Reset. *Off mode* turns the light off. *On mode* turns the light on and lets you select the color (green, orange, or red) and whether the light should be solid or pulse. *Reset mode* returns the light to the default blinking green state (which has a slightly different pattern than pulse).

One way to use the light is for debugging. For example, you can use the light to give a visual indication of how a program is doing—perhaps green if everything is going well and red if there's a problem. You can also set it to indicate which part of a program is executing or whether a condition is met, without impacting the normal operation of your program.

the ColorCopy program

The *ColorCopy* program makes the Brick Status Light show the color that the Color Sensor detects. If you hold a red object in front of the Color Sensor, the light turns red; if the sensor detects green, the light turns green; and if it detects yellow, the light turns orange (the Color Sensor doesn't detect orange, so yellow is the closest match). If the Color Sensor reading is anything else, the light turns off.

Figure 14-11 shows the completed program. The Switch block uses Color Sensor - Measure - Color mode to determine which color is detected. The first three cases match red, green, and yellow and turn the light on with the appropriate color. The Pulse parameter is turned off so that the light stays on as long as the sensor detects the same color. The bottom case is the default case that is used if the sensor reads anything other than red, green, or yellow, and turns the light off.

Run the program. The light should match the color of an object held in front of the Color Sensor. When no object is in front of the sensor, or if the object is not red, green, or yellow, the light should turn off.

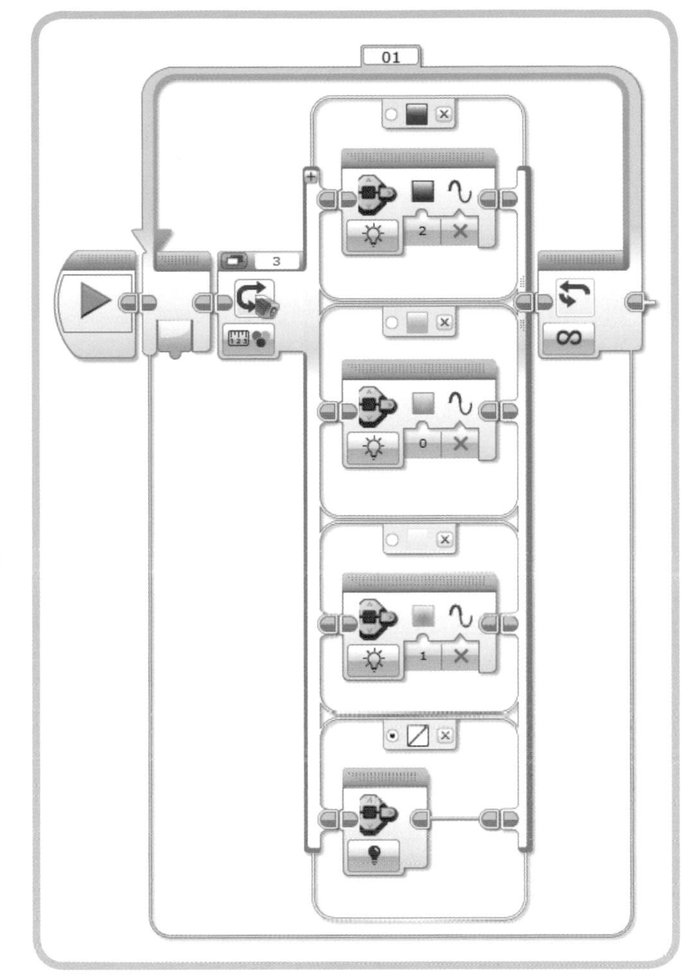

Figure 14-11: The ColorCopy program

CHALLENGE 14-2

Write the *ProximityAlarm* program, which sets the Brick Status Light depending on how close the robot is to an object. Use either the Infrared or Ultrasonic Sensor to determine the distance and show a blinking red light if the robot is closer than the threshold you chose, and a solid green light otherwise.

the display block

The *Display block* has four main modes: Text, Shapes, Image, and Reset Screen. You're already familiar with the Text mode. The *Reset Screen mode* simply returns the display to the normal information shown when a program is running. The Image and Shapes modes require a little more explanation.

displaying an image

Image mode lets you display a picture on the EV3 screen. The EV3 software includes a wide variety of images to choose from, including several faces, arrows, dials, and other objects. You can also use the EV3 Image Editor to create your own image files (select **Tools ▸ Image Editor** from the menu).

Figure 14-12 shows how the Display block looks using Image mode. Click the **File Name** box to select an image. Click the **Display Preview** button to show or hide the preview window, which lets you see how the image will look on the EV3 screen (Figure 14-13).

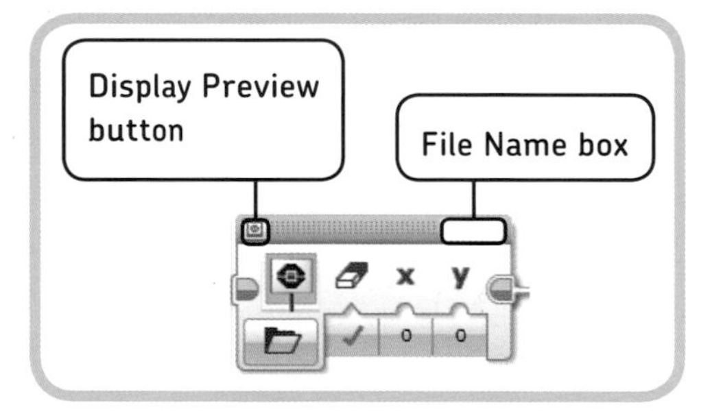

Figure 14-12: The Display block in Image mode

Figure 14-13: The preview window

The X and Y parameters set the location of the top-left corner of the image. The EV3 screen is made up of a grid of dots called *pixels*. (*Pixel* is short for *picture element*.) The screen is 178 pixels wide and 128 pixels high, and the location of each pixel is indicated by an X and Y value. The X value tells you the left-to-right location, with the values going from 0 to 177. The Y value tells you the top-to-bottom location, with 0 at the top of the screen and 127 at the bottom (see Figure 14-14).

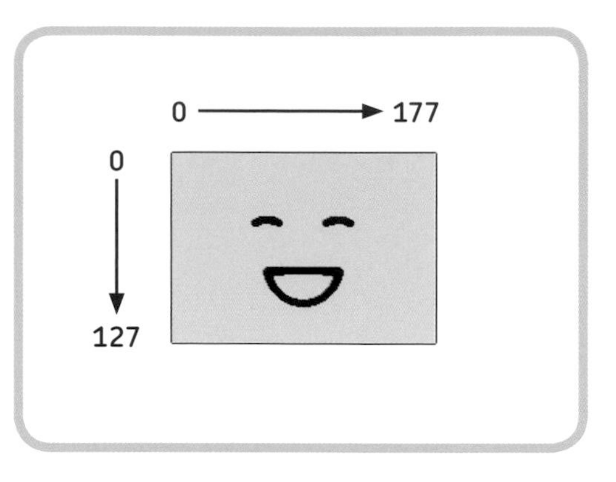

Figure 14-14: X and Y values for the EV3 screen

Set the X and Y values to change where the image is shown on the EV3 screen. For example, Figure 14-15 shows the Big smile image (also used in Figure 14-14) moved up to the top of the screen by setting the Y parameter to -41 (which moves the image up 41 pixels). Depending on the values you choose for X and Y, part of the image might be cut off.

Figure 14-15: The Big smile image at the top of the display

the Eyes program

Showing an image on the display is an easy way to give a robot more personality. The *Eyes* program is a simple example that uses the Random block and six images like the one shown in Figure 14-16 to make the EV3 look a little bored, as if it's waiting for the user to do something. The

Figure 14-16: The Bottom left image

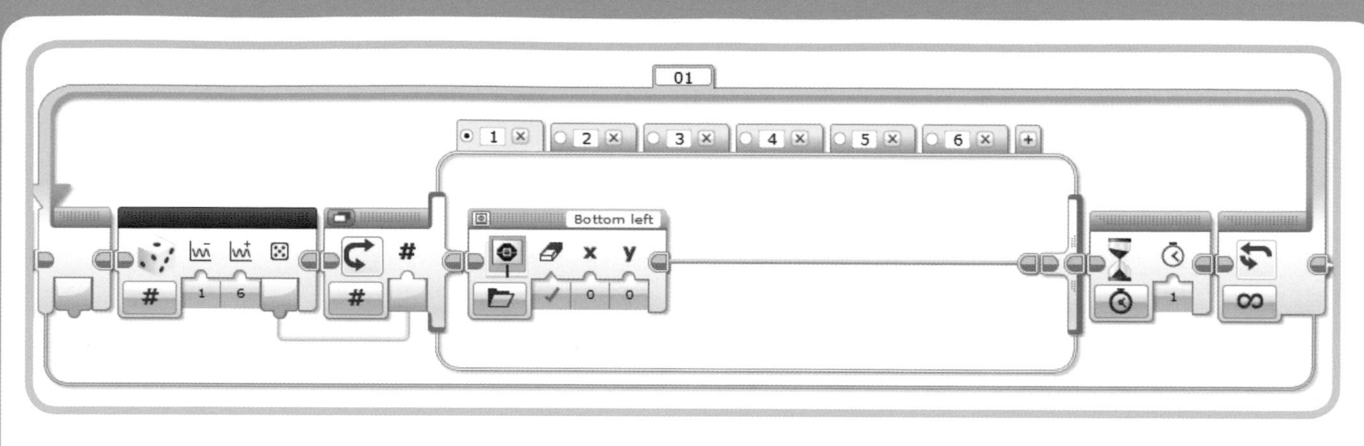

Figure 14-17: The Eyes *program*

only difference between the six images is the direction in which the eyes are looking.

The program is shown in Figure 14-17. The Random block generates a number between one and six and passes it to the Switch block. Each of the six cases of the Switch block contains a Display block showing a different image. The image files for each case are shown in Table 14-1. The Wait block gives a little pause between each image change.

table 14-1: image files for the eyes program

Case	Filename
1	Bottom left
2	Bottom right
3	Middle left
4	Middle right
5	Up
6	Down

Run the program and you should see the eyes randomly looking in different directions.

drawing on the EV3 screen

The Display block's *Shape mode* lets you draw points (single dots), circles, rectangles, and lines. Figure 14-18 shows the Display block in Shape - Point mode. Set the location of the point by entering the X and Y parameters. The Color parameter, indicated by the black box icon, determines whether the point is drawn in black (turning the pixel on) or white (turning the pixel off). If you ever want to clear a screen without drawing anything new, you can simply draw a single white pixel to the screen with the Clear Screen parameter set to true. (The pixel you draw won't show up because all the other pixels are white, too.)

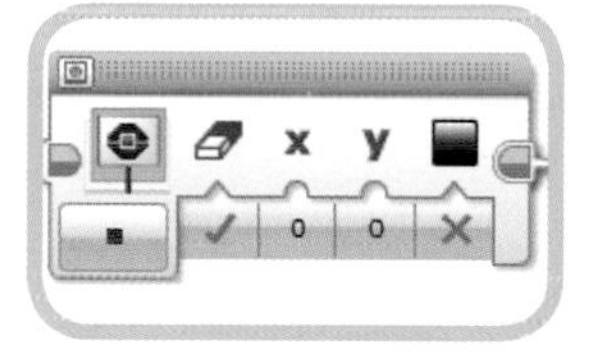

Figure 14-18: Drawing a point

Figure 14-19 shows the Display block with the Shape - Circle mode selected. In this case, the X and Y parameters control the center of the circle, and the Radius parameter lets you control the size of the circle. The Fill parameter (indicated by the paint can icon) controls whether the inside of the circle is filled in or if only the outline is drawn. The preview window shows how the circle will be drawn.

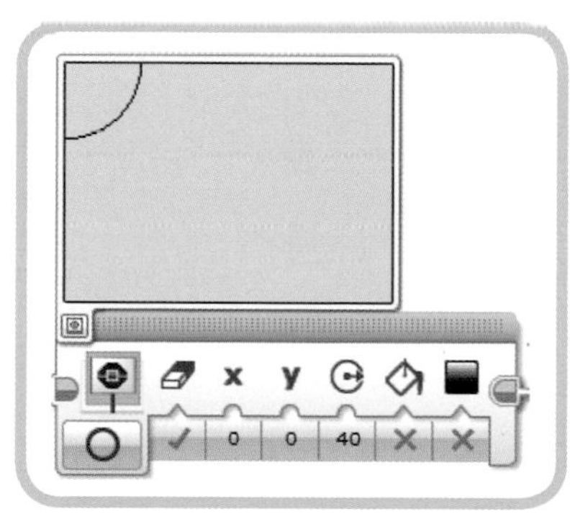

Figure 14-19: Drawing a circle

Figure 14-20 shows the Display block with the Shape - Rectangle mode selected. The X and Y parameters mark the upper-left corner of the rectangle, and the size of the rectangle is defined by the Width and Height parameters.

In Shape-Line mode, the Display block appears as in Figure 14-21. To draw a line, you need one point for each end of the line. The X1 and Y1 parameters set the location of one end of the line, and the X2 and Y2 parameters set the location of the other end.

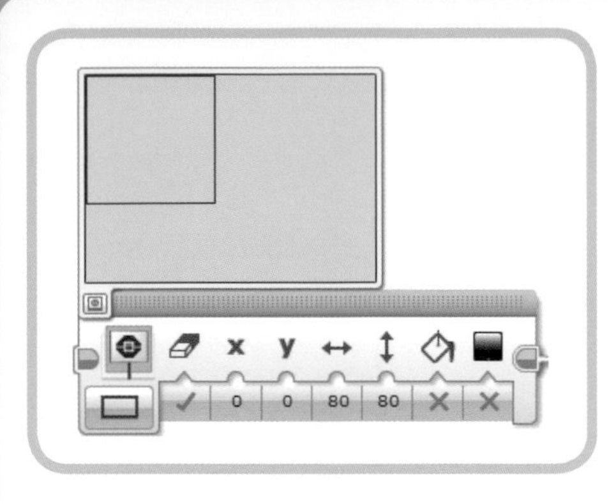

Figure 14-20: Drawing a rectangle

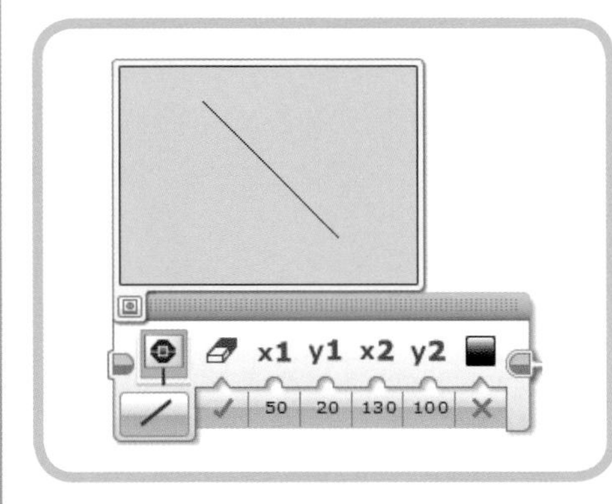

Figure 14-21: Drawing a line

the EV3Sketch program

In this section, you'll create the *EV3Sketch* program, which uses the line-drawing feature of the Display block to turn the TriBot into a sketchpad using the TriBot's two wheels to control where the line is drawn. The basic idea of the program is fairly simple: It repeatedly draws a line from the previous point to the location defined by the current value of the two Rotation Sensors.

The program uses two variables, X and Y, to store the last location used. At the beginning of the program, the EV3 screen is cleared, the variables are initialized to 0, and the Rotation Sensors are reset. The program then enters a loop, which first reads the Rotation Sensors to set a new location (the B motor sets the new X value and the C motor sets the Y value). Then a line is drawn from the old location to the new location, and the values for the new location are stored in the X and Y variables to be used the next time through the loop.

In addition to drawing the line, there should be a way to clear the EV3 screen to start a new drawing. You can do this by setting the Clear option of the Display block if the Center button is bumped. Listing 14-2 shows the pseudocode for this program.

```
clear the EV3 screen
set X to 0
set Y to 0
reset the Rotation Sensors for motors B and C
begin loop
   read the Rotation Sensor for motor B
   read the Rotation Sensor for motor C
   draw a line from X,Y to the point defined by the
      motor B and C positions; if the Center button
      is bumped then set the Clear option
   set X to the motor B position
   set Y to the motor C position
loop forever
```

Listing 14-2: The EV3Sketch *program*

The first section of the program is shown in Figure 14-22. This clears the EV3 screen, initializes the variables, and resets the Rotation Sensors.

Figure 14-23 shows the main part of the program where the Display block draws a line. To draw a line, you need to give the Display block two points: a starting point defined by the values of the X and Y variables, and an end point defined by the values read from the Rotation Sensors for the B and C motors. All of these values are passed to the Display block using data wires.

The Brick Buttons block checks to see whether the Center button has been bumped and outputs true or false accordingly. That value is passed to the Display block and used to control the Clear Screen parameter so that when you press (and release) the button, the Display block clears the EV3 screen before drawing the line.

The final two blocks store the values from the Rotation Sensors in the X and Y variables so that they can be used the next time the loop repeats.

Five blocks supply the settings for the Display block, and because the order of these blocks doesn't matter, I was able to place them so that the data wires don't cross much. (I recommend doing that when possible to make the program easier to read.)

When you run the program, it should start by clearing the EV3 screen. Create a drawing by turning the B wheel to move the virtual pen left to right and the C wheel to move top to bottom. Press the Center button to erase the screen.

The initial drawing starts from the upper-left corner of the EV3 screen. To start a drawing from some other point, move the virtual pen to where you want to start and then press the Center button to clear the screen. You can start a new drawing from there.

Figure 14-22: Initializing the screen, variables, and sensors

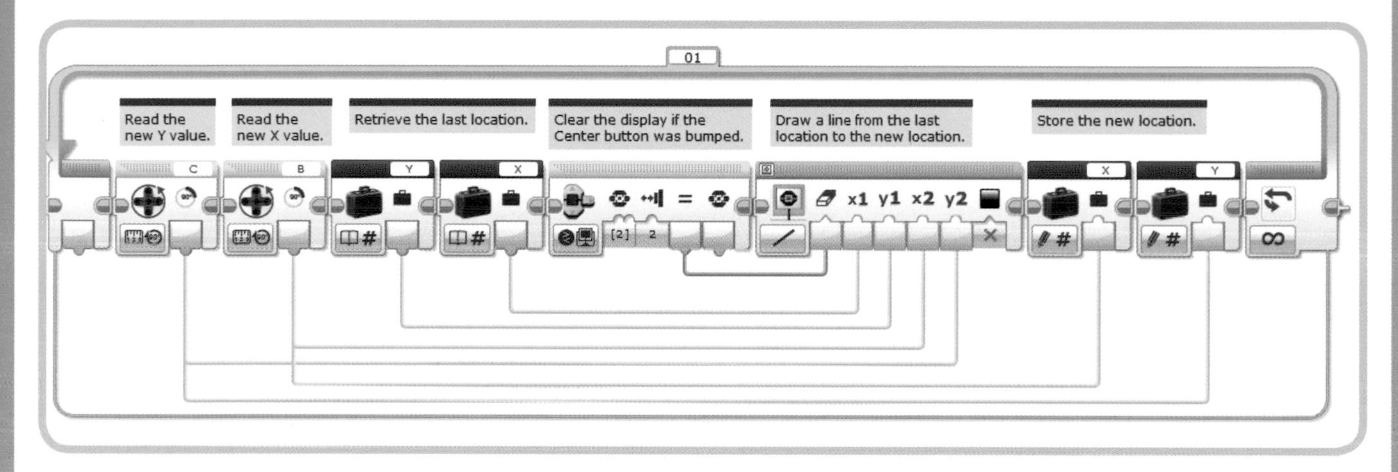

Figure 14-23: Reading the sensors, drawing the line, and saving the new location

further exploration

Here are two more activities to try that use the ideas explored in this chapter:

1. The *PowerSetting* program has a flaw that can be a bit annoying: It doesn't limit the value to the 0–100 range of possible Power values. Modify the program so that the value never becomes less than 0 or more than 100 (in other words, ignore any change that would put the value out of range). To alert users to an invalid value, blink the Brick Status Light red for one second if they try to go below 0 or over 100.
2. Adapt the *CountDown* program from Chapter 13 to show a series of images as the time counts down. You can use one of the groups of images included in the EV3 software to display a dial, progress bar, or timer, or use the Image Editor tool to create your own sequence of images.

conclusion

The EV3 buttons give you a very convenient way to interact with your program. The Brick Status Light block lets you control how the buttons are illuminated, giving you another way to convey information to a user and make your programs more interesting.

The *PowerSetting* program demonstrates how to use the Left and Right buttons to set the value for a variable. The other programs in this chapter show you how to use the Display block to do more than just print text. Using this block, you can display images and create drawings on the screen. Your programs can use these features to take full advantage of the EV3 screen.

15

arrays

In this chapter, you'll learn about arrays, which let you store lists of Numeric or Logic values. Up until now, all the values we've used in our programs (as parameters and in data wires and variables) have been single, individual pieces of data. Using an array lets you store a list of values in a single variable.

After a brief introduction to arrays, we'll create a simple test program to demonstrate how EV3 arrays work. Then we'll create three fairly complex programs that use arrays: one that lets you build simple programs for the TriBot using Brick buttons, one that counts how many times the robot detects a list of colors, and one that implements a memory game.

overview and terminology

An *EV3 array* is an ordered list of values, where each value can be accessed by its position in the list. Each value in an array is called an *element*, and the position of an element is its *index*. In EV3 arrays, the indices start from 0 (which is very common in computer programming). So the first element in an array is at index 0, the second element is at index 1, and so on. The number of values in an array is called the array's *length*. The EV3 software supports Numeric and Logic arrays. A particular array may contain a group of Numeric values or a group of Logic values, but it can't contain a mix of both kinds of values.

As an example, let's say we have an array named SampleValues that contains the colors of objects seen by a program (as detected by the Color Sensor). Say also that the program has seen a blue object, a red object, a white object, and another red object; so the array contains the numbers 3, 5, 6, and 5. Table 15-1 shows the arrangement of indices and values.

table 15-1: indices and values of the SampleValues array

Index	0	1	2	3
Value	3	5	6	5

The length of this array is 4, and the indices go from 0 to 3. Because the indices start from 0, the last index of an array will always be one less than the length of the array. Each element has a unique index, but the values stored in the array don't need to be unique. For example, the SampleValues array has two elements with the value 5 (because two red objects have been seen).

A bit of shorthand is commonly used when writing or speaking about an array element. Instead of saying "the value of the element at index 1 of the SampleValues array," you can say "SampleValues[1]," which is pronounced "SampleValues sub one." (The [1] part is called the *subscript*.)

creating an array

Let's create an array to see what we can do with it. The programs in this chapter all use the Variable block to create an array. The Constant and Array Operations blocks can also create arrays; using the Variable block is more common, however, because you'll usually want to store the array as a variable anyway.

For example, to create the SampleValues array using a Variable block, select the Write Numeric Array mode and set the variable name to SampleValues (Figure 15-1). By default, the Value parameter is set to [ ], an *empty array*. An empty array doesn't contain any elements and consequently has a length of 0. It's not unusual for a program to initialize a variable to an empty array and then add elements to the array as the program runs.

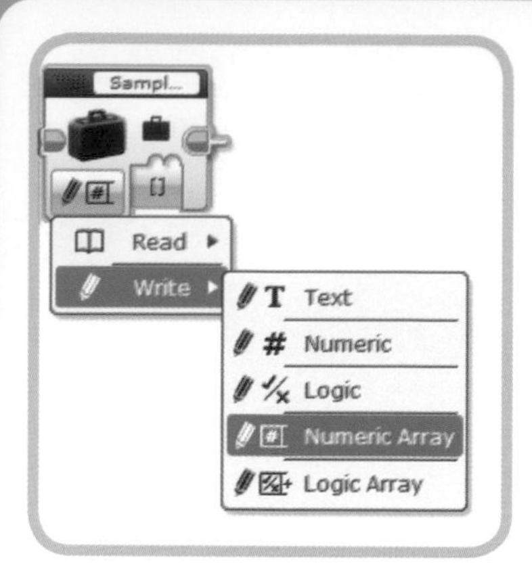

Figure 15-1: Creating an array with the Variable block

Notice that the block's Value parameter box has a double half-circle on top: . This tells you that the parameter takes a Numeric array value. Logic array values are indicated using two triangles instead of half-circles.

To add the four values from Table 15-1 to the Sample-Values array, click the Variable block's Value parameter, and a box will appear where you can add, remove, or edit elements. Click the **+** button to add a new element and then set the value. Figure 15-2 shows the list after adding the first two elements. To remove an element, click the **X** to the right of the element.

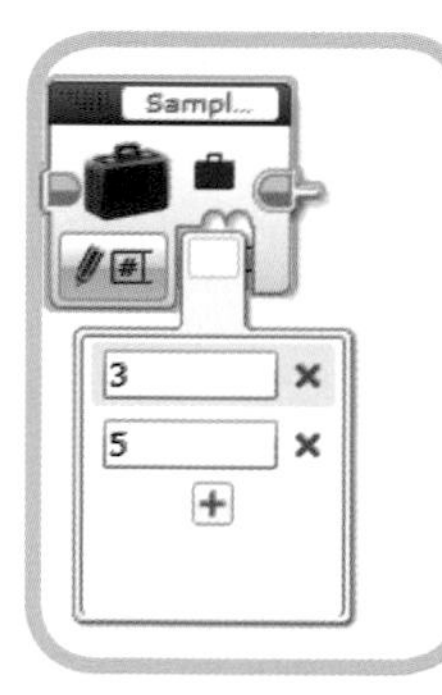

Figure 15-2: Box for setting the element values

Now the block's Value parameter shows the beginning of the array, but most of the elements get cut off because the array is wider than the parameter box. To see all the elements, hover your mouse over the small suitcase icon (as in Figure 15-3).

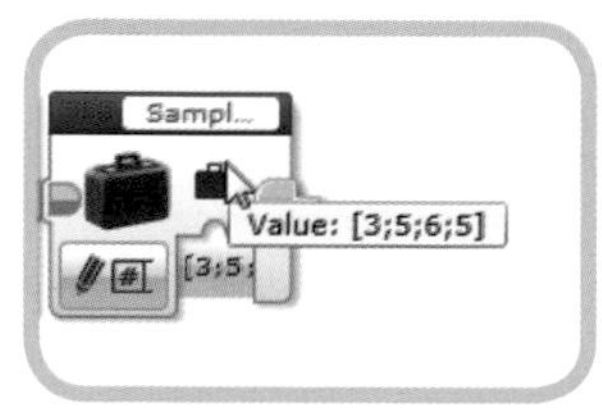

Figure 15-3: Showing all the array elements

Now that you've created an array, let's see what you can do with it using the Array Operations block.

the array operations block

The *Array Operations block* takes an Array value as an Input parameter and allows you to add a new element to the end of the array, read or write the element values of an array, or find out the length (number of elements) of an array. The mode selects which of the four operations to perform and the data type of the array, either Numeric or Logic (Figure 15-4).

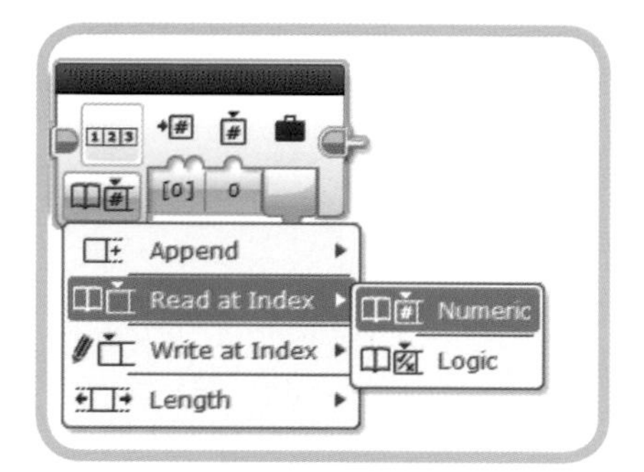

Figure 15-4: The Array Operations block

length mode

The *Length mode* tells you how many elements are in the array. Figure 15-5 shows a program that displays the length of the SampleValues array. In this mode, the Array Operations block has two parameters: an input for an array (usually supplied from a data wire) and an output that gives the length of the array. Notice that the data wire going from the Variable block to the Array Operations block is thicker than the other data wire. This tells you that the data wire holds an Array value.

The first two blocks in the program create the Sample-Values array and put the array on the data wire for the Array Operations block to use. The Array Operations block determines the length of the array and passes this value to the Display block, which shows the value on the EV3 screen. The Wait block gives you time to read the screen before the program ends. Using the SampleValues elements from Table 15-1, this program should display 4.

read at index mode

In the *Read At Index mode*, the block takes an array and an index as Input parameters and gives the element at the specified index as an Output parameter. For example, the program shown in Figure 15-6 displays the value of SampleValues[3] (the element with index 3).

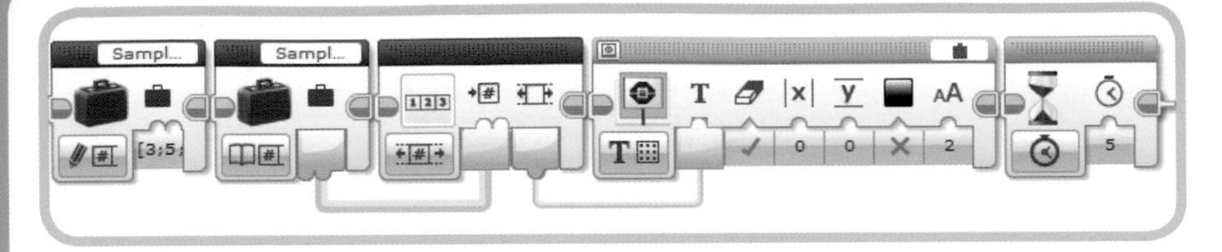

Figure 15-5: Displaying the length of the SampleValues array

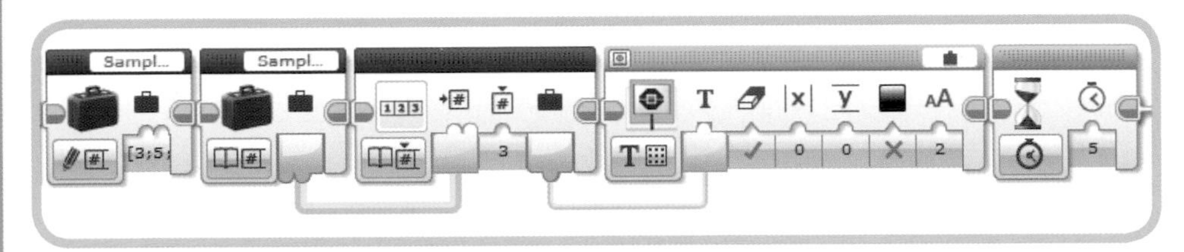

Figure 15-6: Displaying SampleValues[3]

When this program runs, the Array Operations block takes the SampleValues array as an input, reads the value at index 3, and outputs that value on a data wire. The Display block then shows the value, which should be 5. (Remember: we start at 0 so reading index 3 gives us the fourth element in the array.)

Trying to read an element that doesn't exist produces an error. If the index you give the Array Operations block is greater than or equal to the length of the array, your program immediately stops and the EV3 displays a three-corner yield sign that's often used to denote an error or warning.

write at index mode

The *Write At Index mode* lets you change an element's value. Figure 15-7 shows a program that sets SampleValues[3] to 4. In this mode, the Array Operations block takes three inputs: the original array, the index of the element to change, and the element's new value.

It's important to note that the Array Operations block by itself doesn't change the value of the SampleValues variable. To do that, we need to use a Variable block to store the updated Array value. Figure 15-8 shows the updated array being sent out of the second data wire. As you can see, the 5 that was at the end of the array has been replaced by a 4.

If you write a value to an array index that doesn't exist yet, the EV3 software extends the array to the length necessary for the given index, but any other indices that were added this way will be filled with random values.

append mode

The *Append mode* adds a new element to the end of the array, increasing the length of the array by 1. The program in Figure 15-8 adds a new element to the array, with the value 7.

When you run this program, you can see that 7 was added to the end of the array. The length of the array is now 5.

NOTE When examining a data wire, the EV3 software shows only the first five elements of the array.

the ArrayTest program

In this section, I'll present the *ArrayTest* program, which demonstrates some typical array operations. The program creates an empty array, adds the first five multiples of 2 (starting at 0) to the array, and then displays the values on the EV3 screen.

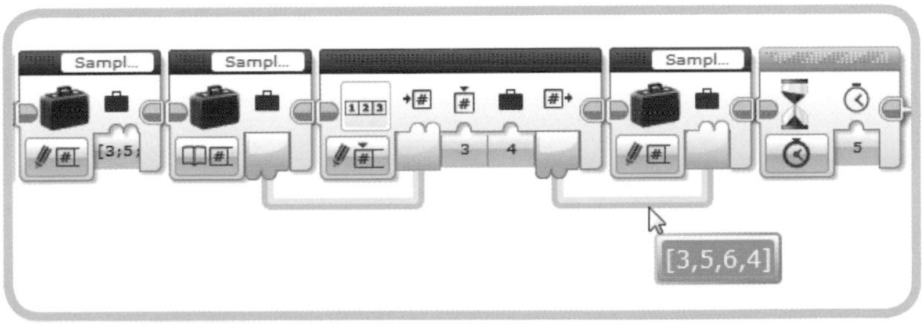

Figure 15-7: Setting SampleValues[3] to 4

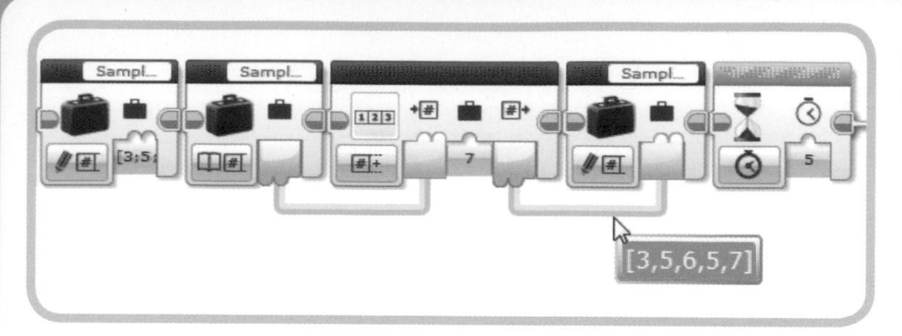

Figure 15-8: Appending a new element to the SampleValues array

The first part of the program, shown in Figure 15-9, begins by creating a Numeric array variable named ArrayValue and setting the value to an empty array. The Loop block repeats five times. Every time the program reads the current Array value (with a Variable block), it appends a new value that's twice the Loop Index to the array (using a Math block and an Array Operations block), and then a second Variable block stores the updated array. When the loop completes, the ArrayValue variable should hold an array containing five values: 0, 2, 4, 6, and 8.

NOTE **To see the array being constructed, add a one-second pause before the end of the loop, and use the EV3 software to examine the value on the data wire coming out of the Array Operations block. The pause gives you time to see the array change after each loop.**

The second half of the program, shown in Figure 15-10, is a bit more complicated, but the basic idea is simple: Go through the array and show each element on the screen. I'll go through and explain each block from left to right.

1. The first Display block clears the screen.
2. The Variable block reads the value in ArrayValue and sends it on data wires passed to the Array Operations blocks.
3. The first Array Operations block checks the length of the array and uses the Length value to control how many times the Loop block repeats.

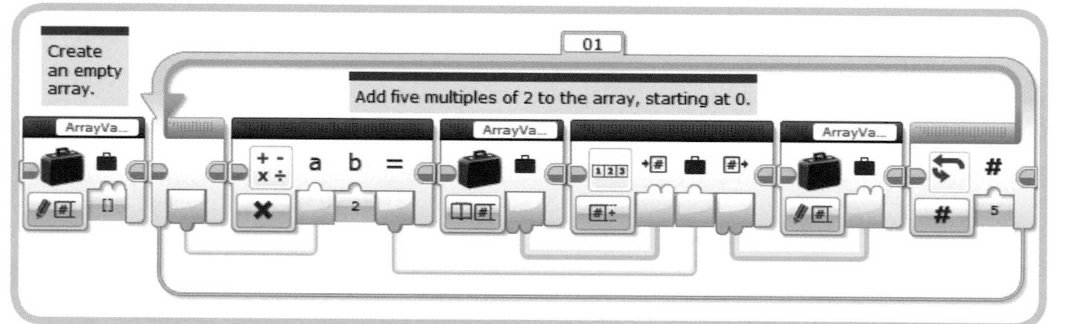

Figure 15-9: The ArrayTest program, part 1

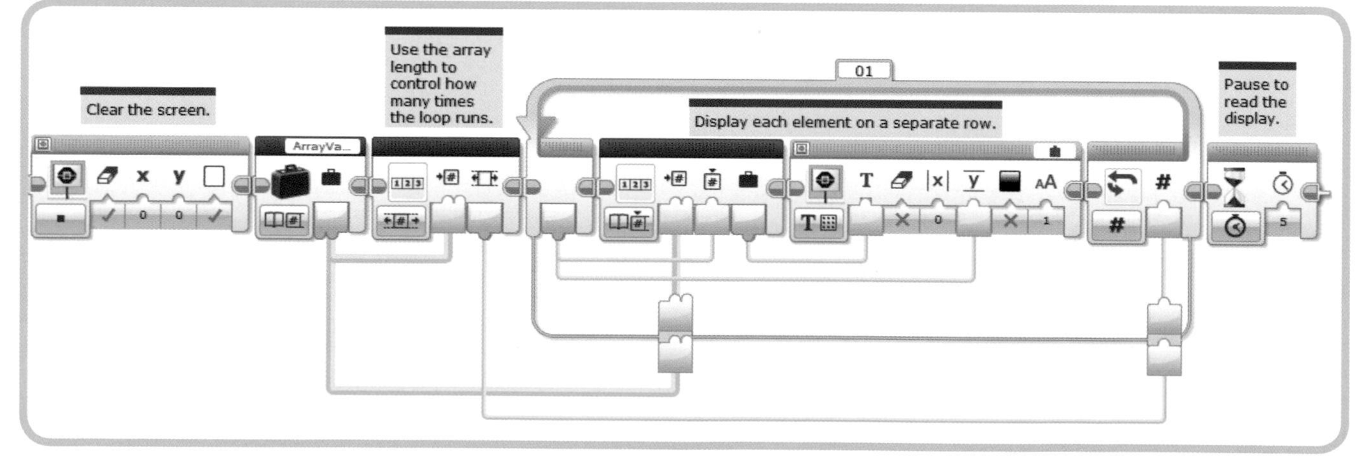

Figure 15-10: The ArrayTest program, part 2

4. The second Array Operations block reads one value from the array, using the Loop Index for the Array Index, and passes it to the Text block. So the first time through the loop it reads ArrayValue[0], the second time ArrayValue[1], and so on.

5. The Text block displays each value on a separate line by using the Loop Index for the Row parameter. The Clear Screen parameter is set to false so that the screen isn't cleared each time through the loop.

6. The Wait block pauses the program so that you can see the result before the screen is reset.

Note that the ArrayValue variable doesn't need to be read inside the loop because the value never changes. Each time through the loop, the data wire going to the Array Operations block contains the same value (the entire array). Only the value used for the Index parameter changes, because it comes from the Loop Index.

Another important point is that this program works because the Array Index, Loop Index, and Display block Row parameter all start at 0. Counting from 0 may seem odd, but as long as it's done consistently, it turns out to be quite convenient.

CHALLENGE 15-1

Displaying the values in an array is often very useful when debugging your programs. Create a My Block that does this based on the code in the *ArrayTest* program. Your new block should take an array as an Input parameter and display each value. Remember that the EV3 screen only has 12 lines, so you should display the first 12 elements, wait for the user to press a button, display the second 12 elements, and so on until you reach the end of the array. Show both the index and value for each element (otherwise it's easy to lose track of the index for longer arrays), perhaps using the DisplayNumber My Block.

HINT If there are more than 12 elements, the Row parameter to the Display block needs to "wrap around" so that the 13th element is shown on row 0. A simple way to accomplish this is to use the Array Index modulo 12 for the Row parameter.

the Button-Command program

The *ButtonCommand* program uses the Brick buttons to give the TriBot a list of movement commands to execute. The first part of the program constructs an array of commands based on the buttons you press, and the second part of the program executes those commands, making the robot move according to your instructions.

We use all five buttons on the EV3 Brick. The Left and Right buttons command the robot to spin a quarter-turn to the left or right, and the Up and Down buttons command the robot to move forward or backward one rotation. We use the Center button to indicate when we're done entering commands and are ready for the TriBot to move.

creating the array of commands

The program uses a Numeric array called CommandList to store the commands to execute. We represent each command using the numbers corresponding to the Brick buttons, as shown in Table 15-2.

table 15-2: button numbers and associated commands

Button	Number	Command
Left	1	Spin a quarter-turn to the left.
Right	3	Spin a quarter-turn to the right.
Up	4	Move forward one rotation.
Down	5	Move backward one rotation.

To construct the command list, the program first creates an empty array. It then enters a loop that waits for you to bump a button, and either adds the button number to the list or exits the loop if you bumped the Center button. As you're creating the list of commands, they are displayed on the EV3 screen.

I'll start with the code to add the commands to the array, and then I'll add the blocks to display each value in the next section. The code shown in Figure 15-11 clears the screen, creates the empty array, and adds commands to the array until you bump the Center button.

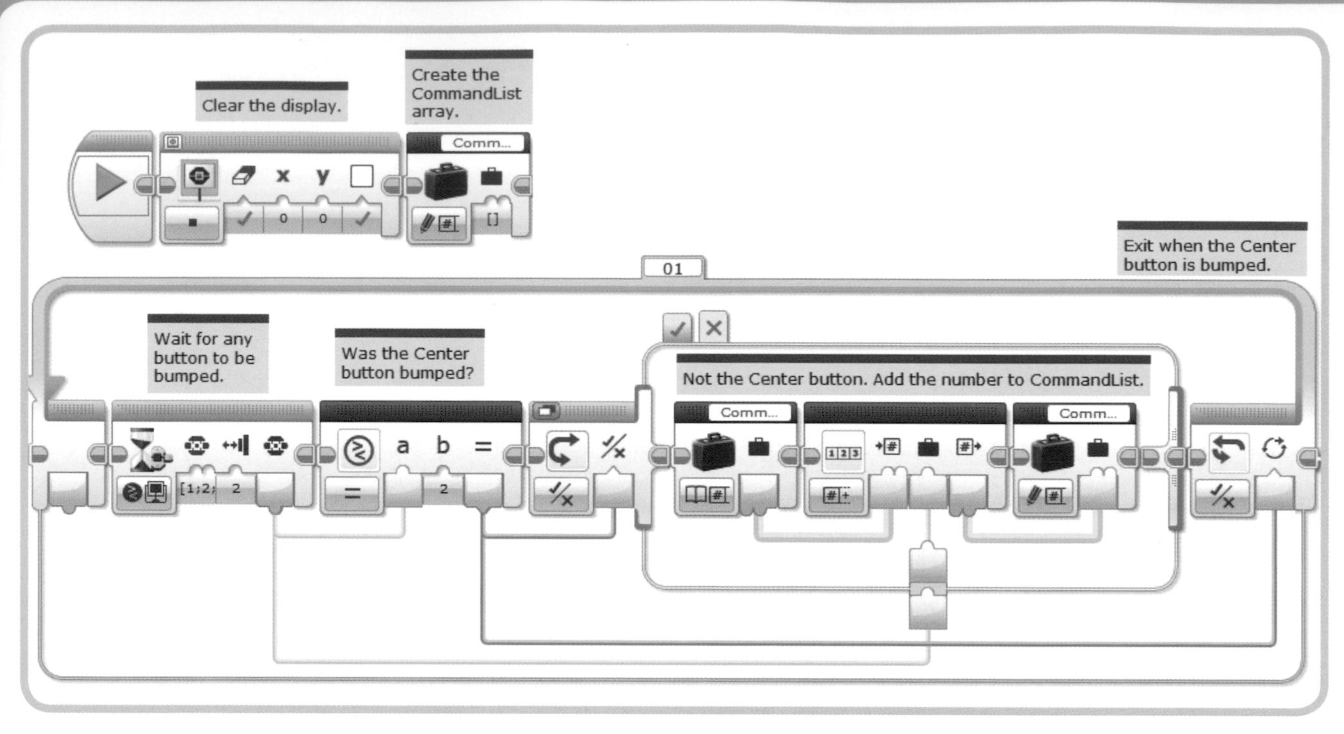

Figure 15-11: Adding commands to the CommandList array

The first two blocks are self-explanatory, but the blocks within the loop require more explanation.

1. The Wait block uses Brick Button - Compare mode to wait for any of the five buttons to be bumped. (All five buttons are checked, as shown in Figure 15-12.) The Button ID Output parameter indicates which button was bumped.

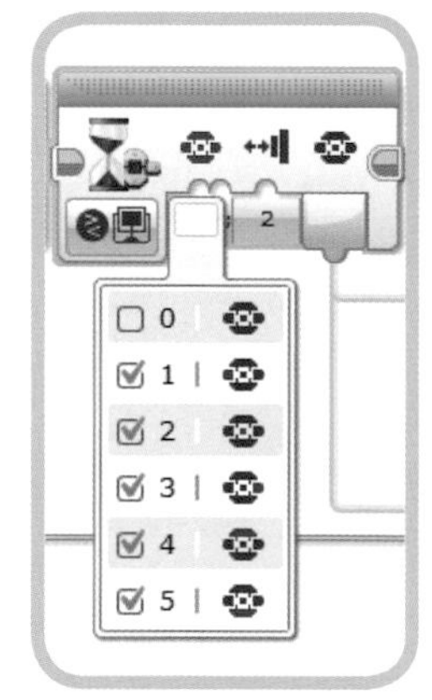

Figure 15-12: Wait for any of the five buttons to be bumped

2. The Compare block checks whether the Center button (number 2) was the one that was bumped. The result is passed to the Switch block and the Loop block.
3. The true case of the Switch block is empty, so if the Center button is bumped, we go straight to the end of the loop. If a different button was bumped, we go to the false case of the Switch block and run the three blocks there.
4. The first Variable block (within the Switch block) puts the current CommandList array on a data wire.
5. The Array Operations block appends the Button ID from the Wait block to the end of the array.
6. The second Variable block stores the updated Array value back in the CommandList variable.
7. The Loop block uses the result from the Compare block to decide when to exit the loop so that the loop exits when the Center button is bumped.

displaying the commands

As each new command is appended to the array, the program should add the number to the screen. So after pressing Up, Up, Up, Right, the display should show "4443". We also want the program to display as many values as possible. If we simply show each value on a new row, that will limit us to 12 values. (Similarly, if we use a new column for each value, that limits us to 22 values.) So for this program, we'll use rows and columns: The first value appears at the top left (row 0, column 0), the next value appears at (0,1) (that is, row 0, column 1) and so on, until we reach the end of the row. Then the next value appears on the row below: row 1, column 0. This lets us fit up to 264 (22 times 12) values on the screen, which should be plenty for this program.

We'll need two simple formulas to compute the row and column for each value based on the Loop Index. (Remember, this Loop Index will match each element's index in the array.) To calculate the row to use, we divide the Loop Index by 22 and round it down (using the floor function) to the nearest integer: floor (Loop Index / 22). When the Loop Index is between 0 and 21, the result of the expression is 0, and the new command is printed on the first row. When the Loop Index is between 22 and 43, the result of the expression is 1, and the command is printed on the second row, and so on.

To determine which column to use, we need to find the remainder of the Loop Index divided by 22: Loop Index % 22. This will give us the number of spaces left over after deciding which row the command should go on. When the Loop Index is between 0 and 21, the result of the expression is the same as the Loop Index. When the Loop Index hits 22, the result of the expression is 0, so the command prints in the first column on the left. The next time through the loop, the Loop Index is 23, and because the remainder of 23 divided by 22 is 1, the command prints on column 1. This is another example of using the modulo operator to make the numbers "wrap around"—this time using the width of the display.

Figure 15-13 shows the Switch block from Figure 15-11 with two Math blocks and a Display block added to display each command in the proper row and column. The Math blocks compute the column and row based on the Loop Index, which is input to the *a* parameter of both blocks. The Display block then uses these values to add the new command to the display. The Display block uses Font 1, so that each character takes up one row and one column, and has the Clear Screen parameter set to false.

You may want to test the program at this point and make sure the list of commands is stored and displayed properly. If you run the program from the EV3 software, you can examine the value on the data wire coming out of the Array Operations block to see if the program works as expected.

executing the commands

The second part of the program (Figure 15-14) reads each element of the array and executes the appropriate command. The structure of this loop is similar to the one used in the *ArrayTest* program. An Array Operations block reads the length of the array and uses that value to control how many times the Loop block repeats. Inside the Loop block, the Loop Index is used to read one value out of the array. This is a pattern that you'll often use with arrays.

The element value is then passed to a Switch block. The cases of the Switch block correspond to the Button IDs for each command, and each case uses a Move Steering block to move the TriBot left, right, forward, or backward.

When you run the program, you can enter commands using the four directional buttons, and the list of commands (Button IDs) should be displayed on the EV3 screen. When you bump the Center button, the TriBot should start moving until it has completed all the commands entered.

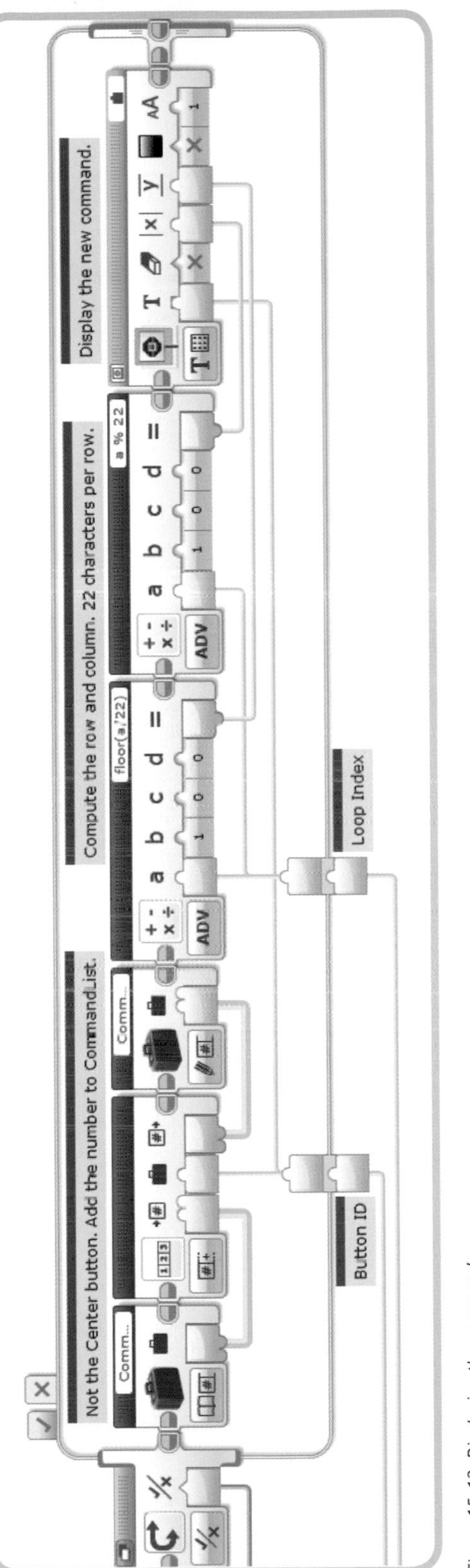

Figure 15-13: Displaying the commands

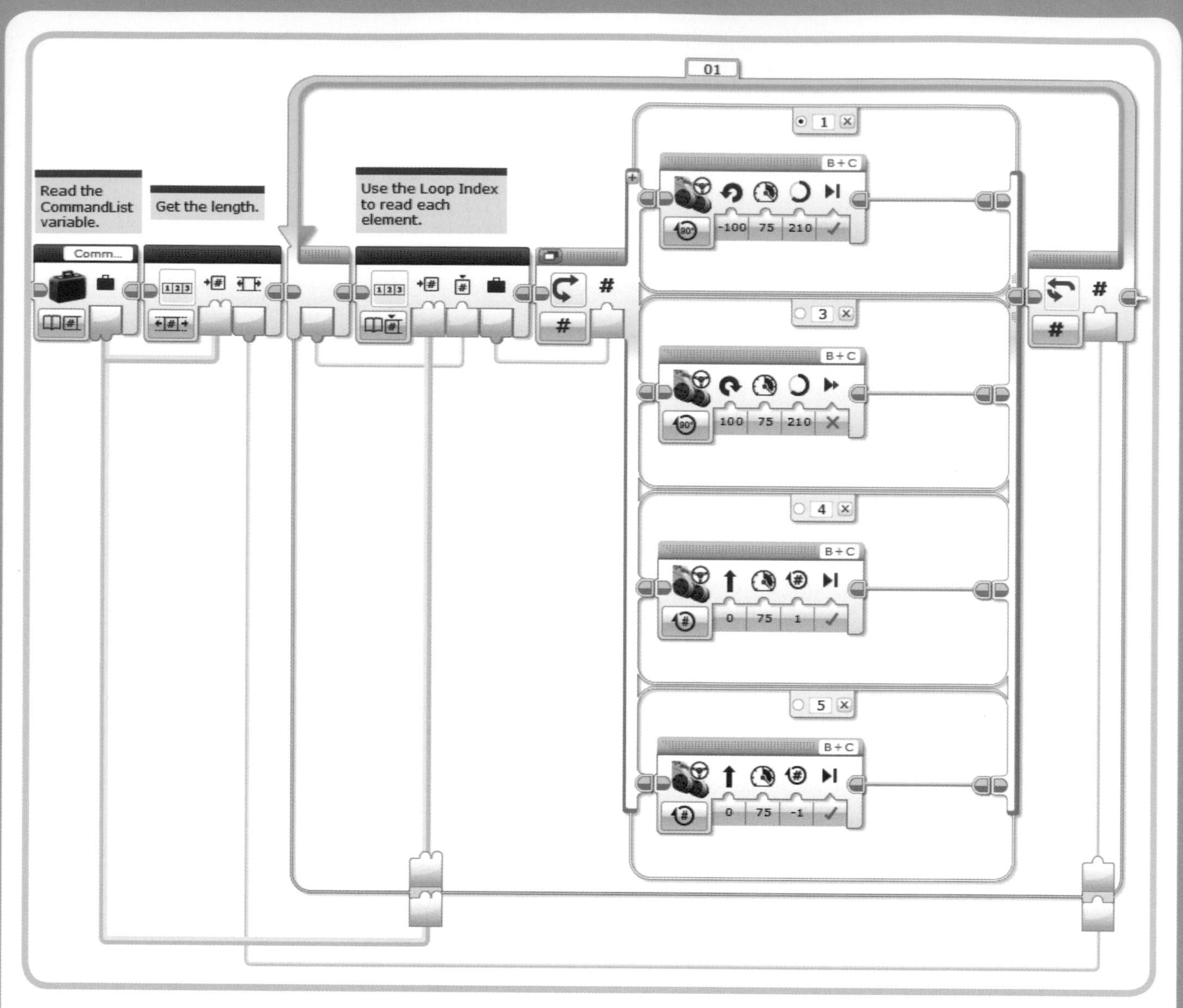

Figure 15-14: Executing the commands

> **CHALLENGE 15-2**
>
> Change the program's actions to make it more interesting. Each case of the Switch statement can contain as many blocks as you want, or use My Blocks for even more complex actions. For example, you could replicate the behavior of the BumperBot with these commands:
>
> 1. Move forward (using the On mode).
> 2. Wait until the Touch Sensor is pressed.
> 3. Move backward 1 rotation.
> 4. Spin for a random number of degrees.

the ColorCount program

The *ColorCount* program counts the number of objects of each color detected by the Color Sensor. The behavior is similar to the *RedOrBlue* program; the program waits for you to place an object in front of the Color Sensor and then says the name of the color and keeps a count of how many objects of each color it has seen. But the *ColorCount* program keeps track of all eight Color values that the Color Sensor can identify (the seven colors plus the No Color value). Recall that the *RedOrBlueCount* program from Chapter 11 could do this for only red and blue objects using two variables. While you could extend the *RedOrBlueCount* program to use eight variables to count the eight possible Color values that the Color Sensor can return, an array is better suited to store those eight values. (When I say eight color values, I'm counting the seven color values plus the No Color value).

To keep track of how many objects of each color the program has seen, we'll need a Numeric array with eight elements, which I'll call ColorCounts. Each time the Color Sensor takes a measurement, it outputs a value from 0 to 7, corresponding to the color it saw, as shown in Table 15-3. Because both the color numbers and the array indices start at 0, we'll store the count of each color number in the matching Array Index. This means that, for example, ColorCounts[2] will hold the number of blue objects, and ColorCounts[4] will hold the number of yellow objects. So for this array, the index tells you not only the position of an element but also the color to which the element refers.

table 15-3: color sensor values

Number	Color
0	No Color
1	Black
2	Blue
3	Green
4	Yellow
5	Red
6	White
7	Brown

Listing 15-1 shows the high-level steps in the program. To make this program easier to write, we'll build a couple of My Blocks. The first one takes a color number as an input and give you back the name of the color as a Text value. This can be useful for both showing the color name on the screen and saying the color using a Sound block. The second My Block adds one to the count for a particular color. This step is straightforward but requires five blocks and takes up a lot of space on the Programming Canvas, so using a My Block will make the main program smaller and easier to read.

```
create the ColorCounts variable as an array with eight 0s
display the eight color names and the starting count (0)
begin loop
   wait for the Center button to be bumped
   read the Color Sensor
   use a Sound block to say the name of the color
   add one to the count for the color
   display the new count for the color
loop forever
```

Listing 15-1: The high-level steps of the ColorCount *program*

the ColorToText my block

We can take the Numeric output from the Color Sensor and produce the corresponding color name using a Switch block. The Switch block is pretty straightforward but will be quite large because it needs eight cases to handle all of the possible Color values. We'll also need to repeat this code twice in the program—all the more reason to make it into a My Block.

To create the ColorToText My Block, I first wrote the *ColorToTextBuilder* program, shown in Figure 15-15. The My Block is created from only the Switch block and the blocks inside it. The Constant block at the beginning and the Variable block at the end are only there to add data wires, which will turn into the ColorToText My Block's parameters.

The Switch block takes a number as input and selects the matching case. Each case contains a Constant block that places the corresponding (text) name of the color on the data wire; the only difference between the cases is the Text value set in the Constant block.

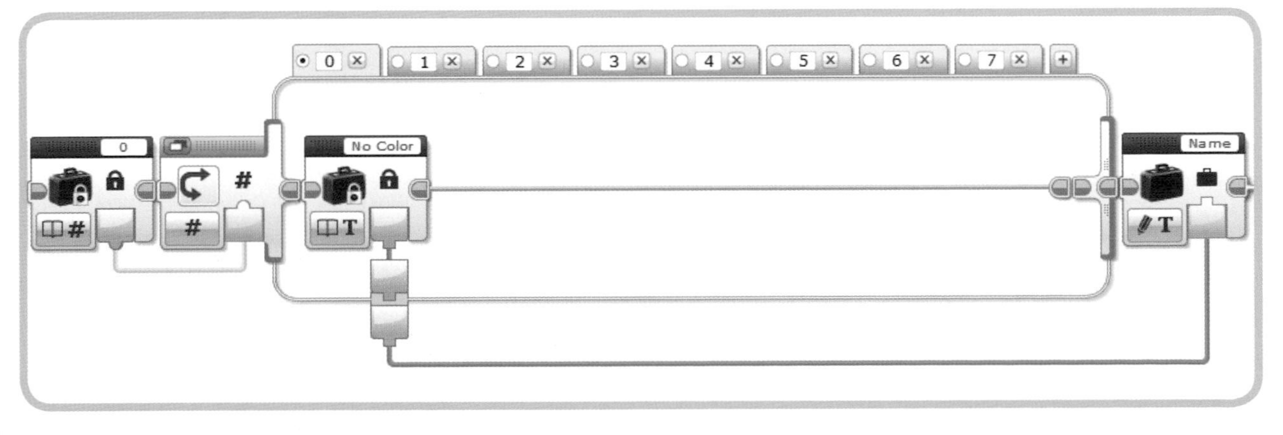

Figure 15-15: The ColorToTextBuilder *program*

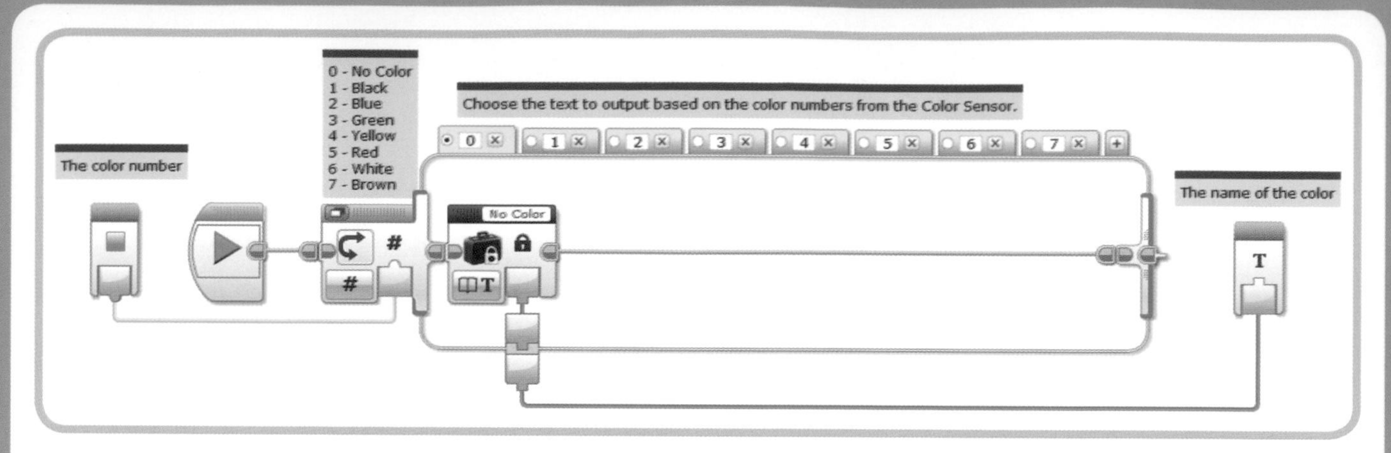

Figure 15-16: The ColorToText My Block

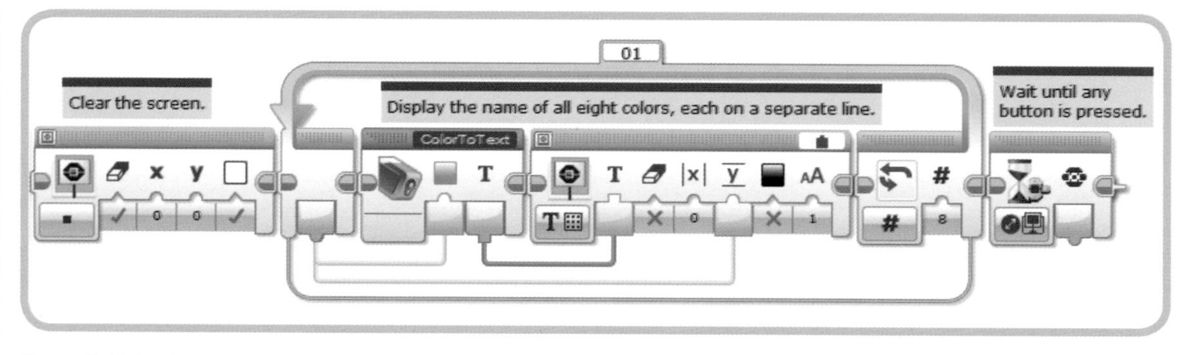

Figure 15-17: The ColorToTextTest *program*

To create the My Block, select the Switch block and click **Tools ▸ My Block Builder** on the menu. Set names and icons for the two parameters. (I used *Color Number* and *Text*.) Figure 15-16 shows the resulting My Block with some comments added.

After you build this My Block, it's a good idea to thoroughly test it to make sure that all the data wires are connected appropriately and that the block outputs the right name for each Color value.

The *ColorToTextTest* program (Figure 15-17) clears the EV3 screen and then displays each color on a separate line. The Loop block repeats eight times, sending the Loop Index as an input to the ColorToText block to get the names of all eight colors, and as an input to the Display block to set the row for each color. The first time through the loop, the Loop Index will be 0, so the output from the ColorToText My Block should be No Color, displayed on line zero (the top of the display). The next time, the Loop Index will be 1 and the program should display Black on line one. The loop repeats until all eight color names are displayed. The Wait block at the end uses Brick Buttons Change mode to wait until any button is pressed, so you can read the display before ending the program.

Run the program, and the EV3 display should show the following:

```
No Color
Black
Blue
Green
Yellow
Red
White
Brown
```

If any color is missing or in the wrong order, edit the ColorToText My Block to fix the problem.

the AddColorCount my block

The ColorCounts variable holds the array, which contains the number of times each color has been seen. Each time a new object's color is identified, we need to add 1 to the corresponding element in the array. To read and update the variable, we need five blocks, which we'll combine into a My Block to make the program shorter and easier to read. (Refer back to Chapter 12 if you need a refresher on creating a My Block.)

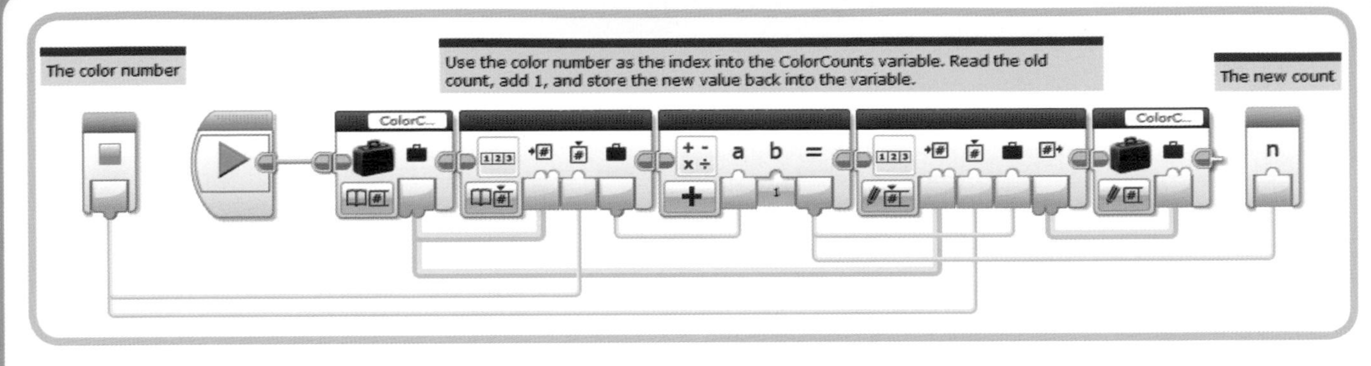

Figure 15-18: The AddColorCount My Block

Figure 15-18 shows the completed AddColorCount My Block, which takes the color number as an input and adds 1 to the corresponding ColorCounts array element. We'll need to use the new value in the main program to update the display, so the My Block supplies that value as an Output parameter.

The first Variable block reads the value of the ColorCounts variable and puts the entire array on a data wire. The Array Operations block takes the array and color number as inputs, and outputs the current count for that color. The Math block adds one to the count, and the following Array Operations block writes the new value back into the array. The final Variable block then stores the new Array value back in the ColorCounts variable.

using a data wire to select a sound file

When our program recognizes a color, it uses a Sound block to say the color name. We can select the sound file to play for each color by passing the color name (from the ColorToText My Block) to a sound block with a data wire (as shown in Figure 15-19). To do this, you need to select Wired where you normally select the filename (Figure 15-20), and then connect a data wire to the Filename parameter.

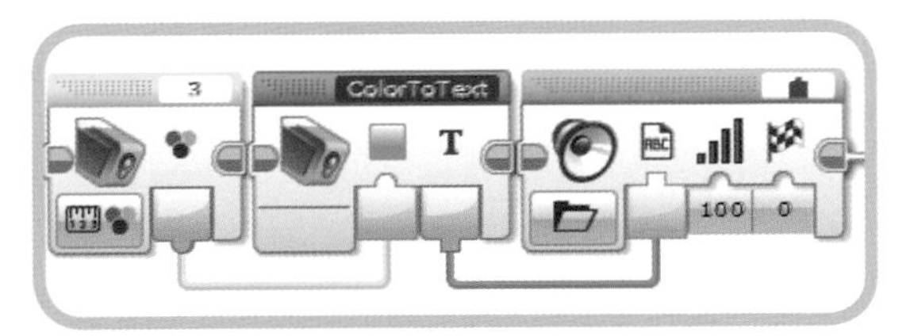

Figure 15-19: Using a data wire to select the sound file

When you select a sound file using a data wire, you need to make sure that the EV3 Brick has the appropriate sound file stored on it. (When you select a sound file manually, the EV3 software handles this automatically.) This means that we have to add the eight files we need to the project and make sure the sound filenames match the color names being sent in on a data wire from the ColorToText block.

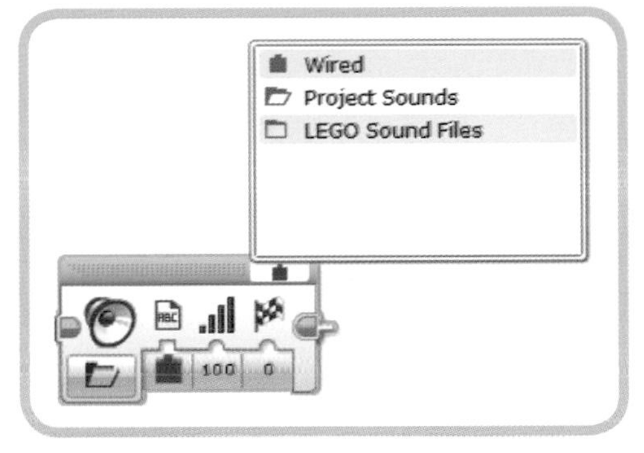

Figure 15-20: The Sound block with Wired selected for the filename

You can see which sound files are included in your project by looking at the Sounds tab on the Project Properties page (Figure 15-21). To add a sound file to the list, all you need to do is add a Sound block to a program and select the desired file. That sound file stays in the list of sounds included in your project, even after you select a different sound file or delete the Sound block.

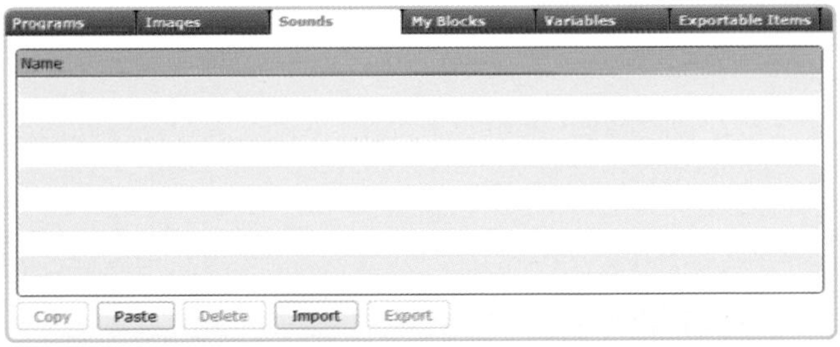

Figure 15-21: The list of sound files on the Project Properties page is empty until you use a Sound block to select a file.

Follow these steps to add the seven color names to the project:

1. Create a new program named *ColorCount*.
2. Add a Sound block to the program.

3. Select **Black** for the filename (under **LEGO Sound Files ▸ Colors**).

4. Select **Blue** for the filename (on the same Sound block).

5. Continue changing the filename until you've selected all seven colors.

The list of sound files should now look like Figure 15-22. Double-check to make sure you have all seven files.

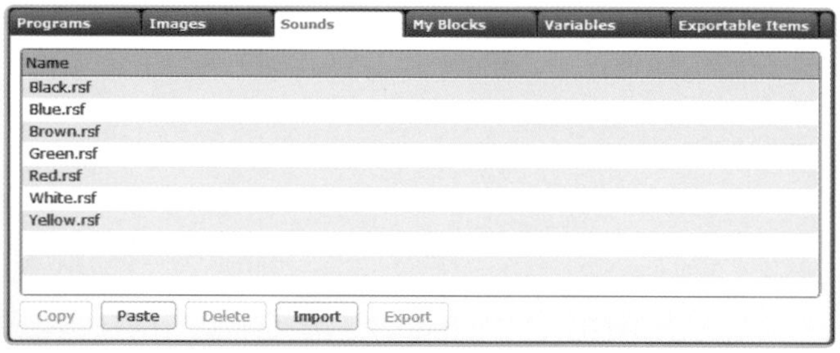

Figure 15-22: Including the seven color sound files

Now we need to take care of the No Color case. There isn't a sound file named No Color, so we'll create one. In the *RedOrBlue* program, we used the Uh-oh sound when the Color Sensor failed to recognize a color. To use that sound file in the *ColorCount* program, we can rename the file to *No Color* so that the sound filename matches the text passed in from the Color-ToText block. Here are the steps for creating a No Color sound file from the Uh-oh sound:

1. Select **Uh-oh** for the Sound block's filename (under **LEGO Sound Files ▸ Expressions**).

2. Open the Project Properties page.

3. Select *Uh-oh.rsf* in the list of sound files.

4. Click the **Export** button. This opens a file dialog allowing you to save the sound file. Change the name from *Uh-oh.rsf* to *No Color.rsf* (be sure to include the space).

5. Click the **Import** button. This will open a file dialog where you can select the file to add to the project. Select *No Color.rsf*.

6. On the Projects Properties page, select *Uh-oh.rsf* and click the **Delete** button.

7. Delete the Sound block from the program.

The Program Properties page should now look like Figure 15-23. Of course, you can use a different sound file if you prefer, such as Boo, Sorry, No, or one of the horn or beep sounds. You can also use the Sound Editor tool to create a sound file of your own. The important thing is that whatever file you choose has the name No Color, which matches the Text value used by the ColorToText My Block.

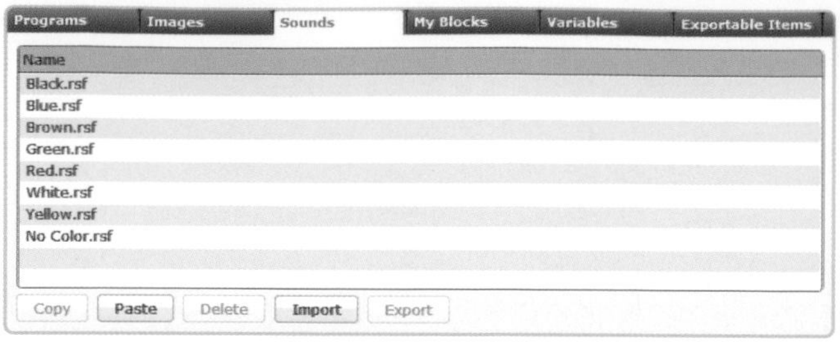

Figure 15-23: The complete list of sound files needed for the ColorCount *program*

initialization

Now we can start writing the program. First, we need to get everything set up by creating the ColorCounts array with the eight elements set to 0, and displaying the color names and the initial count.

The blocks that perform this initialization are shown in Figure 15-24. Here's how this section of code works:

1. The Variable block, set to **Write Numeric Array** mode, creates the ColorCount variable. The Value parameter is set to **[0;0;0;0;0;0;0;0]**. This creates an array with eight values and starts them all off at zero. Refer to "Creating an Array" on page 175 if you're not sure how to set the initial Array value.

2. The Display block clears the screen.

3. The Loop block repeats eight times, once for each color.

4. The ColorToText block takes the Loop Index as input and supplies the corresponding color name as its output.

5. The first Display block shows the color name from the Color-ToText block, using the Loop Index to set the row. The Font parameter is set to **1** so that all eight colors fit on the display.

6. The second Display block shows **0**, also using the Loop Index for the Row parameter. The Column parameter is set to **10** so that the values form a nice column.

After running this part of the program, the EV3's display should show the following:

```
No Color   0
Black      0
Blue       0
Green      0
Yellow     0
Red        0
White      0
Brown      0
```

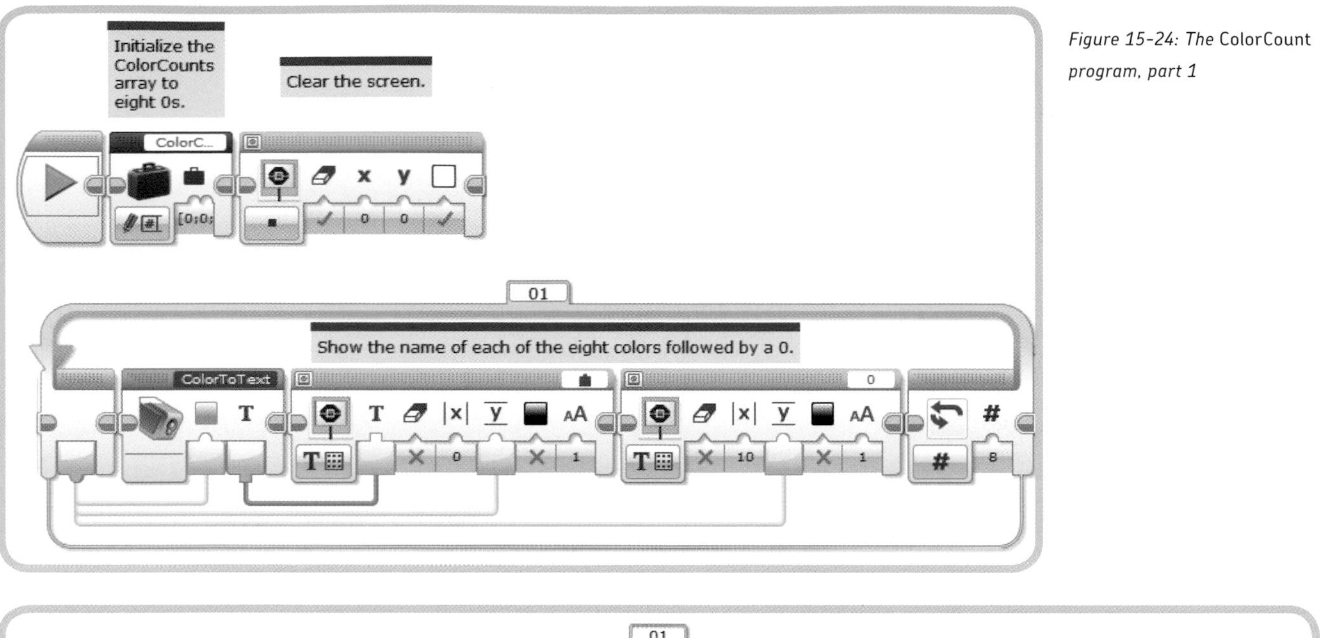

Figure 15-24: The ColorCount program, part 1

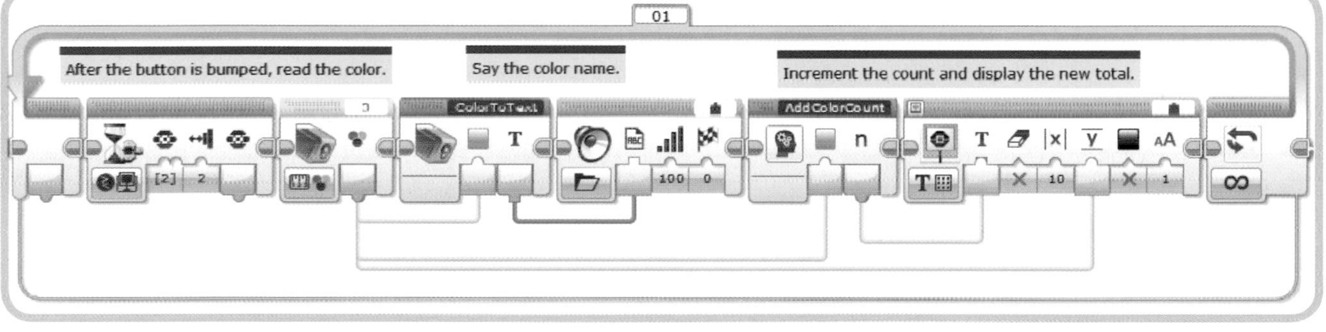

Figure 15-25: The ColorCount program, part 2

Displaying the counts in column 10, rather than immediately after the color name, provides a clear list of information. It also makes it easier to update the display, because we can just update the color counts using 10 for the column and the color number for the row. This requires a lot less code than redisplaying the entire list every time a number is changed.

If you want to test the program at this point, you need to add a temporary Wait block at the end. Otherwise, the program will end and the screen will clear before you can check the result.

counting colors

Now we're finally ready to write the code that does the counting (shown in Figure 15-25). Here's how it works:

1. The Loop block repeats until you end the program.
2. The Wait block uses **Brick Buttons - Compare** mode so your program waits until you bump the Center button.
3. The Color Sensor block in **Measure - Color** mode determines the color of the object in front of the sensor. This value is used in three ways in the following blocks.
4. The ColorToText My Block converts the color number to the corresponding color name.
5. The Sound block uses the color name to select the sound file to play.
6. The AddColorCount My Block adds one to the count for the recognized color. It also provides the new value as an Output parameter, which it sends to the Display block.
7. The Display block shows the new count from the AddColorCount My Block. The color number is used for the Row parameter, and the Column parameter is set to **10**. This will overwrite the previous count for the observed color.

Run the program, and it should start by displaying the list of colors, all followed by 0s. When you place an object in front of the Color Sensor and then press the Center button, the program should respond by saying the color name and changing the count for the color from 0 to 1. Keep testing different objects until you've heard all the colors.

the MemoryGame program

The *MemoryGame* program is a simple Simon Says–style game using the Brick Status Light and the Brick buttons. The Brick Status Light shows a random sequence of colors, and then the user presses buttons to try to repeat the sequence. If the response is correct, the game continues with a longer sequence, and if the response is wrong, the game ends.

The entire program is in a Loop block, which continues until the player enters an incorrect response. We use an Array variable, named Lights, to hold the sequence of lights, using the Brick Status Light values (0 = green, 1 = orange, 2 = red). Each time through the loop, we use a Random block to pick a random value between 0 and 2 to add to the Lights array. The program flashes the corresponding light as the array is filled. Then the user responds by pressing the Brick buttons to match the sequence of lights, using the Left button for green, the Center button for orange, and the Right button for red.

The program starts by turning off the Brick Status Light; otherwise, the flashing green light could cause confusion. Then the main loop of the program starts (Figure 15-26).

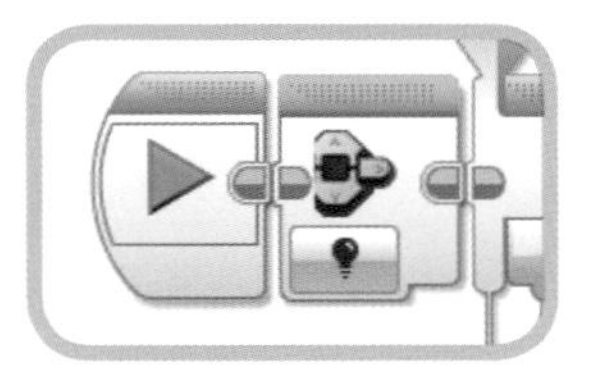

Figure 15-26: Turning off the Brick Status Light before starting the loop

the start of the loop

The first part of the loop is shown in Figure 15-27. The program uses the Loop Index to determine how many values to add to the Lights array. However, because the Loop Index starts at 0 and we want to start the game with one value in the Lights array, we add 1 to the Loop Index and use that for the number of elements in the array.

Each time through the loop, the program uses the Display-Number My Block (from Chapter 12) to show the number of items in the coming sequence as "Level 1," "Level 2," and so on. Then the Lights variable is initialized to an empty array. The program then says "Start" and gives a short pause before showing the sequence of lights.

creating the sequence of lights

The next part of the program fills the Lights array with random values and flashes the Brick Status Lights according to those values (Figure 15-28). The value from the Math block in Figure 15-27 is used to control the number of times this loop repeats, and consequently the number of items added to the array.

NOTE To draw the data wire from the Math block to the Loop block, use the Zoom Out button ( ) on the Toolbar until you can see both blocks.

Each time through the loop, the Random block generates a number between 0 and 2. This value is appended to the Lights array and used to turn on the Brick Status Light. After a short pause, the Brick Status Light is turned off, followed by another pause before the loop repeats or exits. When the loop

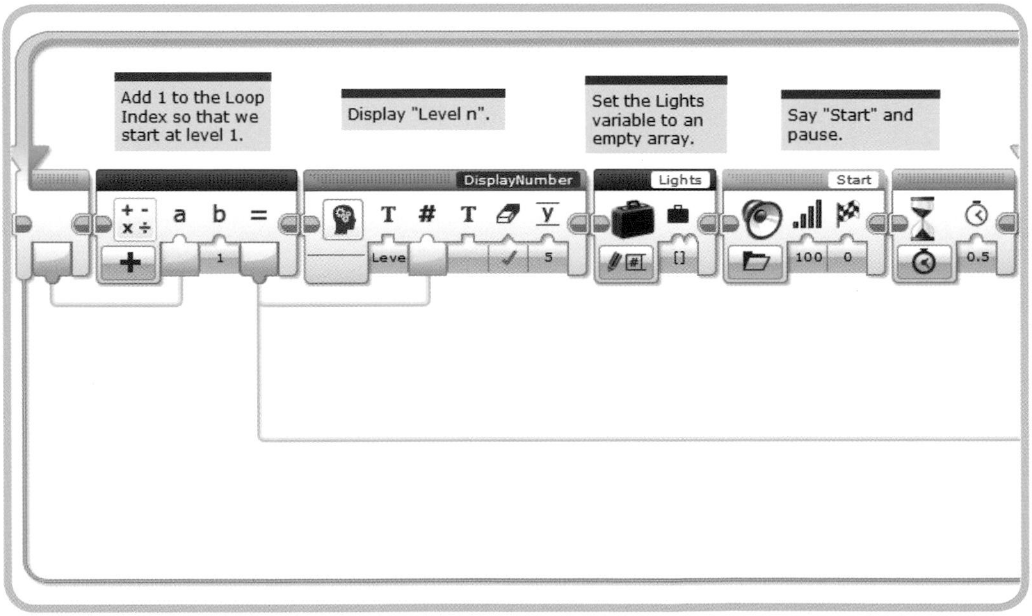

Figure 15-27: Getting ready for one turn

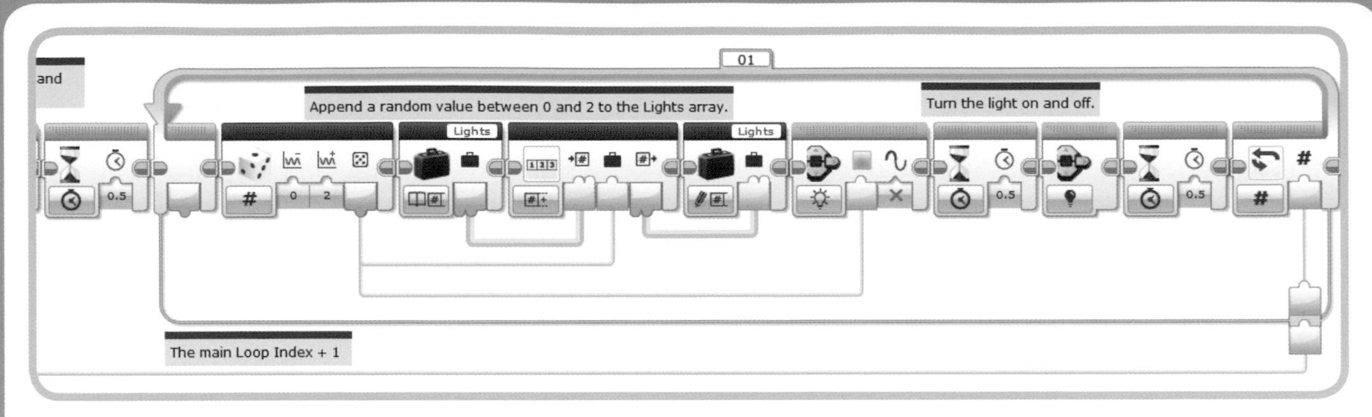

Figure 15-28: Creating the sequence of lights

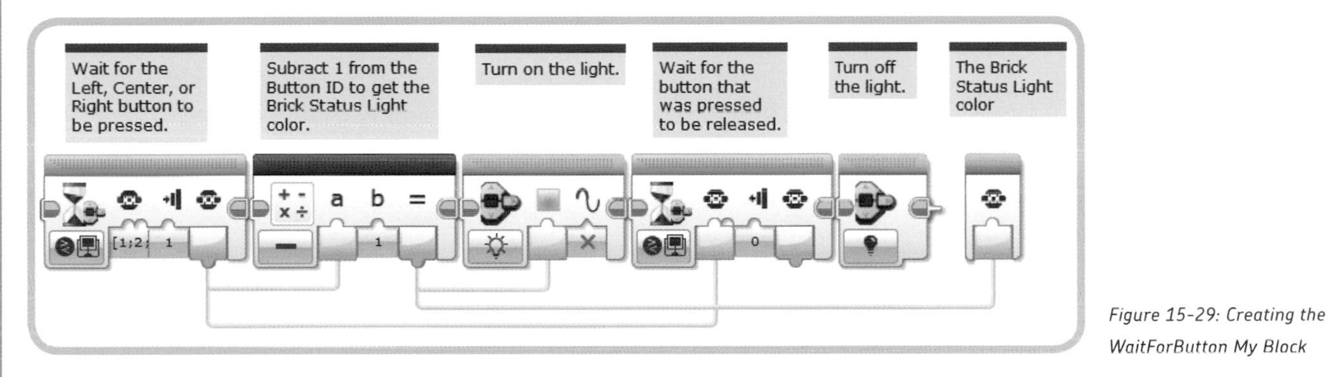

Figure 15-29: Creating the WaitForButton My Block

completes, the Lights array holds all the values that were used to flash the Brick Status Lights. We'll use these values to make sure that the user responds with the correct sequence of button presses.

the WaitForButtons my block

For the final part of the program, we'll create the WaitForButtons My Block, shown in Figure 15-29. When the user tries to repeat the light sequence using the Brick buttons, this My Block shows the matching Brick Status Lights as the buttons are pressed, and outputs the Brick Status Light number for each button press. Figure 15-30 shows how the My Block will appear in your program.

Figure 15-30: The WaitForButton My Block

The first Wait block uses Brick Buttons Compare mode to wait for the Left, Center, or Right button to be pressed. Figure 15-31 shows how the buttons are selected.

Notice that the Button ID numbers we're using go from 1 to 3 while the Brick Status Light numbers go from 0 to 2. The Math block subtracts 1 from the Button ID number to get the correct value to pass to the Brick Status Light block. The light is turned on, and the second Wait block waits until the button that was pressed is released.

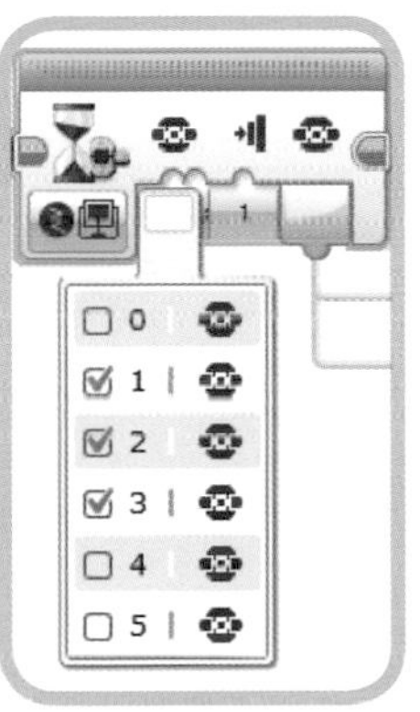
Figure 15-31: Waiting for the Left, Center, or Right button to be pressed

After the button is released, the Brick Status Light is turned off. The My Block returns the number corresponding to the Brick Status Light color (not the Button ID) as its output parameter. This value will be used in the main program to make sure the user pressed the button matching the light that was shown in the first part of the program.

checking the user's response

The final part of the program, shown in Figure 15-32, accepts and checks the user's response. The first thing it does is say "Go," using a Sound block to let the user know that it's time to repeat the sequence. The program then goes through the array and waits for the user to press a button matching the Color value stored in each array element, using the WaitForButton

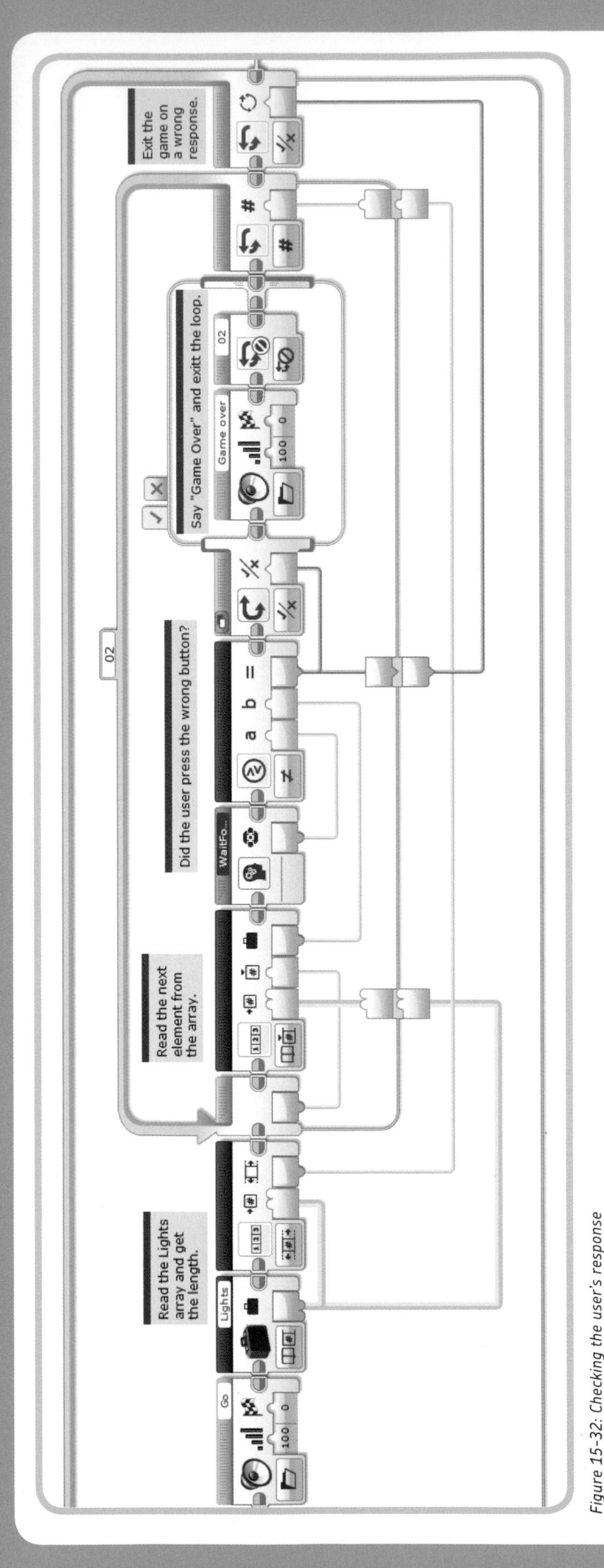

Figure 15-32: Checking the user's response

My Block. The Compare block determines whether the user pressed the wrong button (by checking if the Brick Status Light color associated with the button the user pressed matches the value stored in the array).

If there's a mistake, the code in the Switch block runs and the program says "Game Over," and a Loop Interrupt block is used to exit the inner Loop block that makes up this section of the program. The loop is named *02*, and the Loop Interrupt block is set to match. The result from the Compare block is also used by the main Loop block to determine if it should continue. If there were no mistakes, this value is false and the main loop repeats.

The length of the Lights array is used to control the number of times the loop repeats. When the loop ends, either as a result of the Loop Interrupt block or because the entire sequence was entered correctly, the outer Loop block checks the last output of the Compare block. If the response was incorrect, this value is true, the loop exits, and the program ends. On the other hand, if the response was correct, the value is false and the loop repeats, showing a new, longer sequence of lights.

Run the program. It starts by turning off the normal blinking green light and then flashing the Brick Status Light once. Respond by pressing the correct button, and the program flashes the light twice and waits for your response. The program continues until you make a mistake (or end the program yourself).

In the interest of keeping the program short, I omitted any onscreen instructions. To make the program more user-friendly, you can use Display blocks to add instructions at the beginning of the program and a key showing which button to press for each color.

further exploration

Here are some additional activities for exploring the ideas presented in this chapter:

1. There are two common errors when working with arrays: attempting to read an element that doesn't exist, and writing to an element that doesn't exist. Modify the *ArrayTest* program to see what happens in these cases. For example, after the array is populated, try to read ArrayValue[10]. Then try writing a value to ArrayValue[10] and see how this affects the array length and the values stored in the array. That way, you'll know how the software reacts to these errors, and you'll be able to identify them if they pop up by mistake.
2. The *ButtonCommand* program is limited because it only supports four commands. You can improve this by using two button presses to select the command. In the original program, we used the Button IDs for the command numbers. To use two buttons to specify the command, you'll need a way to turn the two Button IDs into a single command number. One simple way is to take the first Button ID and multiply it by 10 and then add the second Button ID (Figure 15-33). This gives you command numbers between 11 (Left, Left) and 55 (Down, Down). Not all numbers between 11 and 55 will be valid commands, because you'll only be using numbers that contain the digits 1, 3, 4, and 5.

   In this version, you'll need three columns to display each command, two to display the digits, and one for a space between each command.

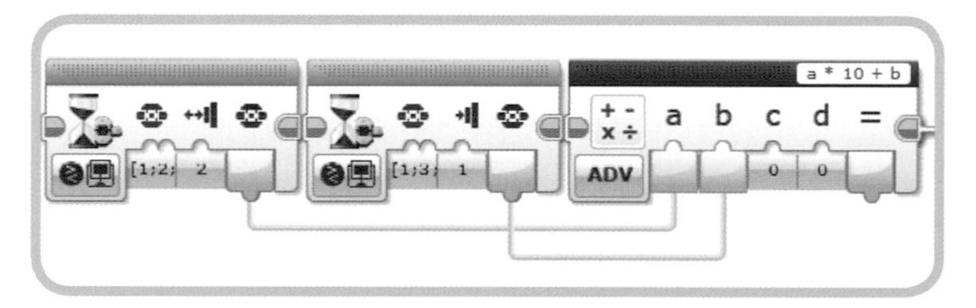

Figure 15-33: Creating a command number from two button presses

3. Add sound to the *MemoryGame* program. Whenever the Brick Status Light is turned on, play a tone to match the color of the light. Do this by creating an array to hold the frequencies to use for each Brick Status Light color. I recommend trying the following set of values: [261.626; 329.628; 391.995]. (These notes make up a C-Major chord. Look up "musical note frequencies" for more information about how tone frequency relates to musical notes.) If we call this array Tones, then Tones[0] would be used with the green light, Tones[1] with the orange light, and Tones[2] with the red light.

conclusion

In this chapter, you learned about EV3 arrays, which let you store lists of values. You used the Variable block to create and store arrays, and the Array Operations block to access the elements and determine the length of an array.

The *ButtonCommand* program allowed you to create a list of commands for the TriBot to execute, essentially creating a program within a program. Using an array allowed you to easily expand the *RedOrBlueCount* program to handle all eight colors that the Color Sensor can detect. The *MemoryGame* program uses the Brick Status Light, Brick buttons, and progressively longer lists of values to test your memory.

Arrays are useful when you want to use a collection of values while your program is running. In the next chapter, you'll learn about files, which you can use to save values from your program, or load values from a different program or from your computer.

16

files

In this chapter, you'll use the File Access block to create and use files in your programs. The information you store in a file is *persistent*, meaning that it's still available after your program ends, even if you turn off the EV3. Using files, you can store information from a program and use it later in the same or a different program.

You'll start by creating a few test programs, and then you'll alter the *MemoryGame* program from Chapter 15 to save the high score in a file. Next, you'll add a menu to the *ColorCount* program so that you can save the object count for each color in a data file and then restore the values the next time the program starts. You'll also learn how to manage your EV3's memory and delete files or transfer them between the EV3 and your computer.

the file access block

The *File Access block* is found in the Advanced palette and has four major modes (see Figure 16-1): read data from a file, write data to a file, delete a file, and close a file. Closing a file tells the EV3 that you're done using it. To demonstrate these operations, we'll create the *FileTest* program, which writes three values to a file and then reads them back and prints them on the EV3 screen.

setting the filename

The File Access block *always* needs to know the name of the file you want to work with. Either click the File Name box in the top-right corner of the block to set the name, or select **Wired** to supply the name using a data wire (Figure 16-2).

Filenames are case-sensitive, and they can be up to 31 characters long. You can include numbers, letters, spaces, and the underscore (_) and dash (-) characters. Try to use meaningful filenames that tell you something about the actual content of your files.

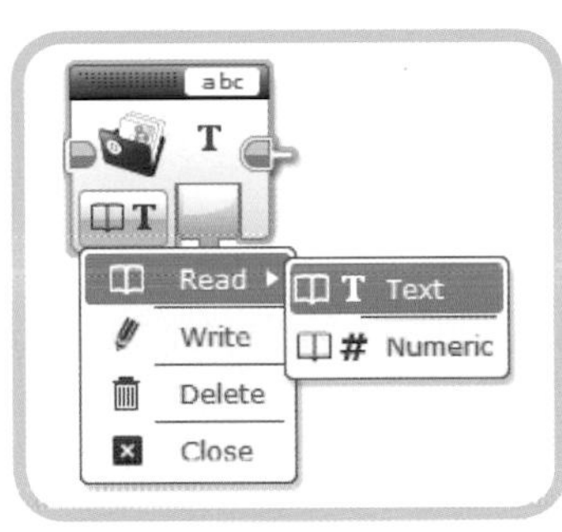

Figure 16-1: Selecting the File Access block mode

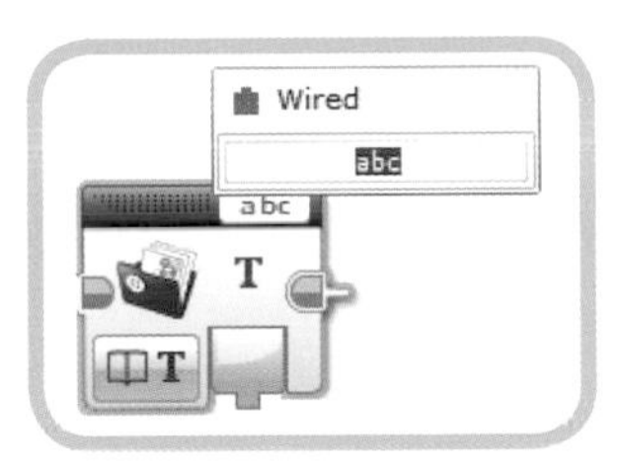

Figure 16-2: Setting the filename

writing to a file

Write mode stores information in an existing file and creates a new file if one doesn't already exist. The File Access block always writes new data at the end of the file, so if the file already exists, the new value is added at the end. If you want to replace existing data rather than add to it, you need to delete the file first.

Figure 16-3 shows the first part of the *FileTest* program, which uses three File Access blocks to write 0, 1, and 2 to a file. We set the filename for all three blocks to *FileTestData*.

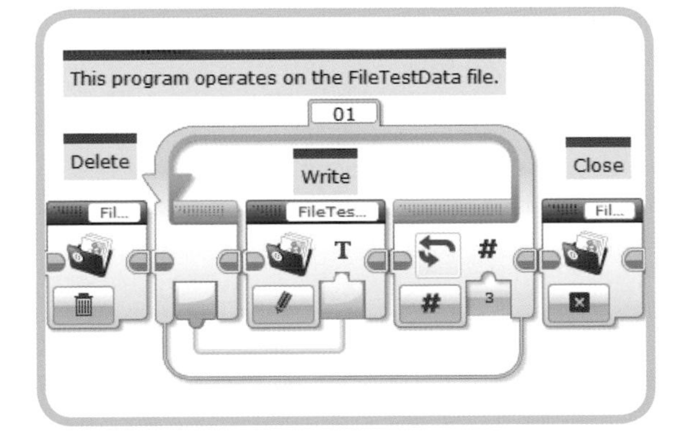

Figure 16-3: Writing to a file

We want to replace the contents of *FileTestData* every time we start *FileTest*, so the first File Access block uses Delete mode to delete the file if it already exists. If we leave this block out, the first time we run the program, it creates a new file and writes the three values. The second time we run the program, it opens the existing file and adds three more values. The third time we run the program, it adds three more values to the file, which now contains nine values.

The second File Access block uses Write mode to write the Loop Index to the file. The File Access block's Input parameter takes Text values, but you don't need to select Numeric or Text in Write mode. Any Numeric inputs to the File Access block are automatically converted to text before being written to the file.

The first time through the loop, the file is created, and the first value (0) is written. The next two times through the loop, the values 1 and 2 are added to the file. The third File Access block uses Close mode to close the file. We need to close the file after writing the values so that when we read the values in the next section of the program, the EV3 starts reading the values from the beginning of the file.

reading from a file

The File Access block's *Read mode* reads Numeric or Text values from an existing file. If the file doesn't exist, your program ends abruptly, and the EV3 shows a File Read Error message.

In Read mode, you need to select the output parameter's data type to use the correct type of data wire. File data is always stored as text, so you can read both letters and numbers as text using Read Text mode. If you know you wrote a number, as you do for the *FileTestData* file, then use Read - Numeric mode to get a Numeric value. This only works if the value in the file is *actually* a number; if it's not, the value will be read as 0, and the EV3 won't tell you that this value doesn't match what's really in the file.

Figure 16-5 shows the second half of the *FileTest* program, which reads the three values from *FileTestData* and prints them on the EV3 screen. The first Display block clears the screen, and the loop repeats three times. Each time through the loop, the File Access block reads a number and displays it. After the loop, the file closes, and the program pauses for five seconds to give you time to read the display. It's good practice to always close a file after you're finished using it, so *FileTest* ends with a File Access block to close *FileTestData*.

Add the blocks in Figure 16-5 to the end of the program from Figure 16-3 to complete the *FileTest* program. When you run the finished program, it should create the *FileAccessTest* file; write 0, 1, and 2 to the file; read the values back; and print them on the EV3 screen.

AVOIDING FILENAME BUGS

Anytime multiple File Access blocks need to use the same file, be careful to enter the correct name into each one. For example, the code shown in Figure 16-3 will fail if all three File Access blocks don't use the same filename. The File Access block usually only shows the beginning of the filename it uses, but you can hover your mouse over the File Name box to display the full name (Figure 16-4). Use this trick to quickly check each of the File Access blocks in your program and make sure they're using the file you expect.

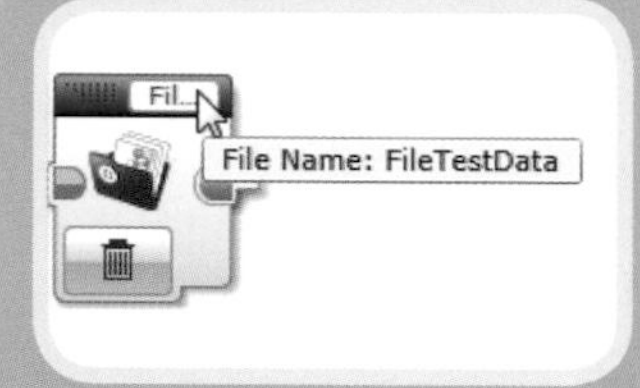

Figure 16-4: Showing the full filename

Other than letters and numbers, the only characters you should use in filenames are the dash and underscore (- and _). You can enter other special characters, such as * and %, in the File Name box, but the EV3 replaces these characters with a space when it uses the file. So if you have File Access blocks using the filenames *Test*One* and *Test%One*, both blocks will use the file Test One, which is likely not what you meant. Remember, too, that filenames are case sensitive; *FileTestData*, *filetestdata*, and *FILETESTDATA* all refer to different files. If you write values to *FileTestData* and try to read them from *filetestdata*, your program will fail.

One way to avoid mistakes when typing a filename is to copy an existing File Access block instead of adding a new one. For example, to build the code shown in Figure 16-3, you could first add the block to delete the file and set the filename. Then, after adding the Loop block, you could copy the first File Access block you made by holding down the CTRL key, clicking and dragging the existing File Access block to create a copy, and placing it in the Loop block. You'd need to change the mode on the copy from Delete to Write, but the filename would already be set correctly.

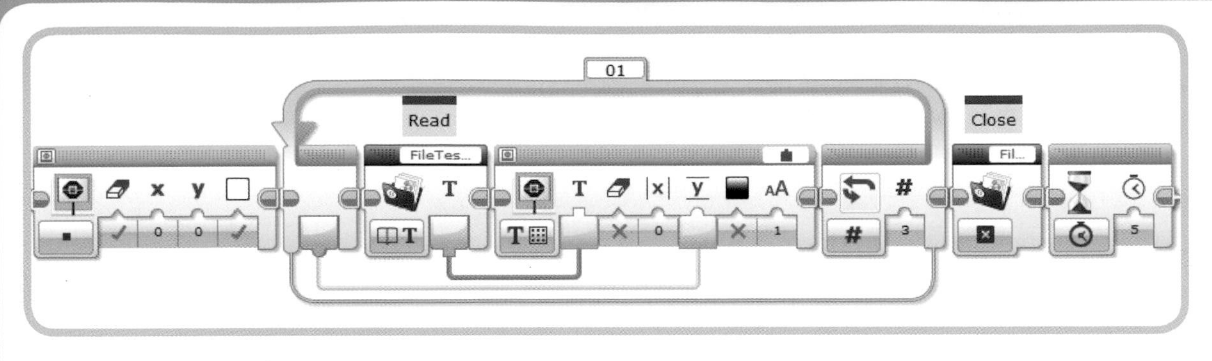

Figure 16-5: Reading from a file

saving the MemoryGame high score

In this section, we'll add code to the *MemoryGame* program from Chapter 15 to save the high score. Recall that this program is contained in a large loop, which ends when the user responds incorrectly. Each time through the loop, the Loop Index is the number of times the user has given a correct response. We'll use the index as the user's score.

The high score is saved in a file named *MG_HighScore*. When the user gives an incorrect response, the main loop exits, and the program compares the user's score with the value stored in the *MG_HighScore* file. If the user beats the high score, the program saves the new high score, prints a congratulatory message, and plays a cheering sound.

We'll place most of the new code at the end of the program, but we do need to make one change at the very beginning of the main loop. To keep track of the user's score, we'll store the Loop Index in a Variable block called *Score* (Figure 16-6). When the loop ends, *Score* will tell us how many correct responses the user gave.

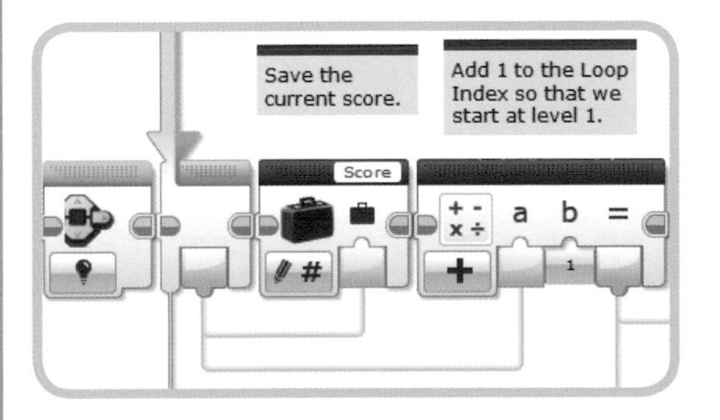

Figure 16-6: Saving the user's score

The rest of the new code follows the main loop. The first part, shown in Figure 16-7, retrieves the previous high score and compares it with the user's score. All of the File Access blocks used by this program work with the *MG_HighScore* file. Let's go through these blocks one at a time.

1. The first File Access block writes a 0 to the *MG_HighScore* file. Adding 0 to an existing file won't affect the program, but this block is at the beginning in case we need to create *MG_HighScore*. Otherwise, we could cause an error by trying to read from a nonexistent file.
2. The second block closes the file so that the next block can read the contents.
3. The third File Access block reads the previous high score from the file. This will either be the 0 written by the first File Access block (if *MG_HighScore* didn't already exist) or the previous high score.
4. The next block closes the file because we're done using it. Notice that the 0 written by the first File Access block is never read if the file already exists.
5. The Variable block reads the user's score.
6. The Compare block checks the user's score. If it exceeds the current high score, the value on the data wire is true; otherwise, the value is false.

After these blocks execute, the two data wires contain the user's score and a Logic value that tells us if this is a new high score. If so, then the next part of the program, shown in Figure 16-8, saves the new high score and congratulates the user. Here's how this part of the program works:

7. If the Logic value from the Compare block is true, the Switch block runs the blocks in Figure 16-8 to update the high score.
8. The first File Access block deletes the *MG_HighScore* file so the new high score replaces the old one.

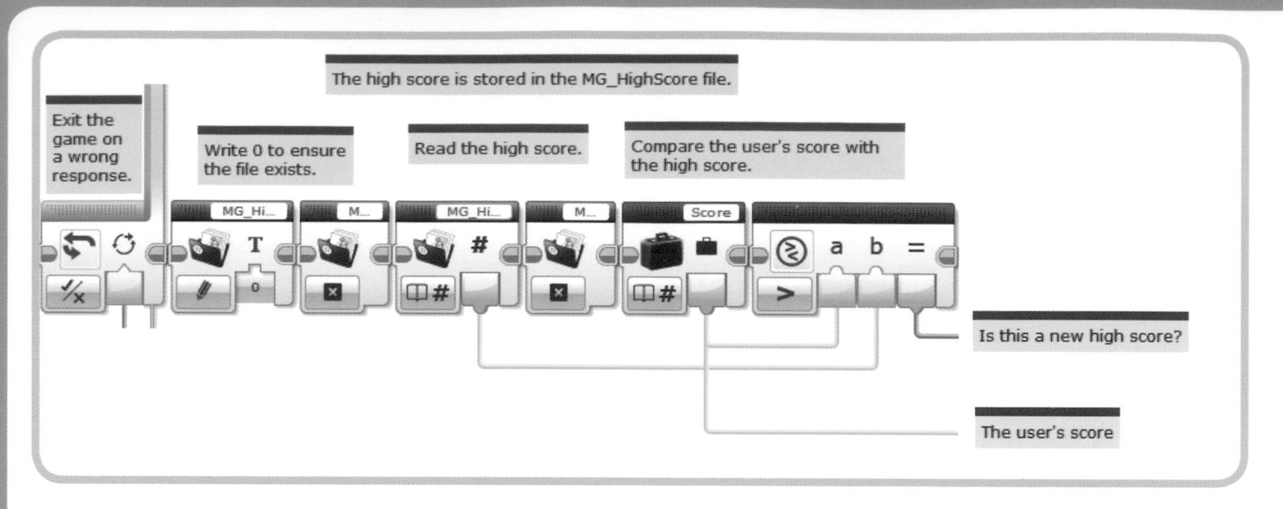

Figure 16-7: Comparing the user's score with the previous high score

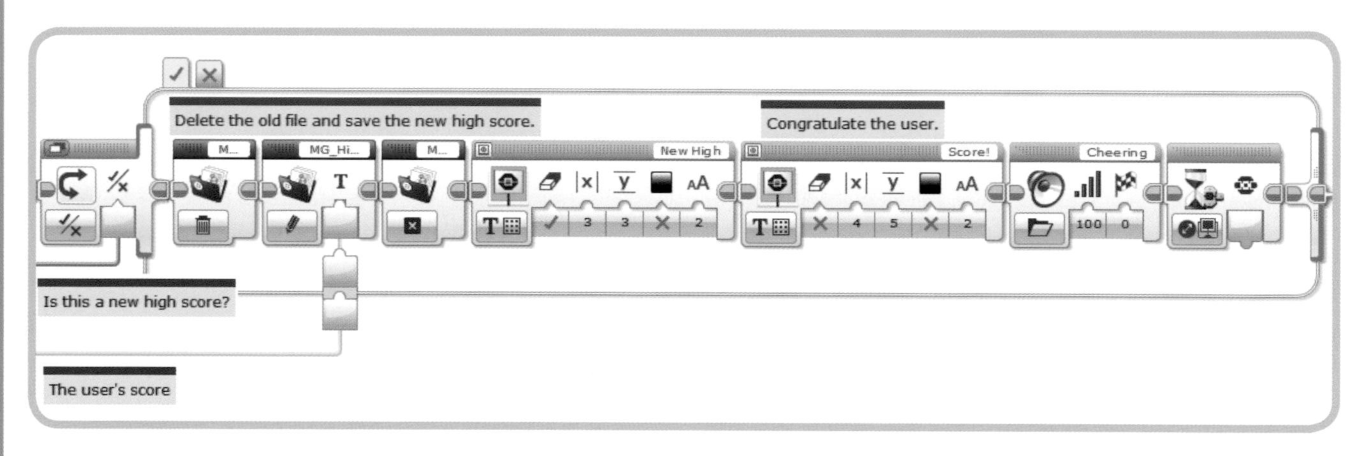

Figure 16-8: Saving the new high score on the Switch block's true case

9. The next File Access block writes the new high score to the file.
10. The third File Access block closes the file.
11. The two Display blocks show "New High Score!" in large letters on two lines near the center of the screen.
12. The Sound block plays the Cheering sound.
13. The Wait block uses Brick Buttons Change mode to pause the program until the user presses one of the buttons.

The first time you run the updated *MemoryGame* program, as long as you respond correctly at least once, you should get a new high score, see the new high score message, and hear the cheers. After that, you should see the message only if you've surpassed your previous best performance.

Remember that before we try to read the old high score, we write an extra 0 to *MG_HighScore* just in case the file doesn't exist. If the user earns a new high score, the file is rewritten and that extra 0 disappears. But after a while, the high score will be harder to beat; we'll just keep adding extra 0s without replacing the file, and the file will grow much bigger than it needs to be. To ensure that *MG_HighScore* only ever contains one value when the program completes, we can rewrite the file every time the game ends (Figure 16-9).

Just add a new data wire to take the original high score and pass it into the Switch block. The three File Access blocks on the Switch block's false case delete the existing file, write the original high score, and close the file.

the FileReader program

Being able to see what your program actually writes to a file can be useful. In this section, I'll present the *FileReader* program (see

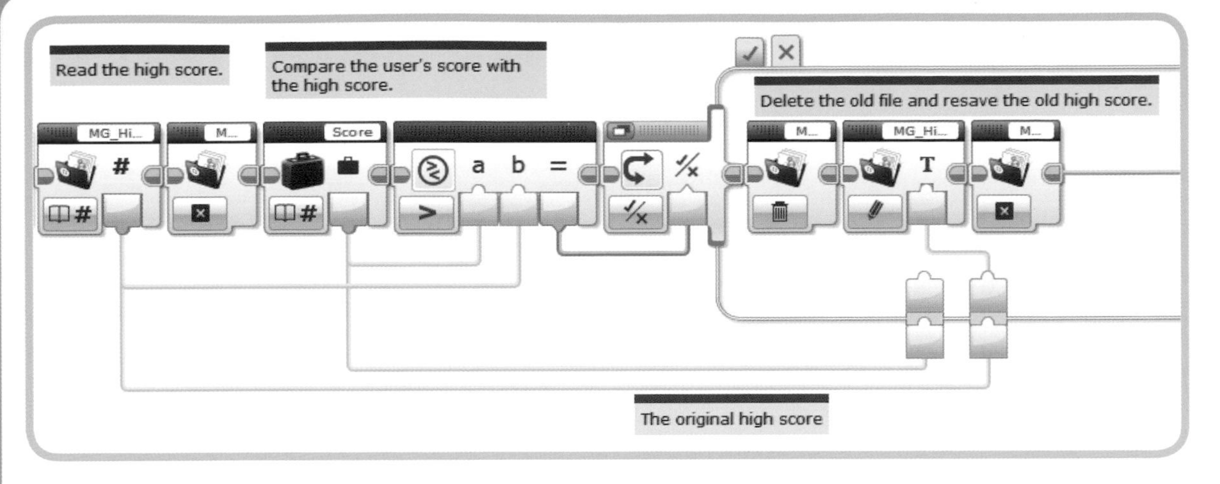

Figure 16-9: Rewriting the high score on the Switch block's false case

Figure 16-10). *FileRead* shows the contents of a file on the EV3 screen, which is convenient when your program uses short files. In this example, the filename is set to *FileTestData*, but you can change the name to any file you want to read.

Each time through the loop, the File Access block reads a value from the file as text. The Loop block is configured to repeat forever because we may not know how many values a file contains ahead of time, but we don't really want the loop to continue forever. We need to know when we've reached the end of the file.

After all the values in a file have been read, a File Access block in Read Text mode returns an empty string. As long as the file doesn't contain any blank lines, we can check for an empty string and use that to tell us that we've read all the data. Compare blocks only work with numbers, so instead we'll use a Switch block in Text mode. The value on the upper case is set to "", which is the empty string. When the File Access block output is an empty string, the upper case runs and the Loop Interrupt block exits the loop. The other case is set as the default, so it'll be used if the File Access block reads anything other than the empty string. The value on this case is automatically set to "False" when you select Text mode.

The remaining two blocks should look familiar; the Display block shows each value read from the file, and the Wait block in Brick Button Change mode pauses so that you can see the values.

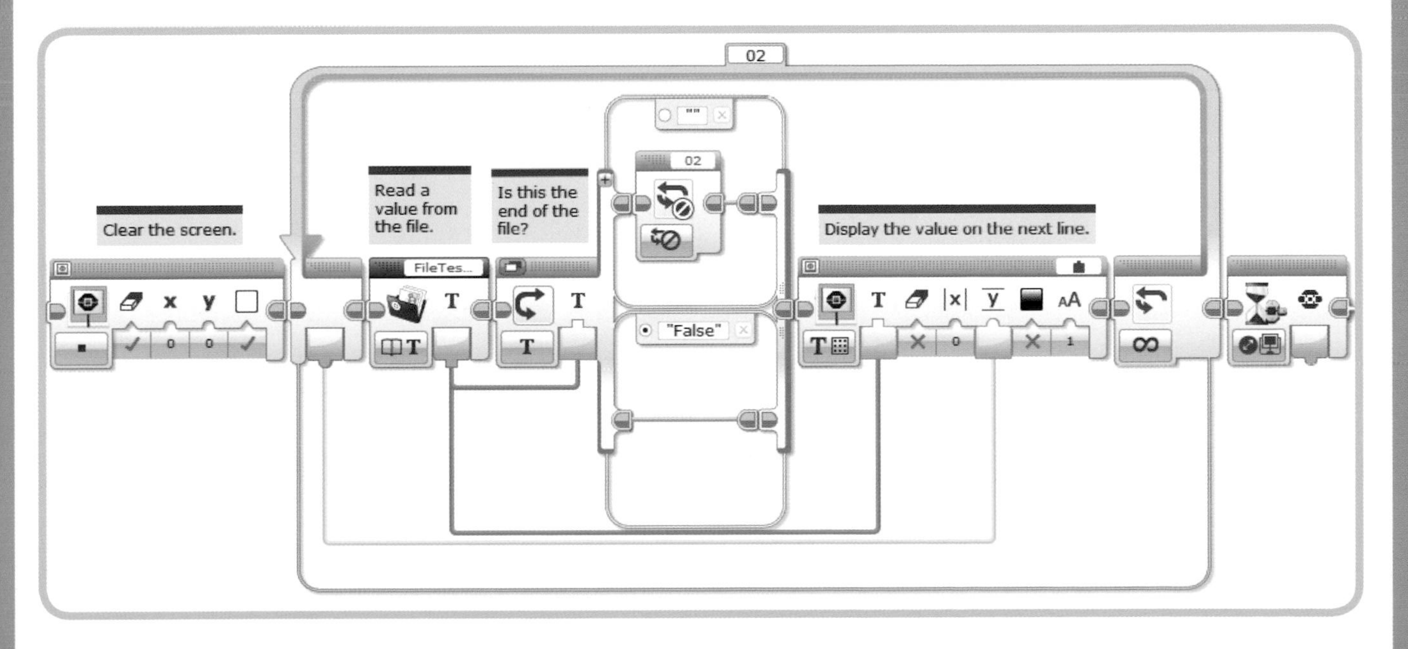

Figure 16-10: The FileReader *program*

CHALLENGE 16-1

The *FileReader* program currently works well only for files with 12 values or less, but you can improve it to work with longer files. Make the *FileReader* program pause and wait for a button to be bumped after reading and displaying 12 values. Then have the program clear the screen and display the next 12 values. This should continue until all the values from the file have been read.

Run this program after running the *FileTest* program, and it should display "0", "1", and "2" on the EV3 screen. Change the filename to *MG_HighScore* to see your best score from the *MemoryGame* program. Change the filename to one that doesn't exist (for example, *NotThere*), and you should see the File Read Error message.

adding a menu to the ColorCount program

Using files can improve many programs! In this section, we'll use a file to store the data collected by the *ColorCount* program from Chapter 15. The code to store the values will resemble the code in the first part of the *FileTest* program, but just saving the eight values at the end of the program won't be very useful. We also need a way to load values from the file. And while we're at it, a way to reset the values to 0 would be nice.

To support these options, we'll add a menu to the *ColorCount* program. When the program starts, it displays a list of choices and waits for the user to select one using the Brick buttons. After performing the selected operation, the program displays the menu again. The menu should show four choices:

Count Start counting objects. This runs the code from the original *ColorCount* program, with just a few changes.
Save Saves the values in the ColorCounts array to the *ColorCountData* file.
Load Loads the values in the *ColorCountData* file into the ColorCounts array.
Clear Sets all eight values in the ColorCounts array to 0.

We'll start by creating two My Blocks—one that displays the four options we need and another that manages a menu given the number of choices. Then we'll use these two My Blocks and a few others to create the new and improved *ColorCount* program. The four menu options described above are specific to the *ColorCount* program, but you could reuse the code that handles the button presses to select options in any program.

FINDING THE END OF A FILE

The *FileReader* program reads all the data from a file as Text values and uses the empty string to see when it reaches the end of the file. But this method doesn't work well if your program needs to read the data as numbers. A File Access block in Read - Numeric mode outputs a 0 after it finishes reading all the data in a file. Depending on the data, 0 could be a valid value, so it often can't be used to indicate the end of the file.

When reading numbers, there are three reasonable ways to tell whether you've read all the data:

1. If you know the number of values ahead of time, you can explicitly control how many values are read. The *FileTest* program always reads exactly three values because we know that's how many should be in the file.
2. When you write the data to the file, you can first write the length of the data. Then the code that reads the file has to first read the length, and then read the values from the file that many times.
3. If you know that a certain value won't be in the data, then you can use that value to mark the end of the file. For example, if you're working with Color Sensor readings, you could use -1 because the Color Sensor never returns a negative number. As the program reads the file, it can check for this *end of file* value, similar to the way the *FileReader* program checks for the empty string. (Because the value is a number, you can check it with the Compare block.)

the CreateMenu_CC my block

When the program starts, it displays the menu options, showing a > to the left of the top option as a selection marker. The display should look like this:

```
> Count
  Save
  Load
  Clear
```

Before we can write the code to manage a menu, we need to display the menu choices. That's the job of the CreateMenu_CC My Block. Figure 16-11 shows the *CreateMenu_CCBuilder* program, from which you can create the My Block. Each Display block prints one menu option, each on a different row, starting at row 0 (the top of the screen). The Column parameter on each block is set to 2 to leave room for the selection marker (>) and a space. The Clear Screen parameter is set to true for the first block, so that the screen is cleared before displaying the menu, and set to false for the remaining blocks.

Here are the steps I suggest using to create this My Block:

1. Create a new program named *CreateMenu_CCBuilder*, and copy the blocks in Figure 16-11.
2. Run the *CreateMenu_CCBuilder* program and the EV3 screen should show the following:

```
  Count
  Save
  Load
  Clear
```

When you're confident that the menu is being displayed correctly, continue with these steps:

3. Select all four Display blocks. Select **Tools ▸ My Block Builder** from the menu to start creating the My Block.
4. Set the name to *CreateMenu_CC* and the description to "Create a menu for the ColorCount program."
5. Select the icon that looks like the one used by the Display block ().
6. Click **Finish** to create the My Block.
7. Press CTRL-Z to restore the *CreateMenu_CCBuilder* program to its original state (restoring the builder program this way just makes it easier to rebuild the My Block if you decide to make changes later).

At this point, you should have a My Block that displays the menu options that we'll use at the beginning of the *ColorCount* program. We'll also use this My Block to build the SelectOption My Block, which manages the menu.

the SelectOption my block

The SelectOption My Block allows the user to select one of the menu options. The Up and Down buttons are used to select an option, and the Center button is used to accept the selection. Here's how the process works:

* After the CreateMenu_CC My Block runs, the display should look like this:

```
> Count
  Save
  Load
  Clear
```

* The SO_Selection variable stores the current selection. It starts at 0 and changes as the user bumps the Up and Down buttons.
* Bumping the Down button should move the selection marker down. If the last option (Clear) was selected before the Down button was bumped, the marker should wrap around to the first option (Count).
* Bumping the Up button should move the selection marker up. If the first option (Count) was selected before the Up button was bumped, the marker should wrap around to the last option (Clear).
* Each time the selection changes, the ">" clears and reprints to the left of the new selection.
* When the user bumps the Center button, the screen clears and the My Block outputs a Numeric parameter indicating the option selected. Options are numbered starting at 0.

The SelectOption My Block has two parameters: an Input parameter to set the number of menu options and an Output parameter to return the user's selection. The block doesn't need to know what the menu options are (because the CreateMenu_CC block already handles displaying the options), but it does need to know how many options there are so that it can wrap the selection marker correctly.

To create the SelectOption My Block, we'll write and thoroughly test the *SelectOptionBuilder* program. This program uses the CreateMenu_CC block to show the menu and a Constant block to set the number of options. At the end of the program, a Display block returns the number of the selected option. In between, we build the heart of the program that we'll

Figure 16-11: The CreateMenu_CCBuilder *program.*

turn into a My Block: the blocks that allow the user to select an option using the Up, Down, and Center Brick buttons.

selecting a menu option

The program uses one variable named SO_Selection to hold the number of the currently selected option. At the beginning of the program, we need to initialize SO_Selection to 0, which represents the first option, and print a ">" on row 0 to mark that selection.

Figure 16-12 shows the initialization section of the program and the code for handling the Center button. The blocks in the upper part of the image display the menu options, put the number of options on a data wire, initialize the SO_Selection variable, and print the first selection marker. This is followed by the blocks on the lower part of the image, which enter a loop and wait for the user to bump the Up, Down, or Center button.

After the user presses a button, the Variable block reads the number of the previously selected option, and the Display block uses this value to print a single space over the ">" that was printed earlier, erasing the mark. The number of the button that was bumped is passed to the Switch block, which has a case for each Button ID (4 for Up, 5 for Down, and 2 for Center). Figure 16-12 shows the case for the Center button, which uses a Loop Interrupt block, named *SO_02*, to exit the loop. This case doesn't use the two data wires that enter the Switch block. (They'll be used by the other two cases.)

NOTE When you create a My Block that ends a loop with a Loop Interrupt block, give it a name that you'd never use in a main program loop to avoid bugs. For example, if you named the loop in the FileMenu *My Block 02* and put that My Block in a main program loop named *02*, the Loop Interrupt block would interrupt the main program's loop as well!

When the Down button is pressed, we execute code specific to the Down button (Figure 16-13), which adds 1 to the current SO_Selection value to move to the next selection. The Math block uses Advanced mode to calculate the new value of SO_Selection, $(a + 1) \% b$, where a is the current selection number and b is the total number of menu options. The modulo operator ensures that the selection wraps around to 0 when we reach the end of the list of options. After the Math block computes the number of the new selection, the Display block prints ">" on the appropriate row, and the value is stored in the SO_Selection variable.

We use the modulo operator again in the code for the Up button, though the math is different because we have to work with negative numbers. The Up button should move through the list in the opposite direction of the Down button, decreasing the selection and wrapping around when we move past the beginning of the list. In this example, if the previous selection was 0, then the new selection should be 3.

But we can't just switch to subtraction because the expression $(a - 1) \% b$ won't give the results we need to wrap the next selection to 3 after we hit 0. When a is 0, $(a - 1)$ will be -1, and -1 modulo 4 is -1, not 3.

The solution is to go forward rather than backward. Moving forward one place less than the total number of options is exactly the same as moving backward 1 place. So in this four-option example, instead of subtracting 1 to move one place back, we add 3 and then apply the modulo operator using the expression $(a + b - 1) \% b$. This should wrap the selection value correctly.

Now that you know the Math block expression to use to move the selection up the list, we can write the code (see Figure 16-14). The only difference between this and the code for the Down button is the expression used by the Math block.

returning the selected option

After moving the selection marker (>) to the desired choice, the user bumps the Center button, which exits the loop. Then the My Block needs to clear the screen (it's good practice to have the block clean up after itself) and return the selected option as an Output parameter. The selected option is stored in the SO_Selection variable, so we need a Variable block to read the value. In the builder program, we show the value of SO_Selection on the EV3 screen, which gives us an easy way to test the program before starting the My Block Builder. Figure 16-15 shows the final four blocks in the program.

When you run the program, it should display the menu choices and move the selection marker properly when you bump the Up and Down buttons. Bump the Center button, and the program should display the number of the selected option.

creating the my block

When you're confident that the *SelectOptionBuilder* program works correctly, you can create the SelectOption My Block from it. Here are the steps to do that:

1. Click the **Zoom Out** button (🔍) on the right side of the Programming Canvas Toolbar until you can see the entire program.
2. Draw a selection rectangle around all blocks except the CreateMenu_CC and Constant blocks at the beginning and the Display and Wait blocks at the end.
3. Select **Tools ▸ My Block Builder**. If you've selected the correct blocks, there should be one Text Input parameter and one Numeric Output parameter.
4. Set the My Block name to **SelectOption** and select the icon for the Brick buttons ( ).
5. Set the name of the first parameter to **Number of Options** and select the icon for the number sign (#).

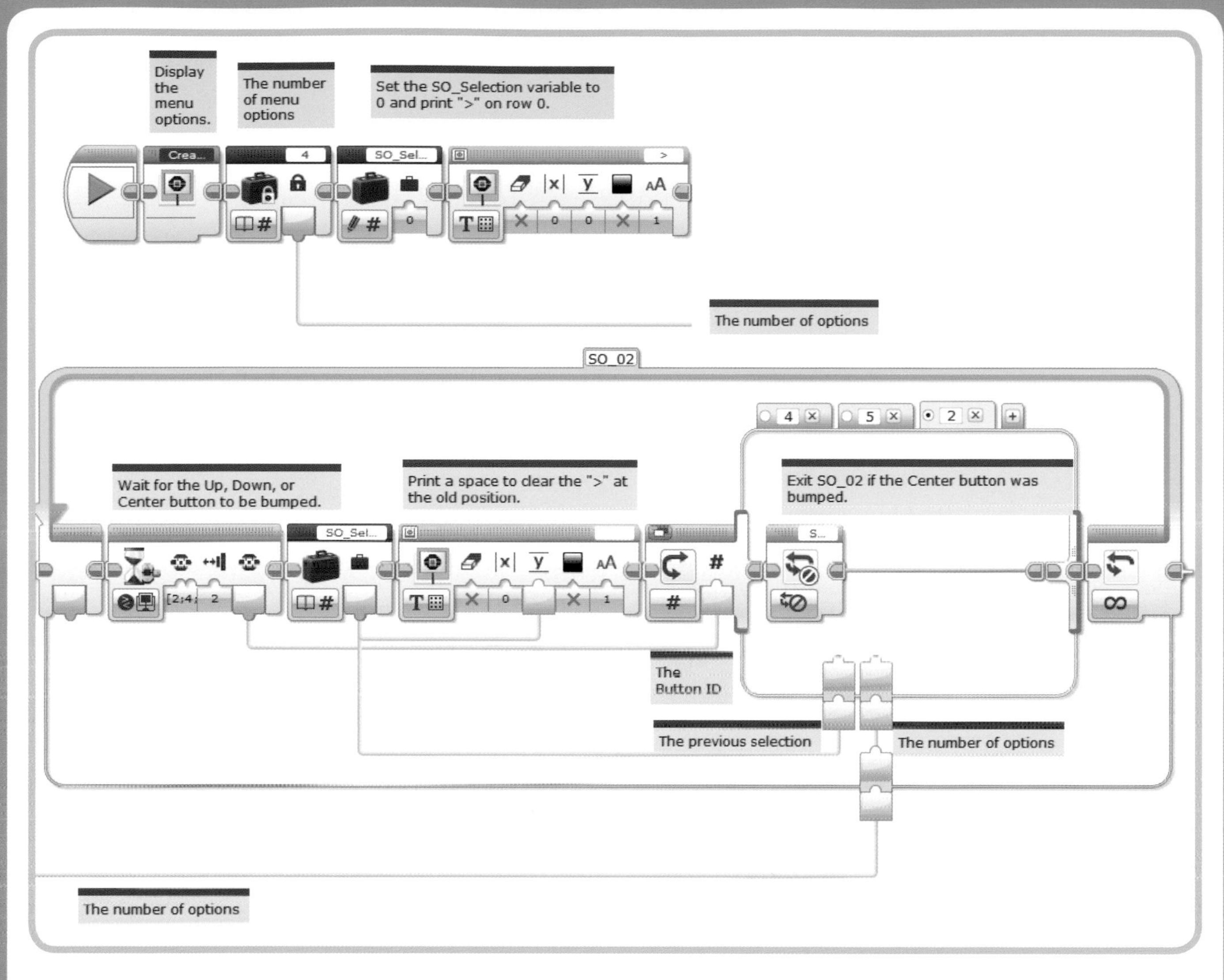

Figure 16-12: Waiting for a button and exiting the loop if the Center button was bumped

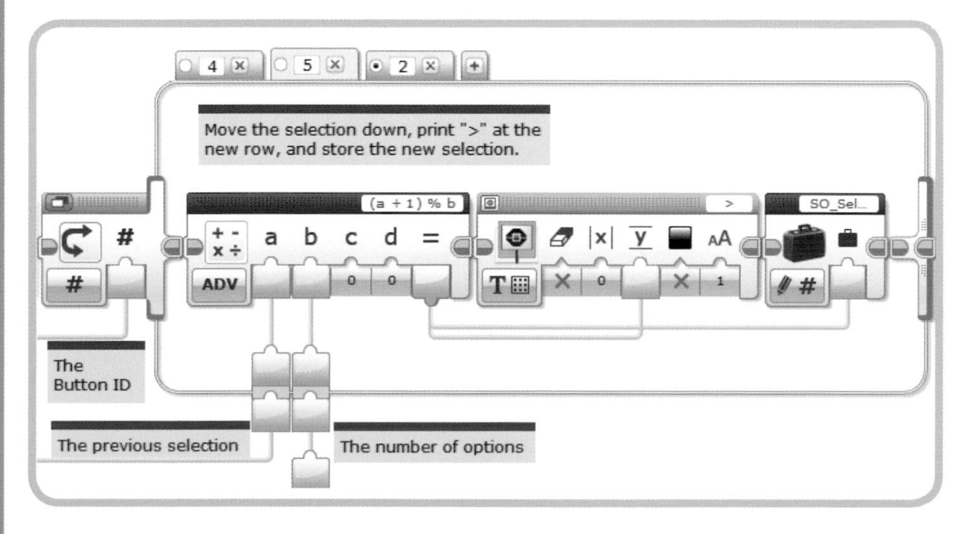

Figure 16-13: Processing the Down button

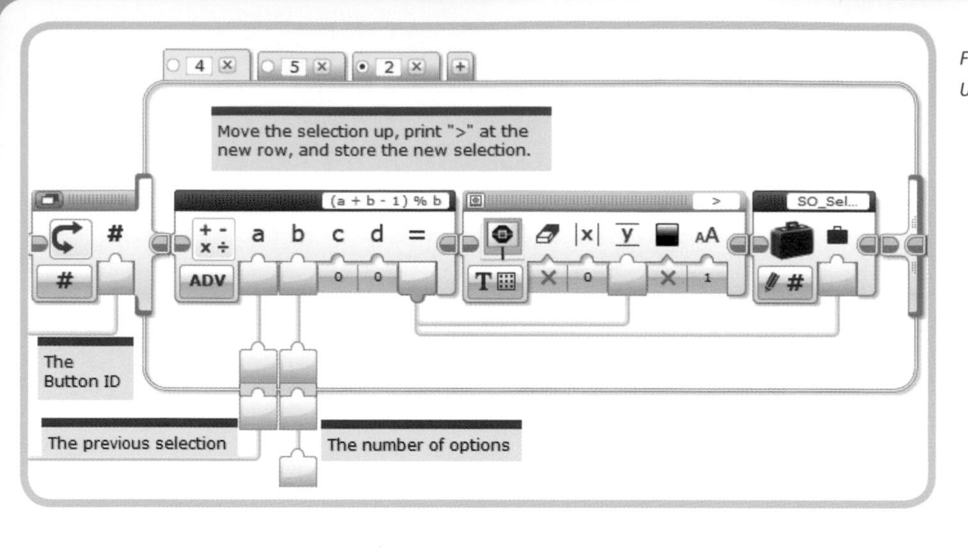

Figure 16-14: Processing the Up button

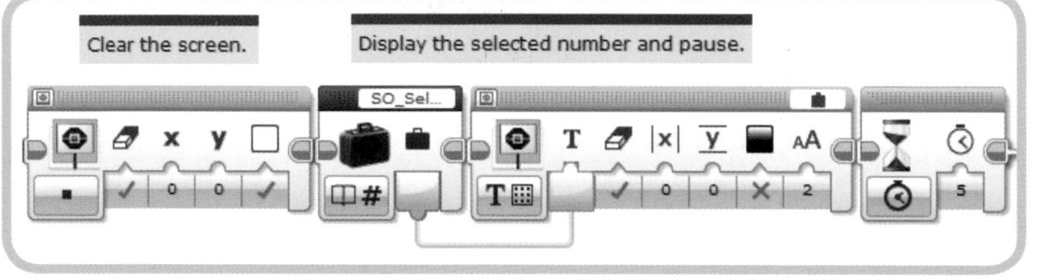

Figure 16-15: Reading and displaying the selected option

6. Set the name of the second parameter to **Selection** and select the icon used for a number (n).
7. Click the **Finish** button.
8. Use CTRL-Z to restore the *SelectOptionBuilder* program.

Now we have a My Block to write the menu options for a program (*CreateMenu_CC*) and another to display the menu and allow the user to select one of the options. The next thing we'll do is build the basic program structure, and then we'll fill in the details of how to accomplish each option from the menu.

the new ColorCount program structure

Programs that contain menus usually have the following structure, and the new *ColorCount* (Figure 16-16) is no exception:

1. Initialize the program data. For the *ColorCount* program, that means setting the ColorCounts variable to an array with eight 0s.
2. Enter a loop that displays the menu and then uses the SelectOption My Block to let the user choose an option.
3. Carry out the selected option using a Switch block. (I like to implement each option in a My Block to keep the code from getting unwieldy.)

The completed *ColorCount* program uses three My Blocks, which we'll create in the following sections, to count the objects and save and load the totals to and from a file. Clearing the totals only requires a single Variable block, so that option is implemented directly.

Create the new *ColorCount* program, leaving out the My Blocks we haven't made yet, and test that the menu display and selection code work. You can test each option as we create the corresponding My Block.

counting objects

The *ColorCount* program from Chapter 15 contains the counting logic we need, so we'll use that as a starting point for creating the Count_CC My Block. Even if you created the new *ColorCount* program from this chapter, copy the original *ColorCount* program from the Chapter 15 project to the Chapter 16 project and change its name to *Count_CCBuilder*.

Count_CCBuilder has two distinct sections: The first displays the name of each color followed by a 0, and the second does the counting. But in the new program, the color count totals won't necessarily be 0 every time we start counting. So in the first section, instead of printing 0s, we need to print the values from the ColorCounts array using a Variable block followed by an Array Operations block to read the count for each color. Figure 16-17 shows the full *Count_CCBuilder* program with the changes we need to make.

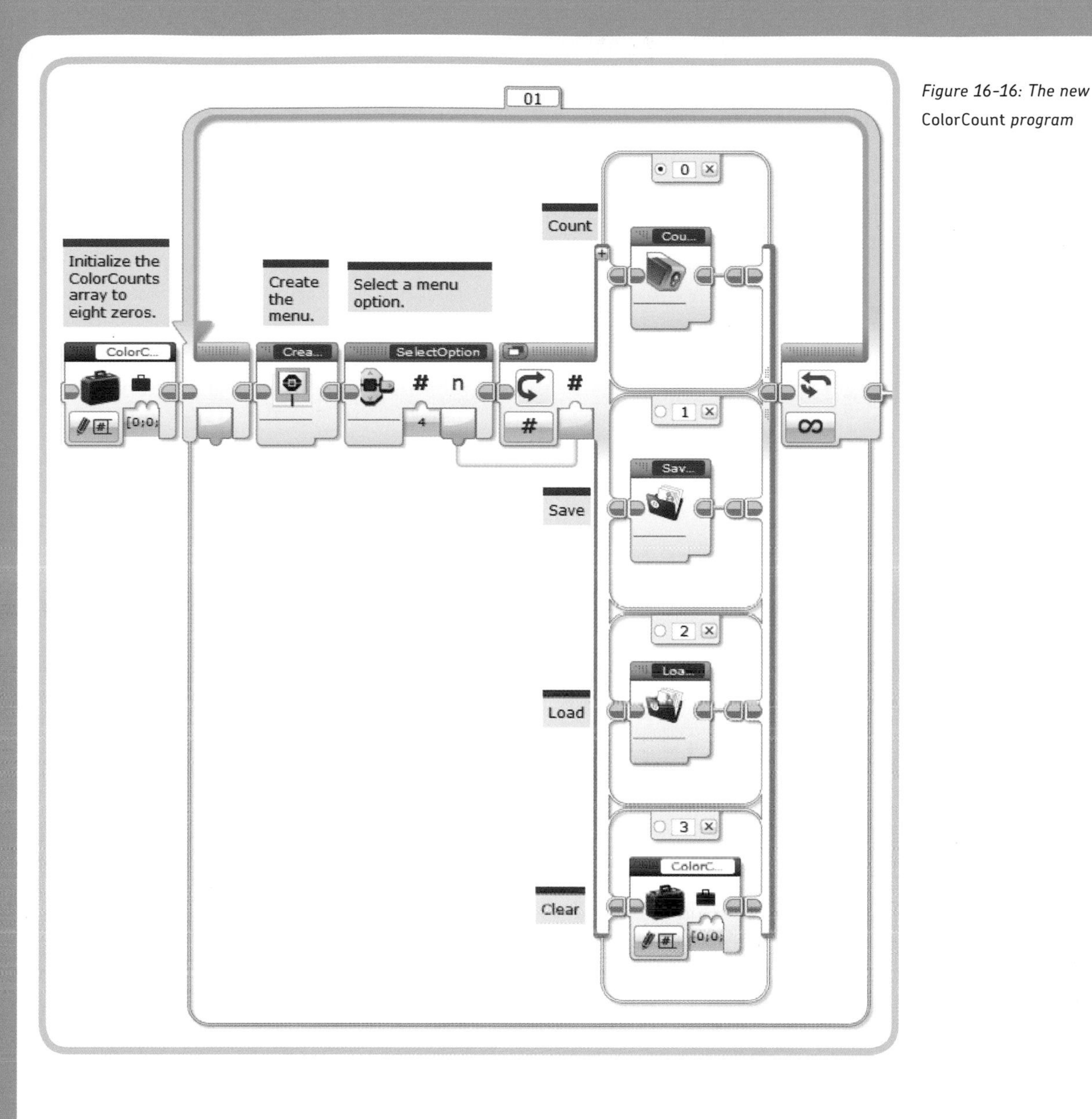

Figure 16-16: The new ColorCount *program*

The second section already correctly identifies each color and stores the totals in the ColorCounts array, but the original loop keeps running until you end it. We need to add a way to end the loop and go back to the menu. The Wait block now waits for either the Center or Left buttons to be bumped. When the Center button is pressed, the program identifies the color and updates the totals. When the Left button is bumped, the Loop Interrupt block inside the Switch block exits the loop, and the main program returns to the menu.

I also renamed the loop from *02* to *CC_02* so that the Loop Interrupt block doesn't accidently end the wrong loop. I don't have a loop named *02* in the *ColorCount* program, but I might decide to reuse this My Block in the future, so it's best to change the name now to avoid a potential naming conflict later.

Test the program to make sure it correctly counts the colors and displays the totals, and create the Count_CC My Block using all the blocks in Figure 16-17, except the Variable block at the beginning. (We don't want the array being reset to 0s each time we run the My Block!) Then, add the new block to the *ColorCount* program. Now when you run the program and select **Count** from the menu, the program should display the color names and totals and then start counting.

Test some objects, watch the totals change, and then bump the Left button. The program should return to the menu. Select the **Count** option again, and you should see the totals from your previous test.

You can also test the Clear option at this point. Select the **Count** option first and test some objects. Select the **Clear** option and then **Count** again, and you should see that the totals have all gone back to 0.

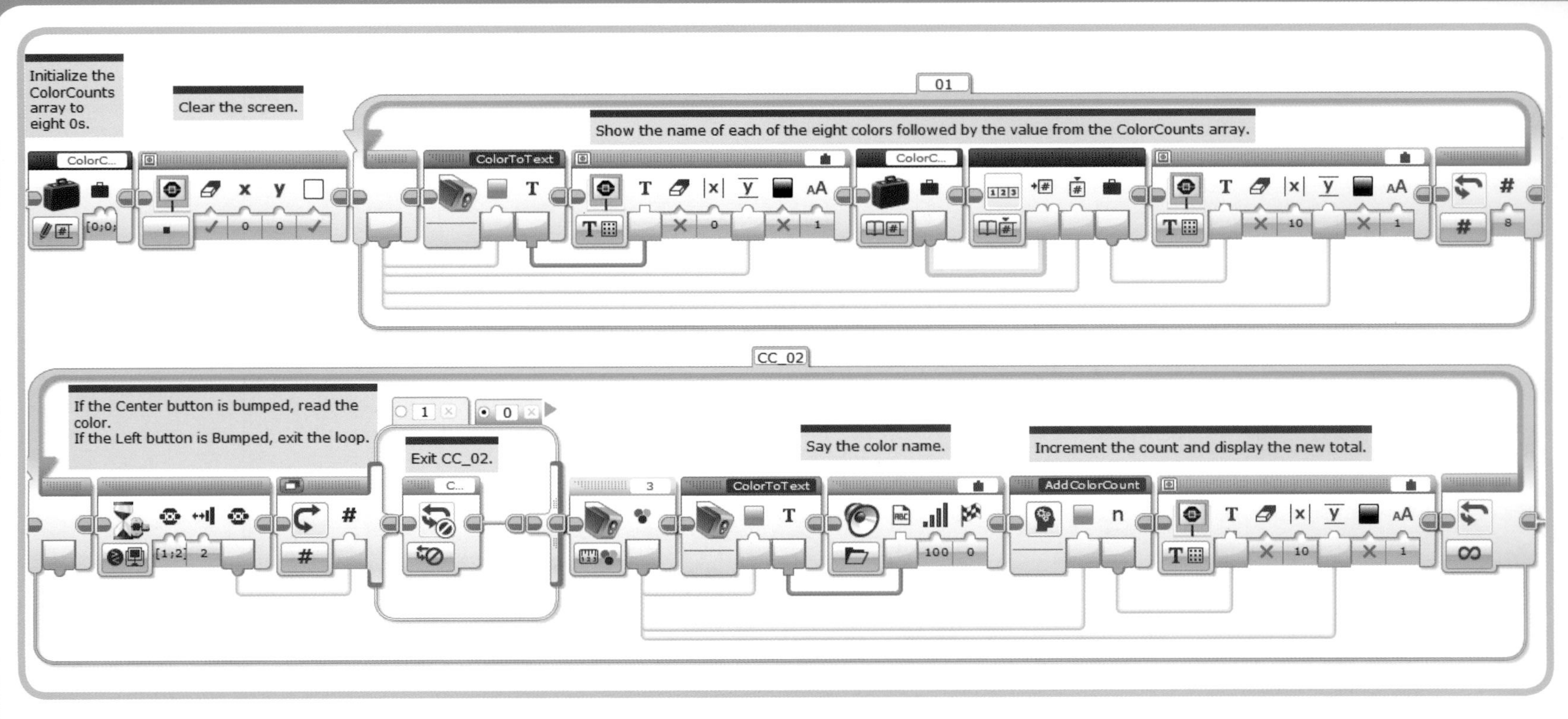

Figure 16-17: The Count_CCBuilder program

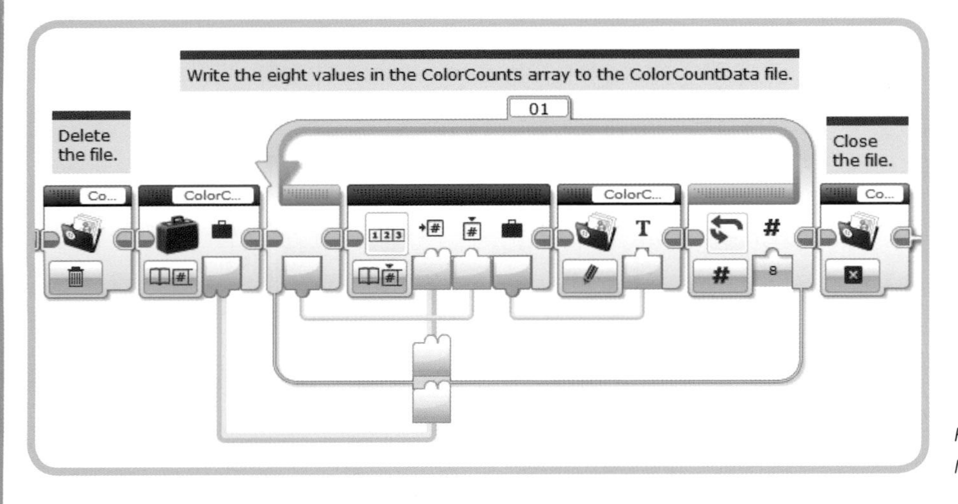

Figure 16-18: The Save_CC My Block

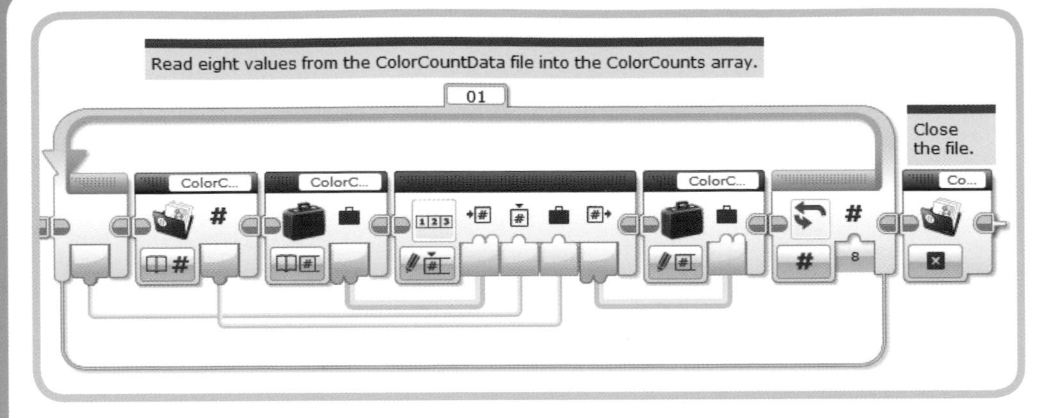

Figure 16-19: The Load_CC My Block

CHALLENGE 16-2

To end the *ColorCount* program, you need to use the Back button on the Brick (or stop the program from the EV3 software), which makes it easy to forget to save the data before ending the program. Add a **Save & Exit** option to the menu that will save the data, and then exit the program's main loop.

saving and loading the count data

The next step is to add the two My Blocks for saving and restoring the data, which use a file named *ColorCountData*. The Save_CC My Block (Figure 16-18) deletes the file, writes the eight values from the ColorCounts array, and then closes the file.

The Load_CC My Block (Figure 16-19) uses a similar structure to read eight values from the *ColorCountData* file and place them into the ColorCounts array. Just make sure you use the Save option at least once before testing the Load option! If the *ColorCountData* file doesn't already exist when you run this My Block, the program ends with an error when the File Access block tries to read the first value.

testing

Add the Save_CC and Load_CC My Blocks to your program as shown in Figure 16-16, and add the **Clear** option if you haven't already done so. Test all four options to make sure they behave as expected. After ending the program, run it again, and use the **Load** option to ensure that it will restore the data from the previous run.

managing memory

All the files on your EV3 (the programs, sound, images, and data files) take up some of the EV3's memory. In this section, I'll show you how to use the Memory Browser tool to see how much memory you've used, delete files to free up more memory, and transfer files between projects or between the EV3 and your computer.

To open the Memory Browser, select **Tools ▸ Memory Browser** from the menu or click the button in the lower-right corner of the Brick Information page (Figure 16-20).

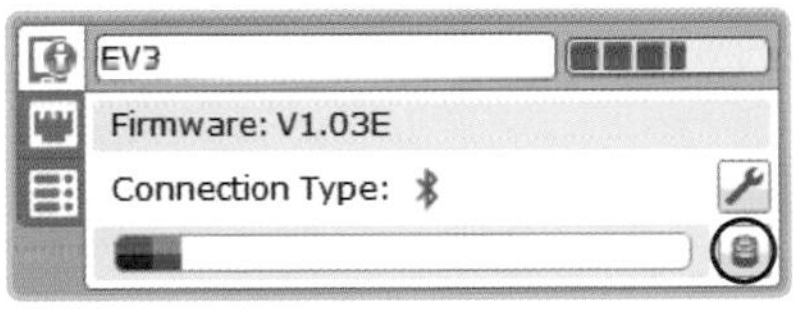

Figure 16-20: The Memory Browser button on the Brick Information page

The left side of the Memory Browser (see Figure 16-21) shows how much free space you have left. You can see from the image that most of the memory on my EV3 is still empty and available.

The right side of the Memory Browser contains a list of the folders and files on the EV3. Working with files on the EV3 is similar to working with files on your computer. Double-clicking a folder opens it so you can see the files it contains. For example, Figure 16-22 shows some of the files in the *Chapter16* folder. This folder contains all the files used by the programs in the

Chapter16 project, including files for each program and My Block, sound files used by the Sound block, image files used by the Display block, and the data files created by the File Access block. Each file's *extension* (the last three letters of the name, after the period) indicates the type of the file. For example, the extension for programs and My Blocks is *.rbf*, and for Sound files it's *.rsf*. The data files created by the File Access block have *.rtf* for an extension.

Figure 16-21: The Memory Browser

Figure 16-22: Some of the files from the Chapter16 *project*

There are five buttons below the list of folders and files:

Delete button Deletes the selected file or folder.
Copy and Paste buttons Allow you to copy files from one project to another. Select the file to copy, and click the **Copy** button. Then select the project you want to copy the file to and click **Paste**.
Upload button Used to copy the selected item from the EV3 to your computer. When you click Upload, a file dialog opens so you can select where to place the file and change the name if you'd like. When copying a data file created by the File Access block, change the file's extension from *.rtf* to *.txt* before trying to open it in your favorite text editor.
Download button Copies a file from your computer to the EV3. When you click this button, a file dialog opens so you can select the file to download. The file is placed in the project that's currently selected in the Memory Browser.

Each project has a separate list of files, and you can't write a file from a program in one project and read it from a program in another project. You can test this important fact about EV3 files with the *FileReader* program:

1. Set the filename to **FileTestData** and run *FileReader* to make sure it works.
2. Create a new project called *Chapter16Test* and copy the *FileReader* program from the *Chapter16* project to the new project.
3. Run the *FileReader* program from the new project.

The program should fail and display the "File Read Error" message because there isn't a file named *FileTestData* in the *Chapter16Test* project. This is actually a good feature because it makes sure that programs in different projects don't accidentally overwrite the same file. If you do want to use a file from one project in a different project, just copy the file using the Copy and Paste buttons.

You'll mostly use the Memory Browser to delete old projects once you've filled up the EV3's memory. But you can also use the Memory Browser to copy data files from the EV3 to your computer if you use the EV3 for data logging, which you'll learn about in the next chapter.

WARNING Downloading new firmware to your EV3 Brick will erase all the programs and any files you've created. Use the Memory Browser to upload any data files you want to keep to your computer before updating the EV3 firmware.

further exploration

For more practice with files, try these two activities:

1. The Save_CC and Load_CC My Blocks are specific to the *ColorCount* program. Create more general purpose My Blocks for saving and loading a Numeric array to and

EV3 TEXT FILES AND WINDOWS

The EV3 runs the Linux operating system and the text files it uses, including the files written by your programs, are written using the Linux text format. Unfortunately, Windows text files are formatted differently. Linux text files use a single special character called a line feed to mark the end of each line. Windows, on the other hand, uses a pair of special characters, a line feed followed by a character called a carriage return, to mark the end of each line. (OS X uses the same text format as Linux, so it doesn't have this issue.)

This difference causes problems when viewing or editing EV3 files using Windows. Some Windows programs, such as WordPad, are aware of the difference and will display an EV3 text file correctly. Others, such as Notepad, will display the entire contents of the file on one line. If you only want to *view* the file, this isn't much of a problem; simply use WordPad instead of Notepad.

Editing a file that you want to download to the EV3 and use in a program is more complicated. No standard Windows tools will write a file in the Linux format. If you save a file in WordPad, each line will end with two special characters, which is one more than the EV3 expects. If the file only contains numbers, then the EV3 will essentially ignore the extra character and read the numbers properly. If the file contains Text values, however, then every value except the first one will start with an extra space because the EV3 converts the carriage return character to a space.

Anytime you want to create a file on your computer, download it to the EV3, and then have an EV3 program work with its contents, you need to remove the carriage returns before downloading so that your EV3 can read the file correctly. There are programs that do this; I recommend *Tofrodos* (found at *http://www.thefreecountry.com/tofrodos/*), which converts files from Windows to Linux format.

Also, remember to save the file in text format—not Rich Text Format—until you're done editing it. Many programs use a file's extension to decide the format to use, and if you save with the *.rtf* extension, those programs might treat your *.txt* file as an *.rtf* instead. So if you want to create a file of commands, you might start off saving it as *commands.txt*. Then, before downloading the file to the EV3, rename it with the *.rtf* extension that the EV3 expects, meaning *commands.txt* would become *commands.rtf*.

from a file. The save My Block should take a filename and an array as input, and the load My Block should take a filename as input and produce an array as an output. When saving the values to the file, first write the number of elements so that when you read the file, you'll know how many values to read.

If the original array is empty, the save My Block should write out the length (0) and nothing else. After the load My Block reads the length, it should check the length (with a Switch block) and only enter a loop to read the values if the length is greater than 0. Setting a loop count to 0 doesn't work because the count is checked only after the loop body runs once.

2. Add a menu to the *ButtonCommand* program that includes options to save, load, create, display, or run the program. The display option is useful so that, after loading a program, you can see the commands it contains, just as you can when creating a program using the buttons.

conclusion

Files allow you to save data from your program onto the EV3. You can use that data later in the program, the next time the program runs, or from a different program. The File Access block contains all the features you need to create a file, write and read data, or delete a file.

The test programs at the beginning of the chapter showed you the basic operation of the File Access block, and then you put this knowledge to use to maintain a high score in the *Memory-Game* program. The changes to the *ColorCount* program were a bit more complex, using My Blocks to provide a menu and store the program data. You can even use the SelectOption My Block you created for this program in other programs that would benefit from a menu.

The Memory Browser tool lets you manage the files on your EV3 (either programs or data files). From this window, you can delete files to make room for other programs and transfer files between projects or between the EV3 and your computer.

17

data logging

In this chapter, I'll show you how to use EV3 features that you've already learned about to log motor and sensor data in files, making your EV3 a *data logger*. *Data logging* is the process of acquiring and recording data.

First, we'll experiment to determine what the Motor Rotation block's Current Power reading really means. Then, we'll examine how the Move Steering block's Steering parameter behaves. We'll wrap up with an experiment to test the reliability of the *LightPointer* program from Chapter 11, which uses the Color Sensor to point the TriBot in the direction of a light source.

I'll only use features common to both the EV3 Home and Education Editions, but the Education Edition has more ways to collect and present data from experiments, making it a great tool for classroom use. If you're working with the Education Edition, it's well worth your time to investigate these features.

data collection and the EV3

Collecting data is critical to any experiment, but collecting data by hand can be tedious and error prone. Most people just aren't that good at recording measurements quickly at precise intervals or over long periods of time. Fortunately, computers excel at this kind of task. The combination of the EV3 Brick computer and sensors makes the EV3 ideal for collecting data.

When designing a program, it can be useful to run some tests to learn how a sensor or motor will react under the conditions you expect your program to experience. The more you know about the motors, sensors, and programming blocks, the easier it is to write working programs, so let's dive right in with an experiment!

investigating the current power reading

The Motor Rotation block has a Measure - Current Power mode, which tells us the motor's *power* as a general expression of strength at a particular moment. The Move block's Power parameter is connected to this mode, and in the following sections we'll write some simple data-logging programs to explore this connection.

the CurrentPowerTest program

For our first experiment, we'll record the Current Power reading of motor B as we adjust the Power parameter of a Large Motor block using the *CurrentPowerTest* program shown in Figure 17-1. The program starts the motor running at 100 percent power and then enters a loop in which the Power parameter of the Large Motor block is stepped down from 100 to 1. It records the Current Power at each step, and each time through the loop, the Power parameter and the Current Power reading is written to the *CurrentPowerTestData* file.

Instead of using a File Access block to write the Power parameter and the Current Power reading separately, the data file created by the program contains one line for each measurement. Each line contains both pieces of information separated by a comma, a format known as *comma-separated values (CSV)*. Files in this format typically use *.csv* for an extension. Spreadsheet programs know how to work with data files in this format, so arranging the data this way will make it easier for us to analyze later.

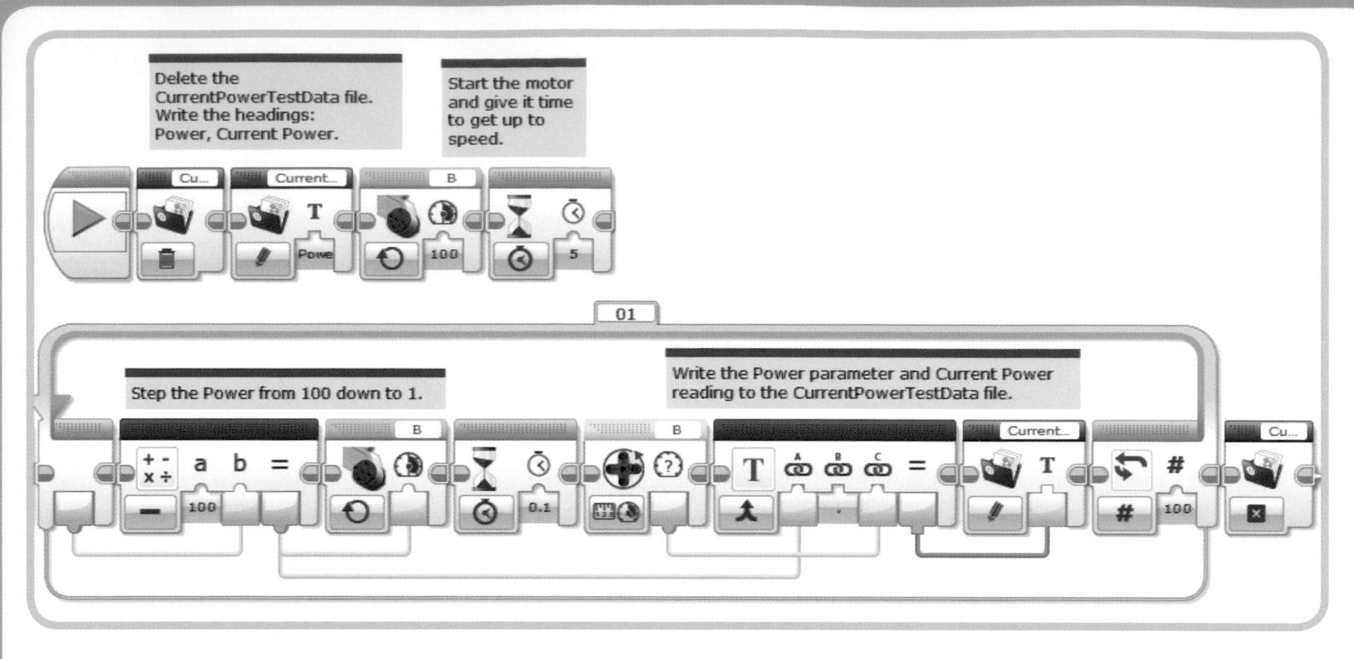

Figure 17-1: The CurrentPowerTest *program*

NOTE In the United States, the values in a comma-separated file are separated by a comma; other parts of the world actually use a semicolon. Change the programs in this chapter to use a semicolon if that's what your spreadsheet program expects.

Let's walk through the purpose of each block in this data-logging program:

1. The first File Access block deletes the file because we want to create a new file with fresh data every time we run the program.
2. The second File Access block writes "Power, Current Power" to the file, to serve as headings for the data.
3. The Large Motor block uses **On** mode to start motor B running at 100 percent power.
4. The Wait block pauses the program for five seconds to give the motor time to accelerate to full power.
5. The Loop block repeats 100 times, one time for each Power parameter from 100 down to 1.
6. The Math block computes the Power parameter by subtracting the Loop Index from 100. The first time through the loop, the result is 100; the next time, it's 99; and so on. The last time through the loop, the Loop Index will be 99, and the Power parameter will be 1.
7. The result of the Math block is passed to the Large Motor block to change the Power parameter.
8. The Wait block gives a small pause to allow the motor to slow down to the new Power parameter.
9. The Motor Rotation block uses **Measure - Current Power** mode to read the Current Power value and put it on the data wire.
10. The Text block combines the Power parameter used by the Large Motor block and the value read by the Motor Rotation block and separates the two values with a comma (,).
11. The File Access block writes the value from the Text block to the *CurrentPowerTestData* file.
12. The final block closes the *CurrentPowerTestData* file.

When you run the program, motor B starts moving at full speed. After five seconds, it slowly ramps down and stops when the program ends, after about 10 seconds. At that point, there should be a file on the EV3 named *CurrentPowerTestData.rbt* that contains the data from this experiment.

Use the Memory Browser (**Tools ▸ Memory Browser** from the menu) to upload the file from your EV3 to your computer. Before you save the file, change the extension to *.csv* so that your text editor or spreadsheet program recognizes the format. It's also a good idea to add a number to the end of the filename so that you can keep data from individual runs of the program (Figure 17-2).

Figure 17-2: Saving the data file

You can open the file in a text editor, word processor, or spreadsheet program. I prefer to use a spreadsheet (such as OpenOffice.org Calc or Microsoft Excel) to analyze data so I can look at the raw numbers and easily create graphs. Table 17-1 shows how the two headings and first 10 measurements should look in a spreadsheet program (the measurement number corresponds to the spreadsheet's row number). The Power parameter starts at 100 and decreases by 1, as expected. The Current Power starts at 76 and stays pretty constant for these first 10 measurements.

table 17-1: the current power for the first 10 power parameter measurements

Measurement #	Power	Current Power
1	100	76
2	99	77
3	98	76
4	97	74
5	96	75
6	95	76
7	94	77
8	93	75
9	92	75
10	91	76

Scrolling down the file to where the Power parameter is 50, I see that here, the Current Power reading more closely matches the Power parameter, as shown in Table 17-2.

When you look at a graph of the entire data set (in Figure 17-3), you should get a better sense of how the Power parameter and the Current Power reading are related.

table 17-2: the current power measurements from power parameters 53 through 44

Measurement #	Power	Current Power
49	53	54
50	52	53
51	51	52
52	50	52
53	49	49
54	48	47
55	47	47
56	46	47
57	45	46
58	44	45

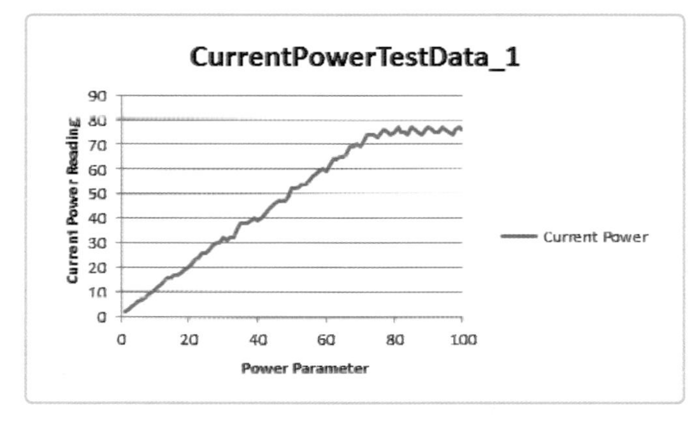

Figure 17-3: Graph of the Power parameter and Current Power reading

From this graph, I can see that when the Power parameter is below 70, the Current Power reading closely matches it. The two values aren't always exactly the same, which is why the graph shows a little wiggle rather than a straight line, but they're also never more than one apart.

The Current Power reading tops out when the Power parameter is about 75 and stays there for higher Power parameters. It turns out that the maximum value of 75 is specific to a particular motor. If I use motor C instead, I get a similar set of data, but with a maximum at 78. Other motors also give slightly different values, all in about the same range.

We'll explore the relationship between the Current Power and the Power parameter further, using the Rotation Sensor to measure how fast the motor is moving at different Power settings, and relate that to the Current Power reading. To make doing this easier, we'll build the LogData My Block to assemble several values into a comma-separated list and write them to a file.

the LogData my block

The *CurrentPowerTest* program (see Figure 17-1) uses a Text block to join the Power parameter and Current Power reading values, separated by a comma. Combining three values together and separating them by commas requires two Text blocks, and each additional value requires another Text block. This code for data formatting can quickly become large, making the program's logic more difficult to see. Creating the LogData My Block stops formatting code from cluttering the main program.

This block also writes the formatted data to a file and adds a *timestamp* to each value, using a Timer block, to indicate when the measurement was recorded. Timestamps are useful in experiments where time is an important factor, and timestamping the data can also help you identify any odd pauses or other timing issues in the program.

Figure 17-4 shows the LogData My Block, which combines a timestamp and up to four Text values. Most values are numbers, but I used Text parameters so that I can use this block to write the headings as well. Here's how this My Block works:

1. The number of the timer to use is passed in on a data wire. The Timer block reads the appropriate timer and passes this value to the first Text block.
2. The first Text block combines the timestamp from the Timer block with one Input parameter, separating them by a comma.
3. The following three Text blocks each take the value from the previous Text block and add a comma and one Input parameter.
4. The final block writes the formatted data to the file whose name is passed in to the My Block as an Input parameter.

The My Block Building window doesn't show which parameter is connected to which block, but you can simply set the name and icon for each parameter, from left to right, and move the Numeric parameter to the left or right as needed. (The order doesn't really matter.) Then, after the My Block is created, you can move the data wires around so that each Input parameter goes to the correct block.

the CurrentPowerTest2 program

The *CurrentPowerTest2* program, shown in Figure 17-5, is based on the *CurrentPower* program and adds a new measurement. Each time through the loop, after changing the Power parameter, the program reads the Rotation Sensor, pauses for one second, and reads the Rotation Sensor again. The difference between the current reading and the previous one will tell us how far the motor has moved during that one-second pause, giving us the average speed in degrees per second. The Current Power reading for the motor is also logged, so once all the data

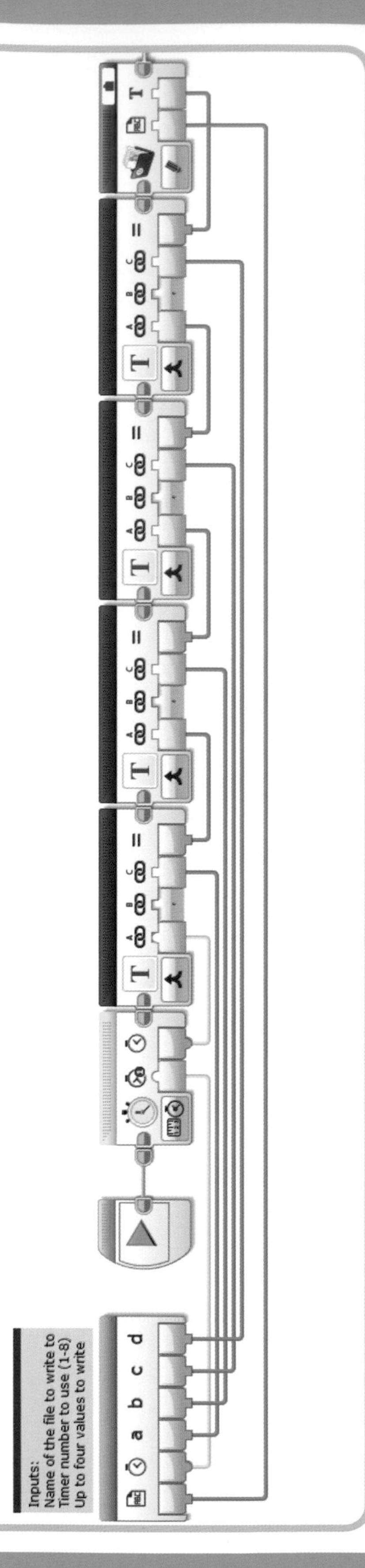

Figure 17-4: The LogData My Block

is collected, we'll be able to see the relationship between the Current Power reading and the motor's actual speed.

The LogData My Block at the beginning of the program writes the headings to the *CurrentPowerTestData* file, and the one in the Loop block writes the Power parameter, Current Power reading, and the computed speed.

This program takes about 100 seconds longer to run because of the one-second pause between the Rotation Sensor readings. After the program completes, copy the *CurrentPowerTestData* file to your computer and examine the data.

Table 17-3 shows the data from my test when the Power parameter is near 100 and when it's near 50. The speed is about 10 times the Current Power reading because the Current Power is in fact the speed of the motor measured in degrees per one-tenth of a second!

table 17-3: data from the currentpowertest2 program

Power	Current Power	Speed
100	79	803
99	80	801
98	79	803
97	79	801
96	79	803
95	79	799
94	78	802
93	79	800
92	78	802
...	...	...
50	51	510
49	48	498
48	48	490
47	47	481
46	47	470
45	44	460
44	43	450
43	43	438
42	42	430
41	41	420
40	40	408
39	39	400

NOTE The Current Power and Speed values in Table 17-3 don't differ by *exactly* a factor of 10 because the pause between Rotation Sensor readings is actually a little longer than a second, which increases the calculated average speed value by just a little. Also, the EV3 constantly adjusts the motor's speed to keep it moving at the correct rate. We could get a more accurate measure of the time between the two Rotation Sensor readings by using Timer blocks to record the time before taking each measurement and then taking the difference between the two timer readings. For the purposes of this program, the simple approach is sufficient.

This data tells us that setting the Power parameter of a Large Motor block to 10 will turn the motor at a rate of 100 degrees per second, and a setting of 50 will turn the motor at a rate of 500 degrees per second. This relationship holds until the Power parameter is about 75; at this point, higher Power values don't move the motor any faster—a condition known as *saturation*. So in practice there's no difference between setting the Power parameter to 80 or 100—both settings make the motor go about 750 degrees per second.

So why doesn't the motor go as fast as you tell it to? The EV3 was actually designed to act this way, which I think is a good decision. Let's say that with the motor used by the EV3 and the expected battery power, the LEGO engineers could guarantee that every motor could move at 700 degrees per second. They *could* design the system so that the Power parameter of 100 translated to a speed of 700 degrees per second, effectively making the relationship between the Power parameter and the speed a factor of 7 instead of 10.

This would preserve the relationship between the Power parameter and the motor's speed throughout the Power parameter's range, but you wouldn't be able to make some motors go as fast as possible. The maximum speed you could specify would be 700 degrees per second, and we know that the motors can actually go a little faster. The system's actual design allows us to

CHALLENGE 17-1

Run the *CurrentPowerTest2* program using the Medium motor to see how the Power parameter, Current Power reading, and actual motor speed are related for that motor. Replace the program's Large Motor blocks with Medium Motor blocks, and change the Port setting on the three Motor Rotation blocks to match. You'll also need to remove the motor from the Lift Arm, or at least disengage the gears.

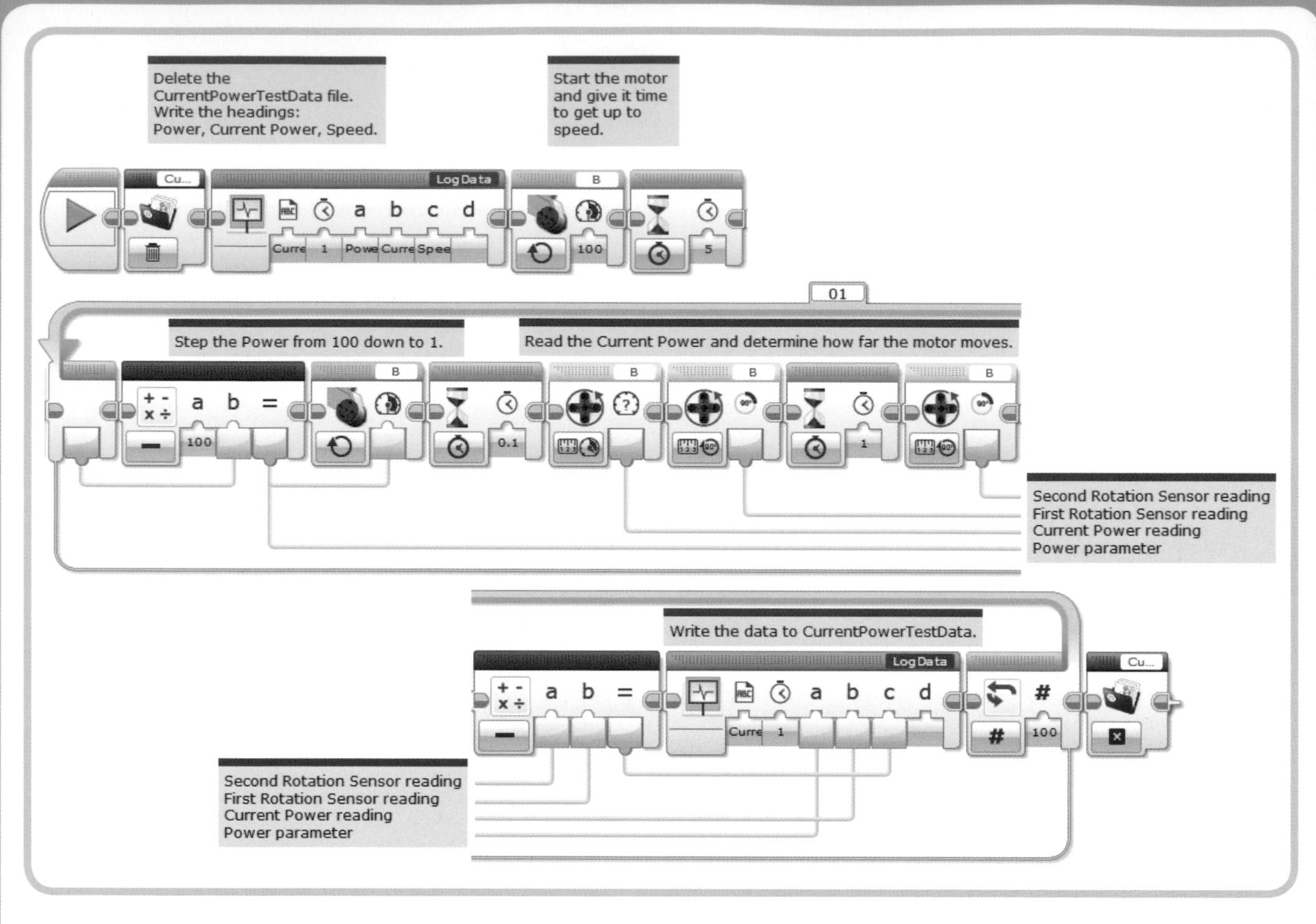

Figure 17-5: The CurrentPowerTest2 *program*

get all the speed we can out of a motor, at the expense of losing the relationship between the Power parameter and the motor speed at the higher end of the spectrum.

testing the current power with the move steering block

The *CurrentPowerTest* program used the Large Motor block to determine the relationship between the Power parameter and the Current Power reading, which revealed how the Power parameter relates to the actual speed of the motor. But in most programs, we actually use the Move Steering block. Does that block hold to the same relationship?

Testing this is just a matter of replacing the two Large Motor blocks with Move Steering blocks and running the program again. Figure 17-6 shows the graph of the data from my test. Below a Power value of 70, the graph shows the same relationship we saw using the Large Motor block. Above 70, there is more variation than the previous tests showed. This happens because the EV3 is trying to keep the two motors moving at the same speed, which is difficult at the top end of a motor's maximum speed.

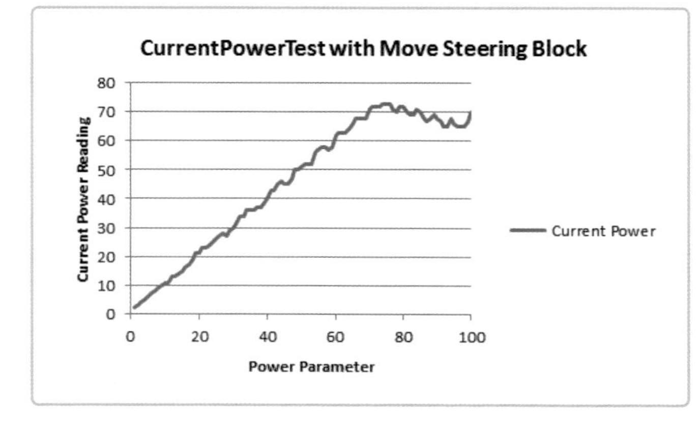

Figure 17-6: Power parameter versus Current Power reading with the Move Steering block

the SteeringTest program

The next program, *SteeringTest* (Figure 17-7), examines the relationship between the Move Steering block's Steering parameter and the speed of the two motors. *SteeringTest* first deletes the *SteeringTestData* file and then re-creates it, writing "Steering, Motor B, Motor C" on the first line as column headings. The program then starts the motors moving with the Steering parameter set to 0. Inside the Loop block, the Steering parameter is stepped up from 0 to 100 for a total of 101 steps because we want to include the values at both ends. At each step, the LogData My block records the Steering value and the Current Power readings from the two motors. The Power parameter for the Move Steering blocks is set to 50, so we are well within the range where the Power parameter should map directly to the Current Power reading.

Run this program, and both motors will spin for about 15 seconds. After the program completes, you should be able to upload the *SteeringTestData* file from the EV3 to your computer.

Figure 17-8 shows a graph of the data. The Current Power for motor B essentially stays constant at about 50 (the measurements vary between 49 and 51). The Current Power for motor C starts at 50 and decreases to -50 as the Steering parameter increases from 0 to 100.

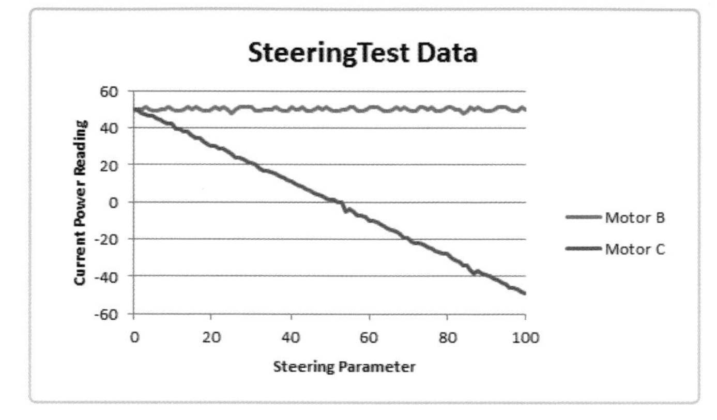

Figure 17-8: The Current Power readings versus the Steering parameter

When the Steering parameter is 0, both motors have a Current Power reading of 50, which makes sense because this Steering value should make the robot move in a straight line. As the Steering parameter increases, motor C slows down, which causes the robot to turn.

When the Steering parameter is at 50, the Current Power of motor C is 0, which means it's not moving at all. With this Steering value, the motor C wheel stays still, but the motor B wheel keeps moving forward, which spins the robot in place around a point centered on the motor C wheel. With the Steering parameter at 100, motor C has a Current Power of -50, so it moves just as fast as motor B but in the opposite direction. This makes the robot spin around a point centered between the two wheels. Any Steering parameter between 50 and 100 causes the robot to spin; the only difference is the point the robot spins around.

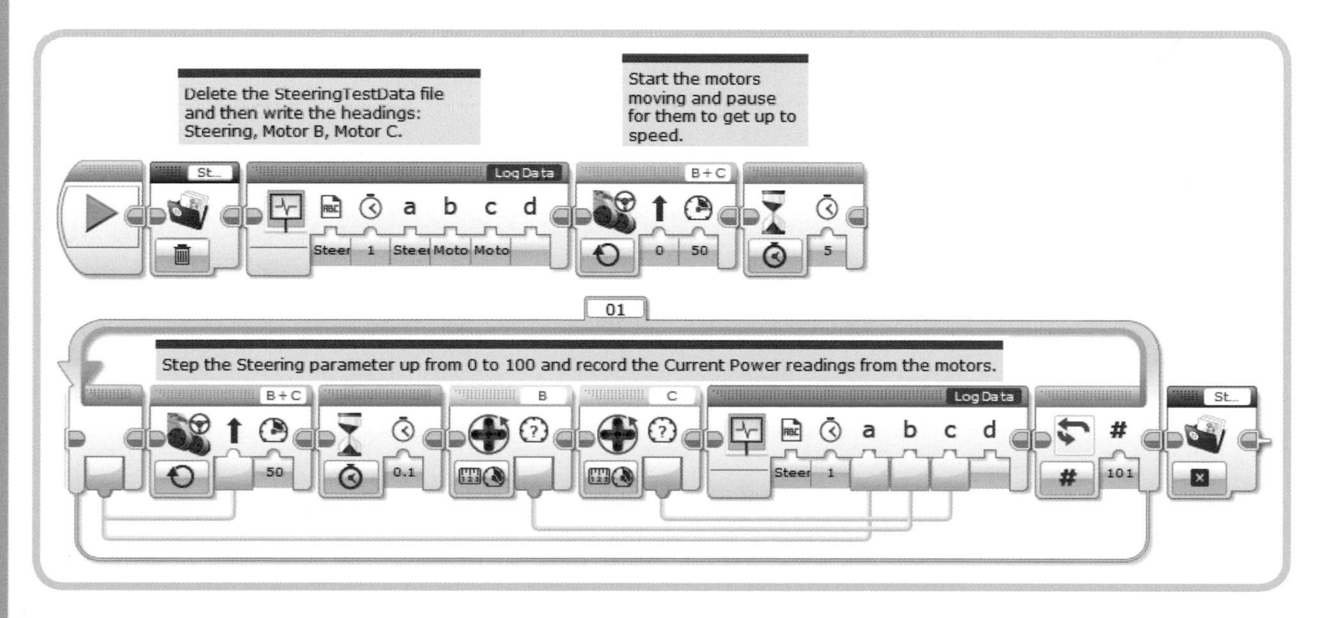

Figure 17-7: The SteeringTest *program*

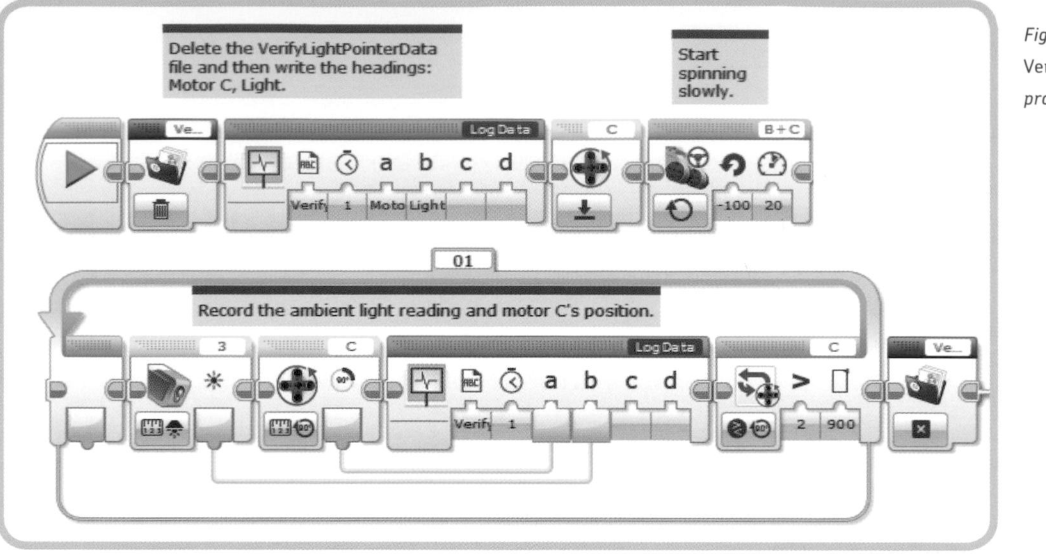

Figure 17-9: The VerifyLightPointer *program*

If you want your robot to keep moving forward as it turns, the Steering value should always be less than 50. Although the range of one side of the Steering parameter is 0 to 100, the useful range when moving forward is really between 0 and about 40. When turning in the other direction, use values between 0 and -40. A value above 40 (or below -40) will make the robot spin or move in a very small circle.

the VerifyLight-Pointer program

The *LightPointer* program presented in Chapter 11 uses the Color Sensor to point the TriBot toward a light source. The TriBot spins in a circle and remembers the position where it detected the brightest light level. After completing a full circle, the TriBot returns to the stored position, which should point it toward the light source.

The *LightPointer* program assumes that the sensor can actually detect the brightest light level while the TriBot is spinning. If this assumption is wrong, the program won't work. For example, the program fails if the robot spins too fast to accurately read the light level or if there is too much ambient light in the room for it to locate the direction of the light source. You can verify this assumption by collecting and analyzing the data from the sensor in an experiment. The *VerifyLightPointer* program shown in Figure 17-9 collects the data.

This program is a combination of the *LightPointer* program and the data-logging programs. First the *VerifyLightPointerData* file is deleted, and then the LogData block is used to write the column headings for the *.csv* file: "Motor C" and "Light." The Rotation Sensor for motor C is reset and then the TriBot starts spinning slowly. Inside the loop, the ambient light reading from the Color Sensor and the position of motor C are written to the file. The loop repeats until the position of motor C reaches a position of 900 degrees (set this value to 700 degrees for the Education Edition). The last block closes the *VerfiyLightPointer-Data* file.

Position the TriBot and a light as shown in Figure 17-10, with the light 90 degrees to the left of the robot, and run the program. The TriBot should slowly spin in a circle and then stop. After the program completes, you should be able to upload the *VerifyLightPointerData* file to your computer and examine the Color Sensor and Rotation Sensor readings.

Figure 17-10: The starting position for the VerifyLightPointer *program*

Figure 17-11 shows a graph of the measurements taken during my test run. The light level increases significantly as the robot turns toward the flashlight and forms a single large peak. The numerical data (Table 17-4) shows that motor C only moves about 1 degree between each Color Sensor reading. Because the motor has to move about 840 degrees to complete a full circle, the TriBot actually rotates very little between each reading, so the chance of the sensor moving past the light without seeing it is very small.

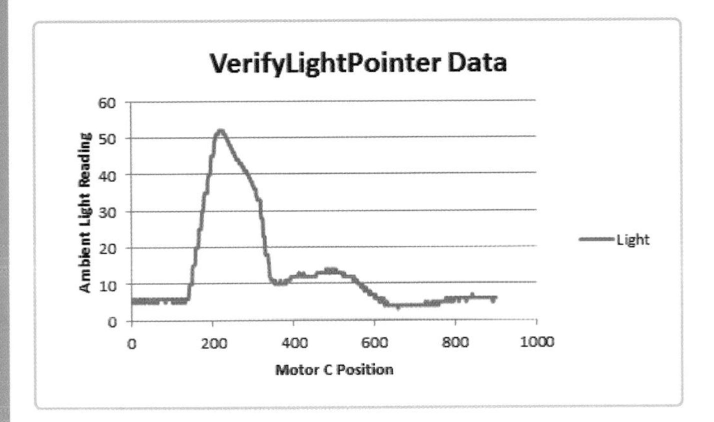

Figure 17-11: The ambient light detected at motor C positions

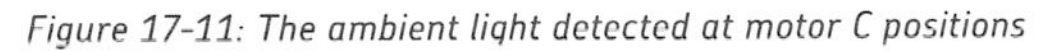

table 17-4: rotation sensor reading

Motor C	Light
2	6
3	6
3	4
4	6
5	6
5	5
6	5
7	5
7	5
9	6
9	6
10	6

Based on this data, it seems that the *LightPointer* program should be able to correctly identify the direction of the light source. If the data had shown only small variations as the robot moved, or showed several peaks, then I'd be less confident that the *LightPointer* program would work correctly.

controlling the amount of data

The *VerifyLightPointer* program collects and records data as fast as possible, creating a large data file in a short amount of time, but for most data-logging programs, you'll want more control over how often the program records data.

Many data-logging programs will be structured like the *VerifyLightPointer* program. After some blocks perform the initial setup, a Loop block will contain the code to collect the data and write it to a file. You can control how often the data is recorded by adding a Wait block at the end of the body of the Loop block.

How long should the Wait block pause? That depends on how long you expect the experiment to take and how often the data you're collecting changes. You need to record the data often enough that you don't miss any important changes, but not so often that you end up with huge data files or run out of memory. Finding the right balance often involves some trial and error, so don't be surprised if you need to change the settings a few times to get them just right.

For example, say you decide to change the *VerifyLightPoint* program so that it takes 20 measurements per second. To do this, you'd add a pause at the end of the Loop block to wait for one twentieth of a second, or 0.05 seconds. Figure 17-12 shows the Wait block added to the program's main loop.

Now the program will pause for 0.05 seconds each time through the loop, which means that the data will be recorded at a rate of approximately 20 readings per second. There may not be *exactly* 20 readings per second because it takes time to control the motors, collect the readings from the sensors, and write the data.

CHALLENGE 17-2

The *VerifyLightProgram* collects sufficient data to identify the direction of the light, represented in the graph by the large peak, because the TriBot is rotating very slowly. Change the Power parameter on the Move Steering block from 20 to 40, and rerun the test to see how much difference that makes. How fast can you make the TriBot spin and still have the graph show a discernable peak?

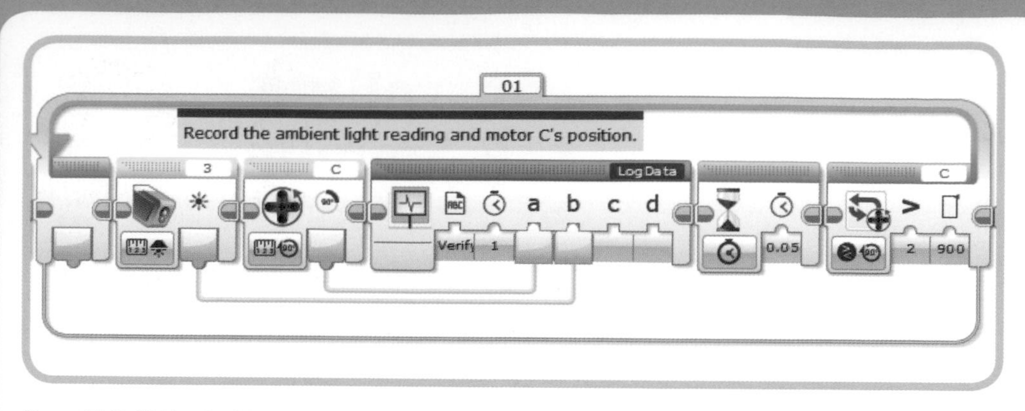

Figure 17-12: Waiting for 0.05 seconds

further exploration

Here are some more activities involving data logging:

1. Create an experiment to show how the Infrared Sensor Proximity reading (from the Home Edition) or Ultrasonic Sensor distance reading (from the Education Edition) changes as the TriBot moves away from an object. Use the position of one of the motors to gauge the true distance from the object. Start very close and have the TriBot slowly back away. Try this with objects of varying colors and textures (for example, a solid wall and a hanging towel may give different results).

   The reading from the Ultrasonic Sensor is a true measurement of distance, so it should track directly with the motor rotation until the point where the sensor stops detecting the object.

2. Write a program that determines how the Ultrasonic Remote's Beacon Heading measurement changes as the angle between the TriBot and the Remote changes. Start with the Remote in front of the robot, and record the Beacon Heading as the robot spins in a circle (this is similar to the *VerifyLightPointer* program).

3. Using the Gyro Sensor's Measure Rate mode, which measures rotation in degrees per second, write a program that helps you to determine how the Power parameter of a Move Steering block relates to the TriBot's rotational speed when spinning (with the Steering parameter set to 100). First, collect data at a slow speed. Then increase the speed to see if the relationship holds and if there's a point at which the TriBot moves too fast to get reliable data from the Gyro Sensor.

conclusion

The EV3 can collect, format, and record data from a variety of sensors, making it a great data logger, and the examples presented in this chapter contain all the steps you'll need in a typical data-logging program. These include creating the data file, collecting sensor data, writing the data to the file with a timestamp, and even controlling the rate of the data collection.

Data logging can help you learn more about the EV3 motors and sensors. For example, the *CurrentPowerTest* program helped uncover the mystery of the Motor Rotation block's Current Power reading and helped you learn what the Move block's Power parameter really does. The closely related *MoveSteeringTest* program collected data that showed how the Steering parameter affects the robot's motion and what range of values is most useful. The *VerifyLightPoint* program used data logging to prove, based on experimental data, that the *LightPointer* program will work as expected.

You can also use your EV3 sensors at home or in a science class for experiments that have nothing to do with robotics! You could use the Color Sensor to compare the brightness of different brands of light bulbs or use the Rotation Sensor to measure area and volume.

For even more experimentation possibilities, get the Temperature Sensor from LEGO Education (*http://www.legoeducation.com/*) or one of the many EV3-compatible sensors from HiTechnic (*http://www.hitechnic.com/*), Mindsensors (*http://www.mindsensors.com/*), or Vernier (*http://www.vernier.com/*).

multitasking

In this chapter, you'll learn how to run groups of blocks in parallel, allowing your robot to perform two or more tasks at the same time, which is called *multitasking*. For example, your program could have one section to control the robot's navigation and another to collect sensor data.

I'll start by showing you how to add a simple odometer to measure the distance traveled by the *AroundTheBlock* program and then move on to add flashing lights to the *DoorChime* program. I'll also discuss the rules of program flow when you use parallel sequences and show you how to synchronize the actions between sequences.

more than one start block

In an EV3 program, a group of blocks connected together is called a *sequence*. All of the programs we've built up to this point have used a single sequence, beginning with the Start block that appears automatically when you create a new program. Multitasking simply puts multiple sequences together in one program.

One way to use multiple sequences in your program is to add another Start block. For example, Figure 18-1 shows a version of the *AroundTheBlock* program (introduced in Chapter 4) that uses two sequences. These *parallel sequences* run at the same time. The sequence at the top of the image moves the TriBot around a square, and the sequence at the bottom continually displays the motor position. When the program completes, the distance traveled by the B motor is displayed. You can use a similar technique to measure how far the *LineFollower* or *WallFollower* programs travel.

Follow these steps to build this program:

1. Create a new project named *Chapter18*.
2. Open the *Chapter4* project and copy the *AroundTheBlock* program to the *Chapter18* project.
3. Drag a Start block from the Flow Control palette onto the Programming Canvas.
4. Add the blocks shown in the bottom part of Figure 18-1. Set the Unit parameter of the Display Number block to **Degrees:** .

When you run this program, the EV3 Brick starts both Loop blocks and then rapidly switches between running the

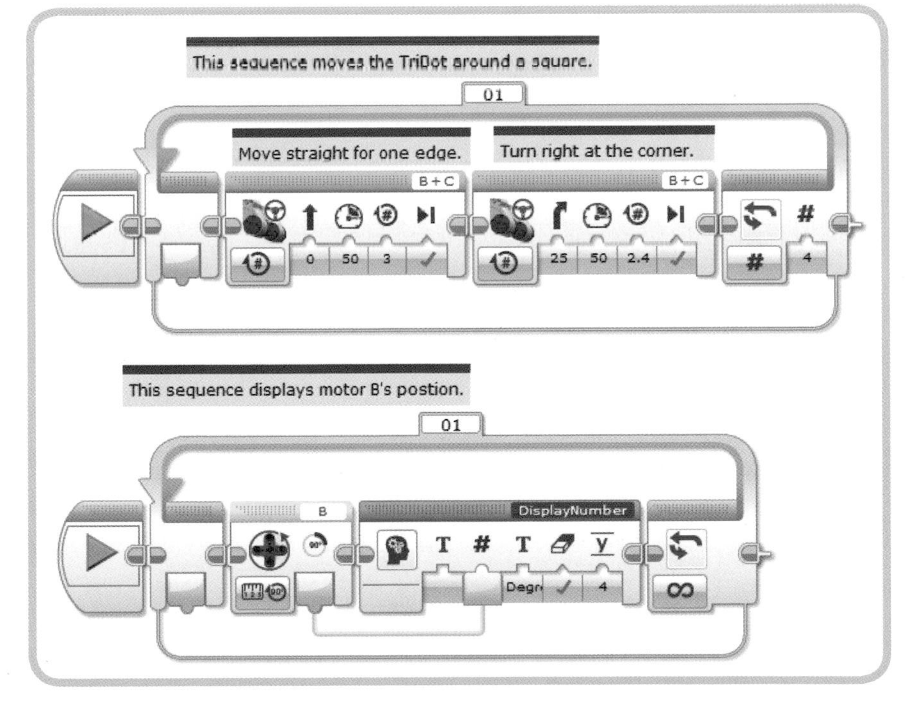

Figure 18-1: Displaying the motor position while moving around a square

code in each sequence. The computer inside the EV3 can't *really* do more than one thing at a time, but it *can* switch between the two tasks quickly, doing a little of one and then a little of the other. This switching happens so fast that you'd never know the difference.

Run the program, and the TriBot should move around the square while the display shows how far the robot has moved. Even after the TriBot has moved around all four sides of the square, the program continues to display the position of motor B. Pick up the robot and turn the motor, and you'll see the position update on the screen.

The original *AroundTheBlock* program ends after the robot moves around the square because the Loop block finishes after repeating four times, and there are no more blocks in the program. When a program has more than one sequence, it continues to run until *all* sequences have ended. The Loop block that displays the motor position is set to run forever, so the program will continue running until you stop it.

the stop program block

Picking up the robot and pressing the Back button to stop a program isn't a huge bother, but there is a better way. The *Stop Program block* (Figure 18-2), found on the Advanced palette, will terminate all running sequences and end the program. Place this block at the end of the upper sequence of the modified *AroundTheBlock* program, as shown in Figure 18-3, to end the program after the TriBot has completed moving around the square.

Figure 18-2: The Stop Program block

In this example, the Stop Program block is placed at the end of a sequence. You can also use this block inside a Switch block if you want to only end the program under certain conditions.

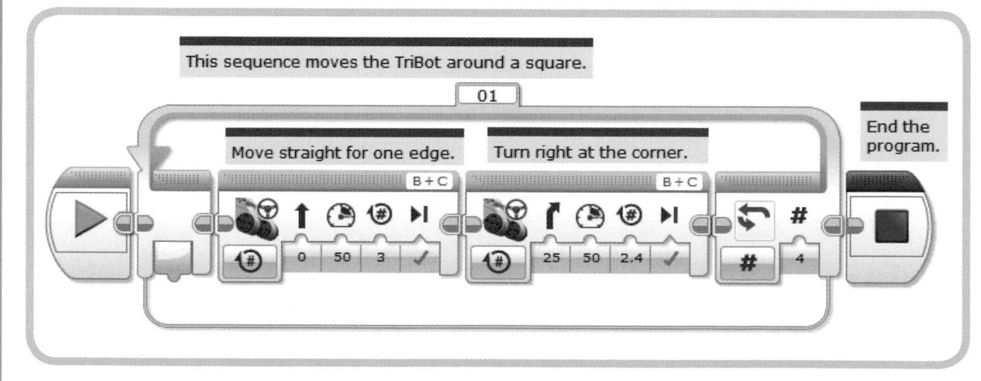

Figure 18-3: Stopping the program after moving around the square

avoiding a busy loop

Programs that depend on a quick response to sensor readings, such as the *LineFollower* program, can be adversely affected by adding a second task. Figure 18-4 shows the odometer code added to the *LineFollower* program from Chapter 13.

Before adding the second task, I could set the Power parameter of the Move Steering block to 50, and the program would work very well. After adding the odometer code, the highest I can set the Power and still get the same reliability is 35.

The code to display the motor position is an example of a *busy loop*, one that repeats as fast as possible and, consequently, uses a large portion of the EV3's processing power. This causes the line-following code to run more slowly, so the robot doesn't react as quickly when the line curves. We can solve this problem by slowing down the display loop, letting the EV3 devote more time to making sure the robot follows the line. Adding a one-second pause to the Loop block (Figure 18-5), to update the display only once a second instead of as fast as possible, allows me to set the Power parameter back to 50. The displayed motor position should still be accurate enough to be useful, even with the delay.

adding lights to the DoorChime program

The new *AroundTheBlock* program uses two Start blocks to run two independent tasks for the entire program. In this section, we'll place multiple sequences in the middle of a program instead. This approach is useful when you want to run two

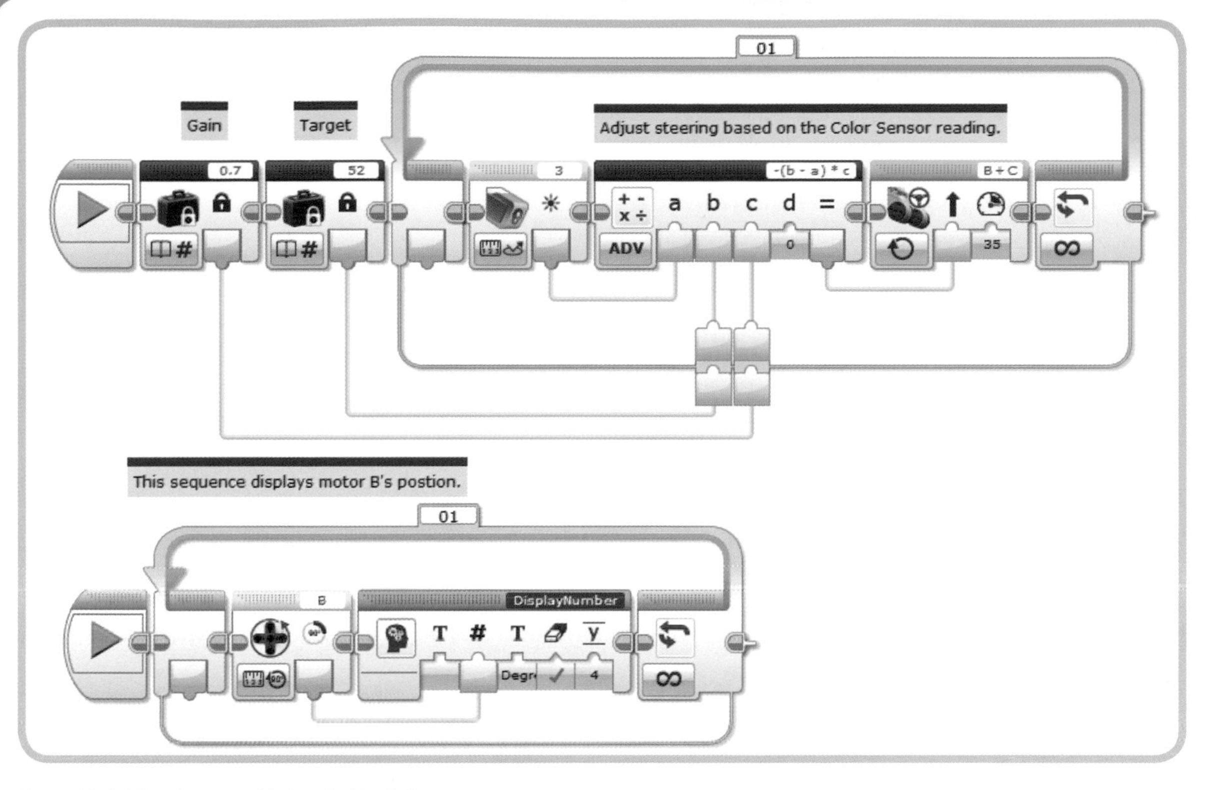

Figure 18-4: The odometer added to the LineFollower *program*

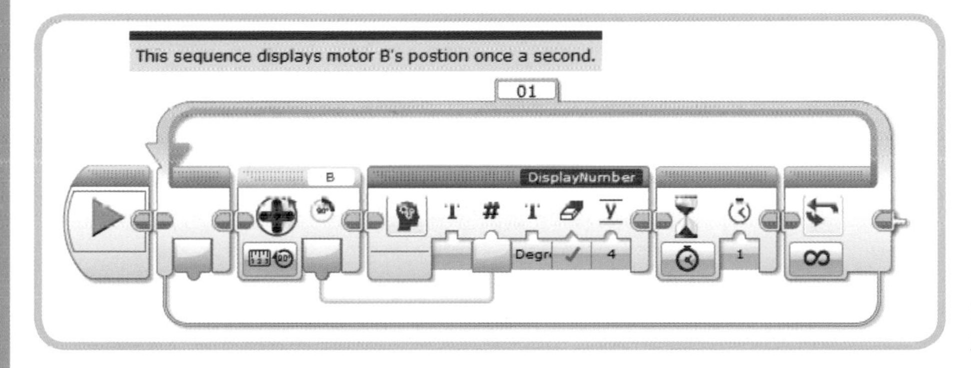

Figure 18-5: Slowing down the busy loop

tasks at some points in the program, rather than the entire time the program is running.

We'll modify the *DoorChime* program from Chapter 12 (Figure 18-6) to flash the Brick Status Light while playing the chime. Because the light should flash only when the chime is playing, we need to add a second sequence at that point instead of creating an entirely new sequence by using a Start block.

Once you make the changes shown in Figure 18-7, the program first turns off the Brick Status Light and then enters the loop and waits for a person to walk by. At that point, the Chime My Block starts running, as do the blocks on the other sequence. The blocks on the second sequence turn on the Brick Status Light and cycle through the three colors, with a quarter-second pause between each change. The Loop Index modulo 3

CHALLENGE 18-1

For simplicity, the odometer for the *AroundTheBlock* program measures distance in degrees of motor rotation, rather than in inches or centimeters. Determine the conversion factor you need to go from degrees to either inches or centimeters and add a Math block to convert the reading to a more useful measurement.

HINT Look at the discussion for the *ThereAndBack* program in Chapter 4 if you're not sure how to convert from degrees to inches or centimeters.

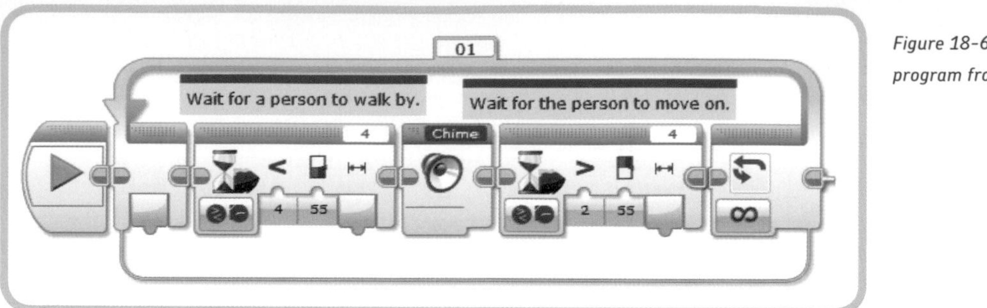

Figure 18-6: The DoorChime *program from Chapter 12*

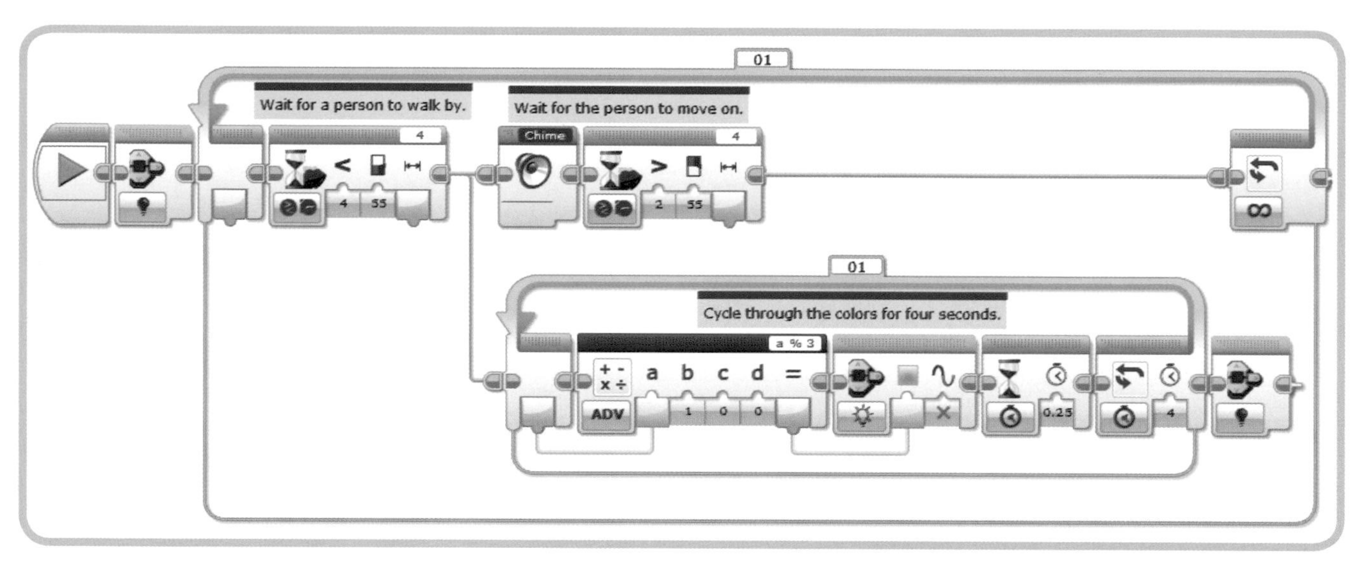

Figure 18-7: Flashing the Brick Status Light while playing the chime

sets the color used by the Brick Status Light block so the value cycles through 0, 1, and 2. The Chime block takes four seconds to play all the notes, so to make the Brick Status Light flash for the same amount of time, the Loop block is set to repeat for four seconds. When the loop completes, the Brick Status Light is turned off.

The thin gray wire that connects the blocks in your program is called a *sequence wire*. It connects the *sequence plug exit* on the right side of one block to the *sequence plug entry* on the left side of another block, as shown in Figure 18-8.

Adding a new sequence is a simple matter of dragging a new sequence wire from the sequence plug exit of one block to the sequence plug entry of another block. Follow these steps to create the new *DoorChime* program.

1. Copy the *DoorChime* program from the *Chapter12* project to the *Chapter18* project.
2. Select the Loop block and drag the handle in the middle of the bottom edge to give you room to add the other sequence (Figure 18-9).

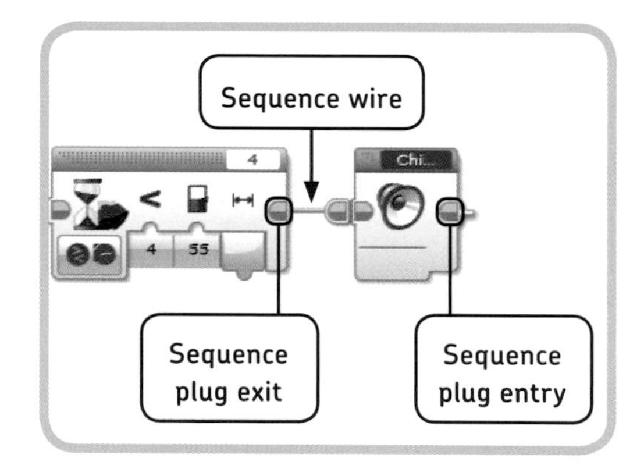

Figure 18-8: The sequence wire and plugs

3. Drag a new Loop block into the existing one and place it below and a little to the right of the Chime My Block (Figure 18-10). The new block appears faded out because it isn't connected to the program yet.
4. Click and drag the sequence plug exit on the right side of the Infrared Sensor block to create a new sequence wire (see Figure 18-11).

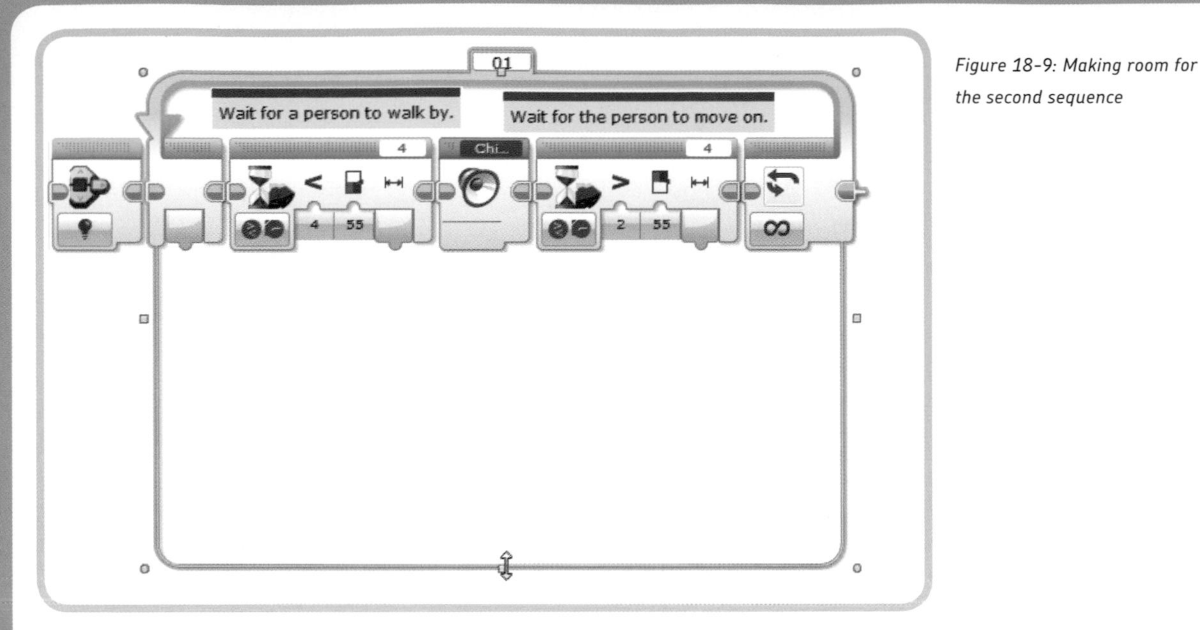

Figure 18-9: Making room for the second sequence

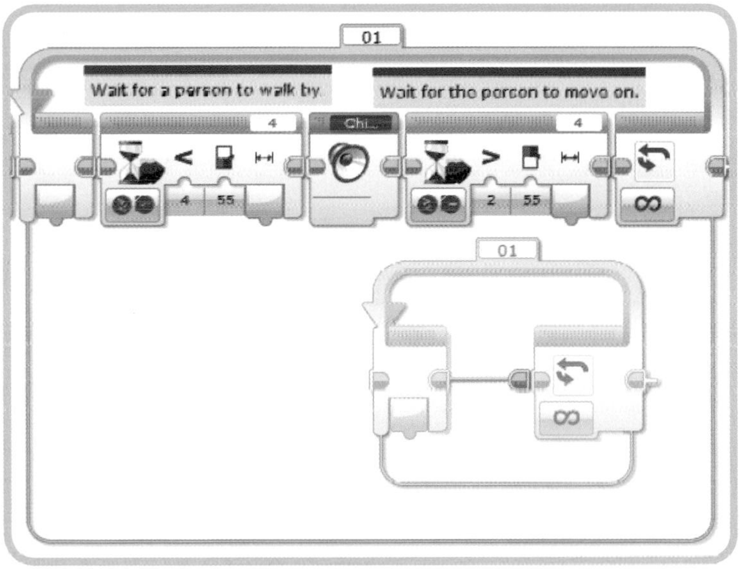

Figure 18-10: Adding the new Loop block

5. Connect the sequence wire to the sequence plug entry on the right side of the new Loop block. The Loop block should no longer look faded (Figure 18-12).
6. With the Loop block in place and connected to the program, add the remaining blocks and data wires to match Figure 18-7.

Now run the program. When someone walks by the TriBot, you should see the Brick Status Light cycle through the three colors while the chime is played.

understanding program flow rules

Using multiple sequences complicates program flow in several ways. For example, you've already seen that a program won't end until it reaches the end of *all* sequences or is ended by the Stop Program block. In this section, I'll discuss some

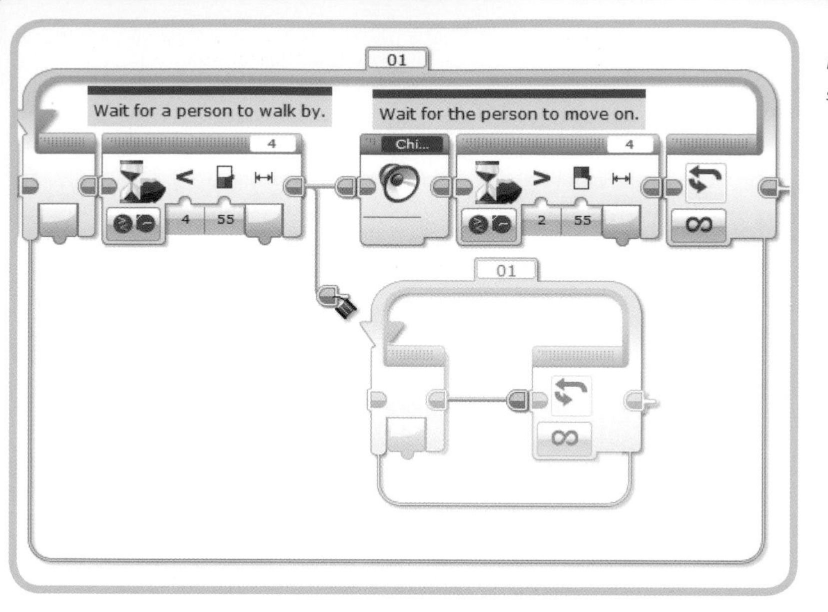

Figure 18-11: Dragging the sequence wire

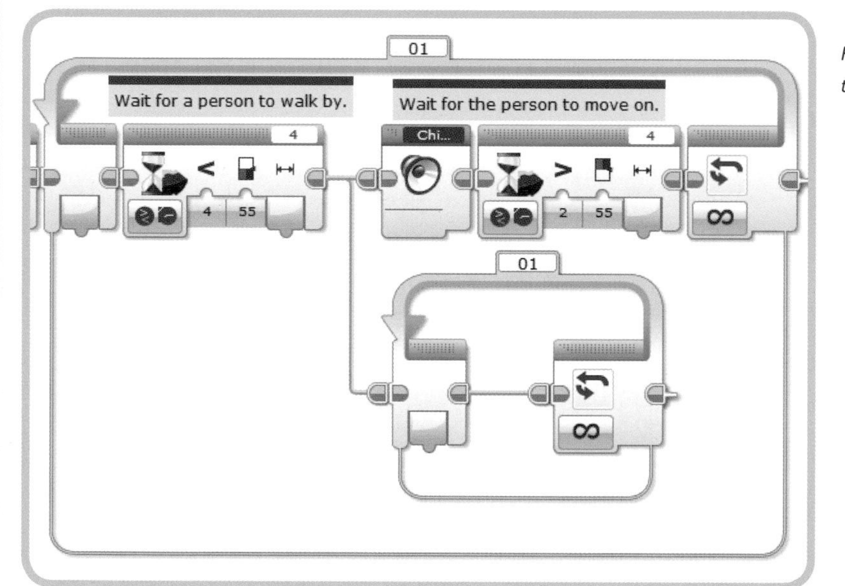

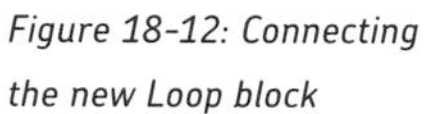

Figure 18-12: Connecting the new Loop block

other program flow rules that are affected by using multiple sequences and show some simple programs that demonstrate these effects.

starting blocks and data wires

A block can only start running after there are values on all the data wires attached to it, as demonstrated by the *BlockStart-Test* program (Figure 18-13). The Display block on the top sequence displays 1 and the Constant block writes 2 to the data wire attached to the bottom Display block. The Display block on the bottom sequence won't start running until after the Constant block puts the 2 on the data wire. When you run this program, the display will show "1", and then after a one-second pause, "2" is added to the display.

Although the normal flow of blocks and data wires is from left to right, there is no left-to-right ordering imposed between blocks on different sequences. The Display block on the lower sequence is to the left of the Constant block but won't run until after the Constant block runs. You can move the Display block to the right, and this would make the operation of the program more visually apparent, but keep in mind that it's the data wire and not the placement of the block that controls when it runs. In your own programs, it's always a good idea to arrange the blocks and sequences to reflect how the program behaves.

Loop and Switch blocks follow the same rule, as demonstrated in the *LoopStartTest* program (see Figure 18-14). The Loop block can't start until after the Constant block puts a value on the data wire, even though this value isn't used until

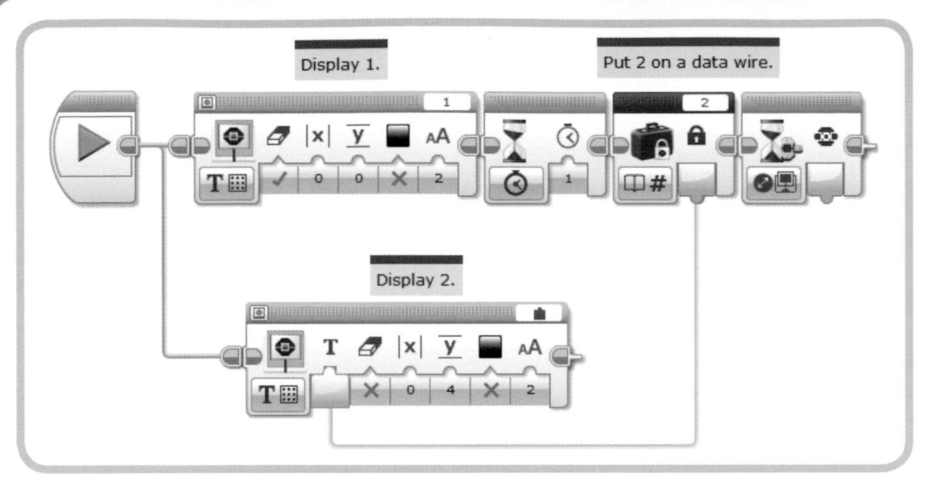

Figure 18-13: The BlockStartTest *program*

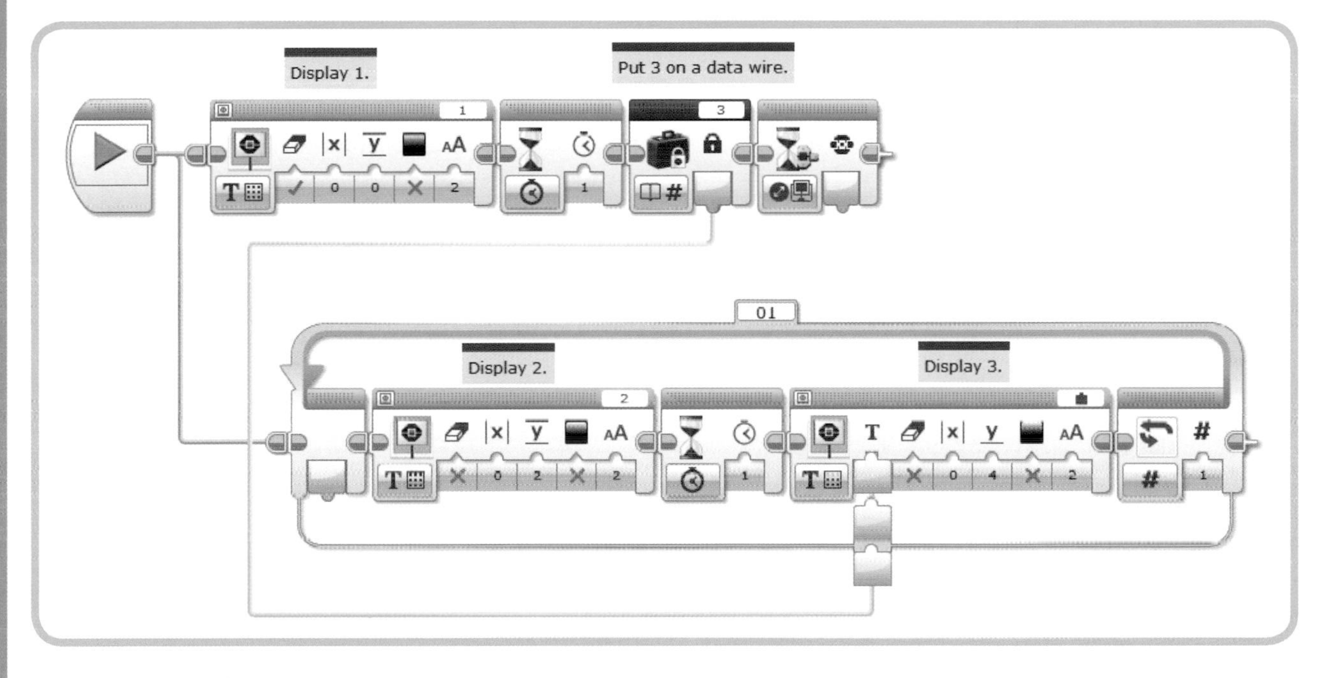

Figure 18-14: The LoopStartTest *program*

the second Display block within the Loop block. When you run this program, it displays "1" (from the Display block on the top sequence) and then pauses for one second. After the Constant block puts a value on the data wire, the Loop block starts. The program displays "2" and then waits for an additional second before displaying "3".

The *BlockStartTest* and *LoopStartTest* programs are deliberately designed to clearly show how the data wire affects the order in which the blocks are run (rather than making the robot do something interesting). Although using data wires to pass values between sequences can be useful, you should be very careful of the dependencies that this imposes. Multiple sequences are most appropriate when the tasks they perform are independent.

using values from a loop or switch block

A data wire that starts inside a Loop block and connects to a block outside the Loop block has a value only when the Loop block finishes. The *LoopCountTest* program in Figure 18-15 shows this rule in action. The Loop block repeats five times, pausing for a total of five seconds. The Display block on the bottom sequence shows the value written to the data wire from the Loop block's Loop Index plug. Because the Display block is outside the Loop block, only the last value (4) is passed to it on the data wire, and only after the Loop block has finished. When you run this program, it pauses for five seconds and then displays "4". This rule also applies to the

Switch block; data wires that leave a Switch block have a value only after the Switch block completes.

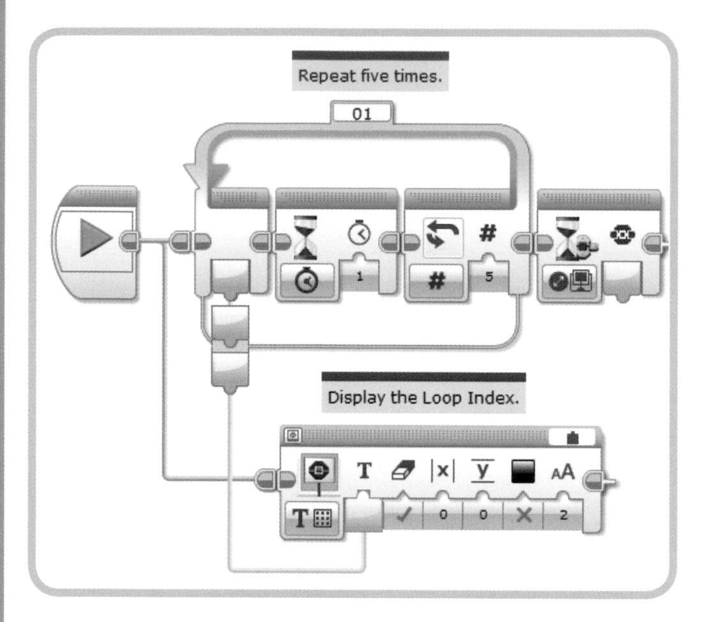

Figure 18-15: The LoopCountTest *program*

using my blocks

Only one copy of a particular My Block can run at the same time, as I'll demonstrate with the DisplayCount My Block (Figure 18-16). This block prints 0, 1, 2, and 3 on the display at one-second intervals, using the row set as an Input parameter.

The *MyBlockTest* program (Figure 18-17) uses two instances of the DisplayCount My Block. When you run this program, it displays "0", "1", "2", and "3", with a one-second pause between each number, on either row 4 or row 0. It then displays the four numbers on the other row. In my tests, the numbers always show up on row 4 first, but that may be a coincidence. In any case, one of the two My Blocks will run to completion before the other one starts.

This rule applies only to My Blocks, so if you replace the two DisplayCount blocks with loops containing Display blocks, the two groups of numbers will be displayed simultaneously. This rule also only applies to two copies of the same My Block. Two DisplayCount blocks won't run at the same time, but a DisplayCount block and a DisplayNumber block will.

NOTE If you really need two copies of the same My Block to run in parallel, create a copy of the My Block and name it something different from the first My Block. You should be able to run the renamed copy and the original at the same time.

This behavior really only becomes apparent with My Blocks that wait for something to happen. Most other My Blocks start and finish quickly enough that you won't be able to observe the rule's effects. For example, you won't notice if two DisplayNumber blocks don't run simultaneously.

synchronizing two sequences

You can control when the second sequence runs based on where you draw the sequence wire. You can also pause the task on one sequence until the task on the other sequence completes by setting a variable on one sequence to be read on the other.

For example, in the *DoorChime* program, the chime and the loop to flash the Brick Status Light both take four seconds. But that time really depends on how the Sound blocks within the Chime block are configured. If you were to add more Chime blocks or change the amount of time that each note plays, the two tasks might not finish together.

Figure 18-18 shows a solution to this problem, using a Logic variable named Done. At the beginning of loop, the variable is set to false. After the sensor detects a person walking by, the Chime block and the loop on the lower sequence start. The loop continues to run until the Chime block completes and the following Variable block sets the Done variable to true. When that happens, the next time the variable is read, the loop exits. It's important to note that the loop doesn't exit immediately when the variable is set, but only after the Variable block reads the value and passes it to the Loop block.

Another alternative that would work for this program is to use a Loop Interrupt block in the top sequence to exit the loop in the bottom sequence, as shown in Figure 18-19. You would just need to make sure to rename the loop on the bottom sequence; otherwise, the Loop Interrupt block will exit the main loop.

keeping out of trouble

Using multiple sequences affects almost every aspect of EV3 programming, including variables, data wires, My Blocks, and program flow. Adding a second sequence allows you to write some incredible programs, but it also increases the number of ways that things can go wrong. Here are some tips to help you avoid the most common problems:

* **Use a second sequence only when it's really necessary.** If possible, find a solution to your problem that requires only one sequence. Don't make your program more complicated than it needs to be!

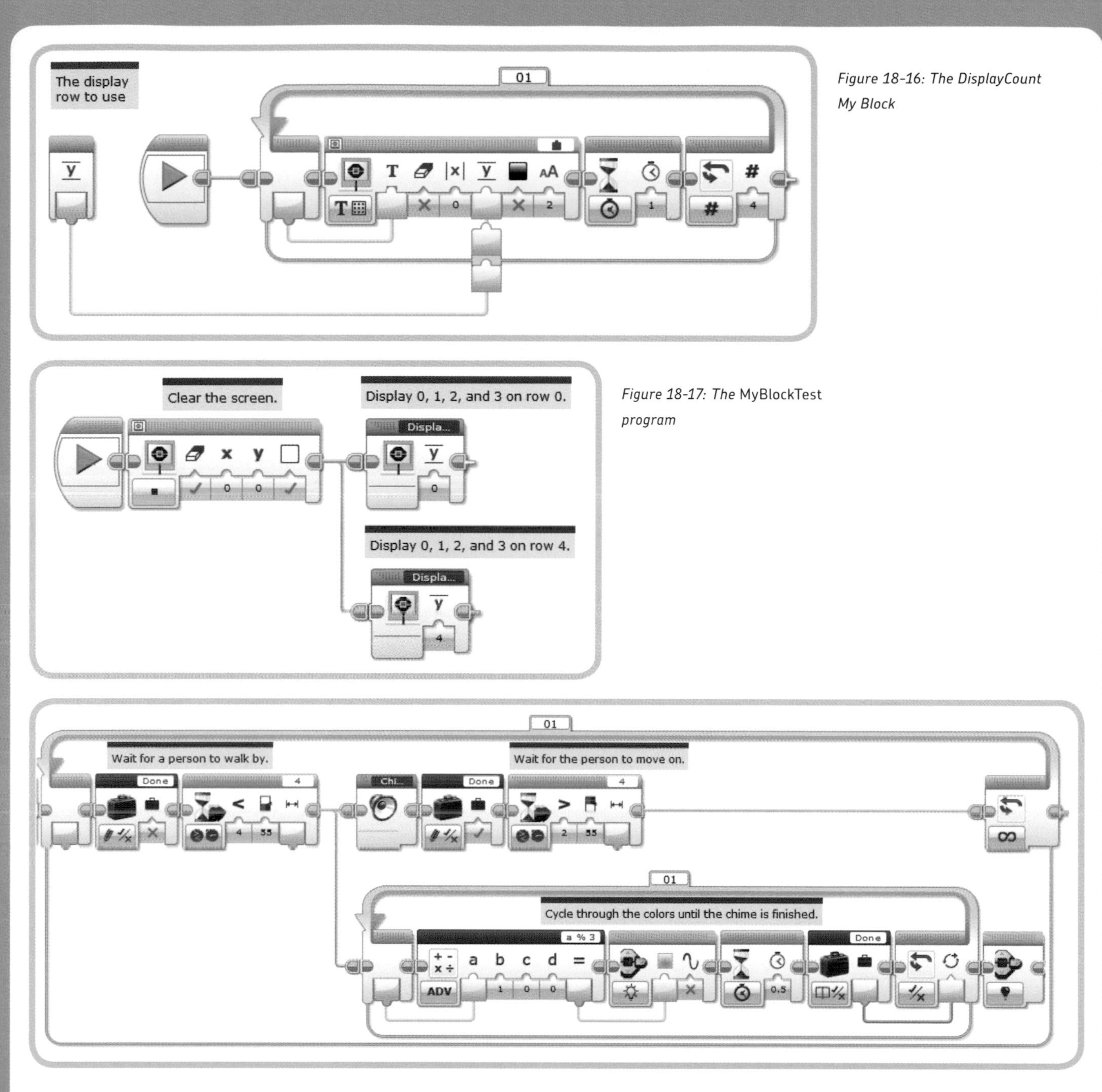

Figure 18-16: The DisplayCount My Block

Figure 18-17: The MyBlockTest *program*

Figure 18-18: Synchronizing the sequences in the DoorChime *program*

* **Edit the program slowly.** The EV3 software tends to get confused much more easily when editing programs with two or more sequences, especially when drawing data wires.
* **Avoid trying to control the same motor or sensor from more than one sequence.** Using any resource (motor, sensor, timer, and so on) from multiple sequences is fraught with peril and very difficult to do correctly. If you want your robot to have different behaviors based on sensor values, consider using nested Switch blocks rather than different sequences.
* **Use variables instead of data wires to pass information between sequences.** This often makes your program easier to understand.
* **Be especially careful with data wires that pass into or out of Loop blocks and Switch blocks.** Reread "Understanding Program Flow Rules" on page 223 if you're not sure what to watch out for.

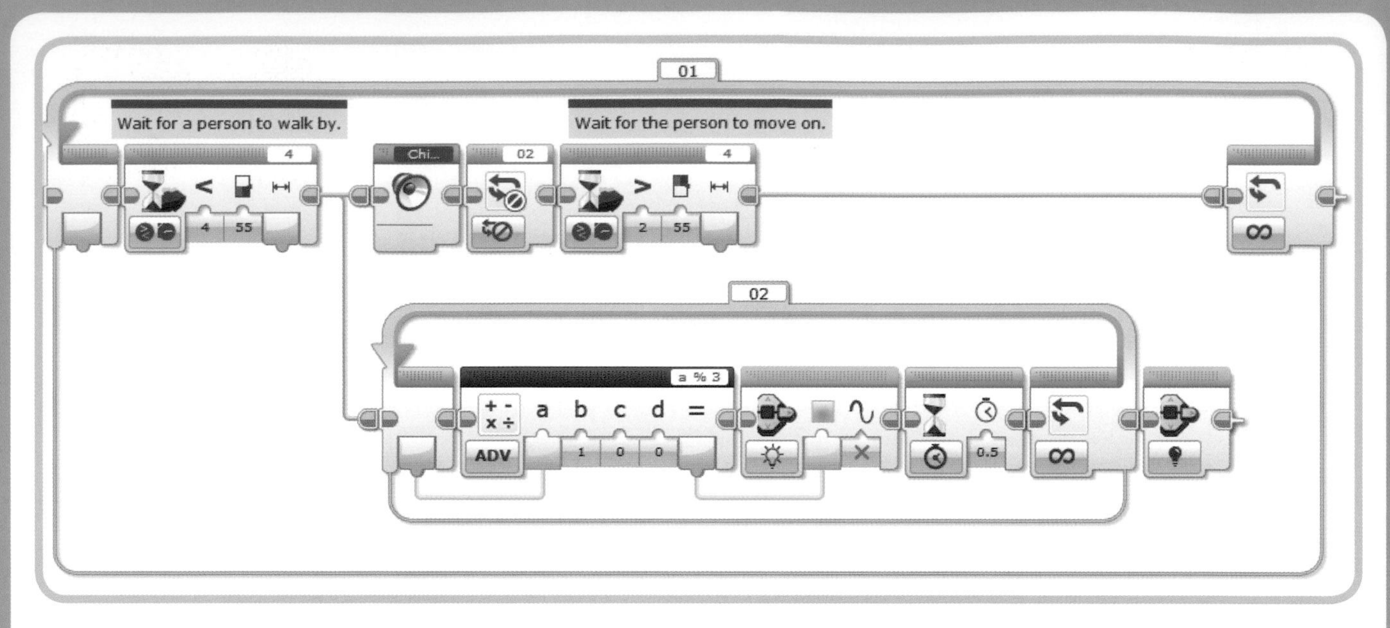

Figure 18-19: Synchronizing the sequences using a Loop Interrupt block

further exploration

Try these activities to learn more about using parallel sequences:

1. The Loop Interrupt block will end any loop with a matching name, even if the loop is on a different sequence. Write a test program that proves this. What happens if the loop is in the middle of running a Move Steering block?

2. Take the *SpiralLineFinder* program from Chapter 10 and rearrange it so that it uses two sequences, one for driving the TriBot in a spiral and the other for detecting the line.

3. Add remote speed control to the *BumperBot* program using the Infrared Remote. First use a variable to hold the Power parameter for the Move Steering block that moves the TriBot forward. Then add a second sequence that uses the buttons on the Remote to adjust the value in the variable.

4. Add a countdown timer to the *MemoryGame* program, using a new sequence in the part of the program that accepts and checks the user's response. As the user is giving a response, the time remaining should be displayed, and if the time reaches zero, the game is over. The amount of time the user has to give the correct response should depend on the number of items in the list; perhaps one second for each item.

conclusion

Using multiple sequences allows your program to perform more than one task simultaneously, a form of multitasking. The changes made in this chapter to the *AroundTheBlock* and *DoorChime* programs demonstrate two simple ways to enhance a program by adding a second sequence.

Although multitasking is a very useful programming technique, it adds complexity to the program flow rules you're familiar with. For this reason, multitasking works best with small, independent tasks.

19

a PID-controlled LineFollower program

Programming your robot to follow a line is an interesting challenge because there are so many possible solutions, ranging from quite simple to highly complex. You've seen a simple three-state controller that works well at slow speeds on a line with gentle curves, and a more complex proportional controller that allows the robot to go faster and make tighter turns. In this chapter, we improve the program further by implementing a full proportional-integral-derivative (PID) control algorithm. Besides adjusting the TriBot's steering according to its current distance from the line (like our proportional controller), this program also calculates a *derivative* value, which detects whether the robot is going along a curve or a straight line by comparing recent measurements, as well as an *integral* value, which detects any drifting by the robot over time.

For the programs in this chapter, mount the Color Sensor on the front of the TriBot as shown in Figure 19-1.

Figure 19-1: Mounting the Color Sensor for line following

The chapter begins with a discussion on line following and the proportional control algorithm to encourage a deeper understanding of how (and why) the *LineFollower* program from Chapter 13 works. Then we'll make small improvements to the existing program so that it uses files and a configuration program to collect and save the sensor limits, and variables to make adjusting the program settings easier. Then I'll show you how to use the more advanced PID control strategy for following a line. By the end of the chapter, you'll have an extremely reliable line-follower program, and you'll understand the principles behind the sophisticated PID control algorithm, which can be used to tackle all sorts of tasks and challenges in robotics.

the PID controller

The *proportional-integral-derivative (PID)* controller is a very common and useful method of controlling all types of machinery, including robots. The ideas behind the PID controller are about 100 years old and are used to control all kinds of mechanisms in everything from ship navigation and printers to musical instruments.

Like the proportional controller introduced in Chapter 13, the PID controller uses a sensor reading, called an Input variable, to adjust a Control variable. For our line follower, the Input variable is the Color Sensor reading, and the Control variable is the Move Steering block's Power parameter. The controller compares the Input variable with a Target value to calculate an Error value. The Error value is then used to determine how the Control variable should be changed.

The proportional controller makes the Steering value proportional to the Error value (the difference between the Color Sensor reading and the Target value). When the TriBot is close to the edge of the line, the Steering value is small, and when it's farther from the edge, the Steering value is large. A proportional controller gives a significant improvement over the previous three-state controller version of the program from

Chapter 9 because it uses more information about how well the robot is following the line. The three-state controller uses the Color Sensor reading to choose among three possible Steering values (turn left, turn right, or go straight). The proportional controller uses the size of the Error value to respond with a much wider range of possible Steering values, which enables the TriBot to do a better job following the line. Although the proportional controller is an improvement over the three-state controller, it still has some significant limitations.

The proportional controller reacts to how far the sensor is from the edge of the line at a given moment, but it doesn't notice how the line changes over time, or whether the robot is gradually drifting in one direction or another. Because it only reacts to the Error value at one instant in time, without any memory of past Error values, it can have problems reacting to sudden changes in the line or accounting for any source of a constant error. A PID controller addresses these factors by adding two additional terms to the expression used to calculate the Steering value: the integral and derivative terms. As you'll see, these two terms use previous Error values to allow the *LineFollower* to go faster, make sharper turns, and follow a straight line more smoothly.

One feature of the PID controller that makes it so popular for controlling automatic systems is that it's a general solution that can be easily adjusted for the particular problem at hand. In this chapter, we'll use a PID controller for a line-following program, but the controller part of the program can be applied to a wide variety of tasks. For example, the controller can be used with the Infrared or Ultrasonic Sensor to make the TriBot follow you at a certain distance, or with the Gyro Sensor to make a robot balance on two wheels. When you understand how a PID controller works and how to tune it for a particular application, you'll be able to use the controller code again and again.

proportional control

Before we start adding PID control to our line-following program, let's take a closer look at how sensor readings change as the robot approaches and goes over the line, and think about how our existing proportional control program reacts to those changes.

The *LightTest* program shown in Figure 19-2 uses the data logging technique from Chapter 17 to collect the reflected light readings as the TriBot moves across the line. This gives us a simple way to record the sensor readings at various distances from the line, and later we'll use a similar program to detect and record the minimum and maximum sensor readings to use for our line follower.

The data is stored in a file named *LightTestData*. The Loop block is configured to run until the motor B Rotation Sensor reads greater than one rotation, which will move the TriBot far enough to collect all the data we need.

Start with the TriBot facing the line, with the Color Sensor about 2 inches (10 cm) away from the line, as in Figure 19-3. Run the program, and the robot should move forward slowly and stop after the Color Sensor passes fully over the line. Ideally, the line should be about halfway between where the Color Sensor was when the program started and where it is when the program completes.

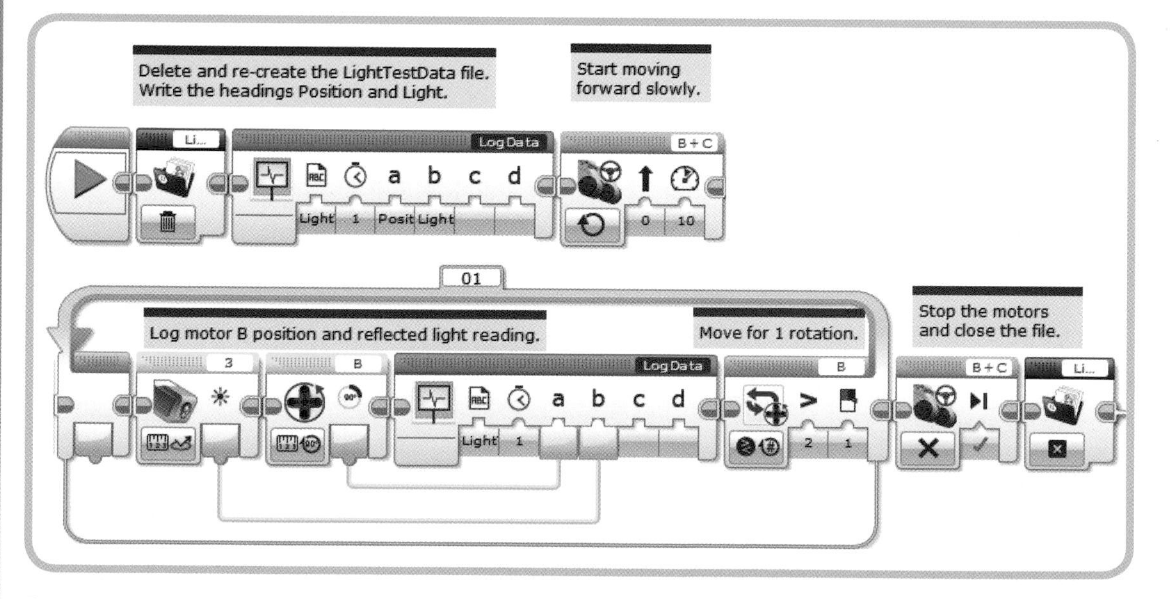

Figure 19-2: The LightTest *program*

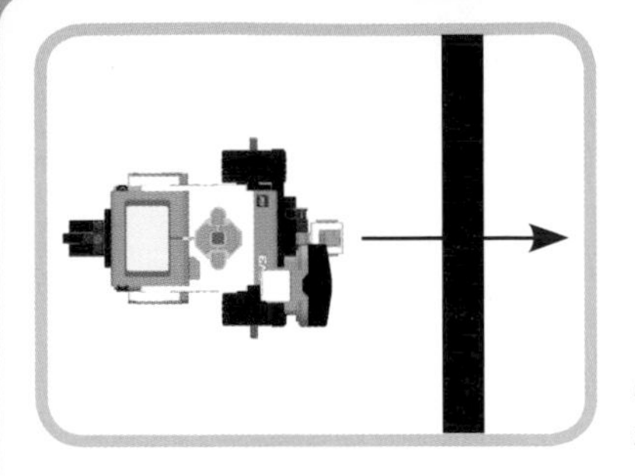

Figure 19-3: Starting position for the LightTest *program*

the raw data

After running the program and uploading the *LightTestData* file to my computer, I created the graph from the data using a spreadsheet program (see Figure 19-4). I also superimposed the position of the line for clarity. When the program starts, the sensor reading is about 62 and stays near 60 as the robot approaches the line. Just before the robot reaches the line, the sensor readings start to decrease and continue to decrease in a straight line until the sensor is fully over the line, at which point the reading is about 5. The reading stays around that value for a small distance, until the sensor starts to approach the other edge of the line. The readings then increase again, in a straight line, until the sensor is past the edge of the line, where the readings again stay around 60. At the very end of the graph, the reading climbs a little to 65.

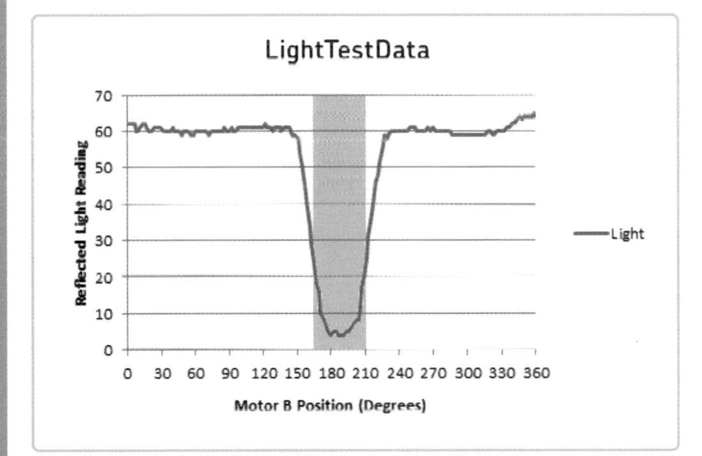

Figure 19-4: The reflected light reading as the TriBot moves over the line

NOTE The values shown here are different from those used in the earlier chapters because I'm using a different test line with a darker background. Other than a slight difference in the background, the test line is the same as shown in Chapter 6: a simple oval made from black electrical tape on a white poster board.

From this graph, we can conclude that the sensor reading will be about 60 when the TriBot isn't close to the line and about 5 when the sensor is directly over the line. When the sensor is near the edge of the line, the reading will be somewhere between 60 and 5, and we can use the reading to judge how far the sensor is from the edge. The nice straight line formed by the readings as they change from 60 to 5 tells us that the reading is proportional to the distance from the edge, which is why we can use those readings as a reliable basis for the proportional line follower.

Notice that the graph is symmetrical, meaning that the left and right halves are mirror images. This symmetry is why we don't try to make the robot follow the center of the line. We can determine if the sensor is at the center of the line (or close to it) by seeing if the value is very close to 5. However, if the reading isn't close to 5, we can't tell which way to move the robot to get it closer to the line. For example, if the sensor reading is 20, we can't tell if the sensor is over the left or right edge of the line.

the good and bad zones

Now let's take a closer look at the sensor readings close to the line and think about how a line-follower program will behave. In Chapter 13, we chose the Target value to use by taking the midpoint of a reading with the sensor off the line, and another reading with the sensor over the middle of the line. As a starting point, we can do the same thing here, using the maximum and minimum values from the *LightTestData* file, which are 65 and 5, respectively. This gives us a midpoint of 35. Figure 19-5 shows a modified view of the data, with a green horizontal line added where the sensor reads 35. To make the following discussion simpler, the x-axis has been shifted and scaled to show the distance in centimeters from where the sensor reads 35.

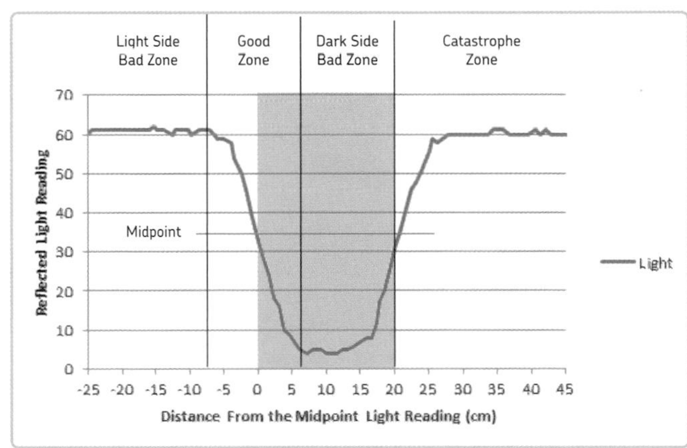

Figure 19-5: Sensor readings near the edge of the line

The graph has been divided into four zones based on where the sensor is relative to the midpoint position. How well the program is able to follow the line depends on which zone the sensor is in.

the good zone

With the midpoint as our target, the program will try to keep the sensor at position 0 on the graph, putting the robot squarely in the middle of the Good Zone, where the sensor reading will be roughly proportional to the distance. As long as the robot stays within about 7 cm from the edge of the line, the sensor reading will give us a good indication of how far the robot is from the target position, and a proportional controller will be able to use the reading to steer the robot toward the edge of the line.

the light side bad zone

On the left side of the graph is the Light Side Bad Zone, which is more than 7 cm from the midpoint position, moving away from the line. When the robot is in this zone, the sensor reading tells us that the robot is too far to the left, but we can't tell how far. The sensor reading is essentially the same whether the robot is 10 or 20 cm away from the edge.

There are three ways the robot can find itself in this area, shown in Figure 19-6. In this figure, the red circle represents the location of the sensor, and the blue line shows the path of the robot as it moves forward.

* When the line curves to the left (Figure 19-6a), the sensor starts crossing the line. If the program overcompensates by using a Steering value that's too large, the robot can move too far to the left.
* When the line curves to the right (Figure 19-6b), the sensor values increase. If the program is too slow to react, using a Steering value that's too small, then it can end up too far to the left before it recovers.
* Lastly, if the gain is too high as the robot follows a straight line, it might have too much side-to-side motion, called *oscillation* (Figure 19-6c); then it can also find itself too far from the line.

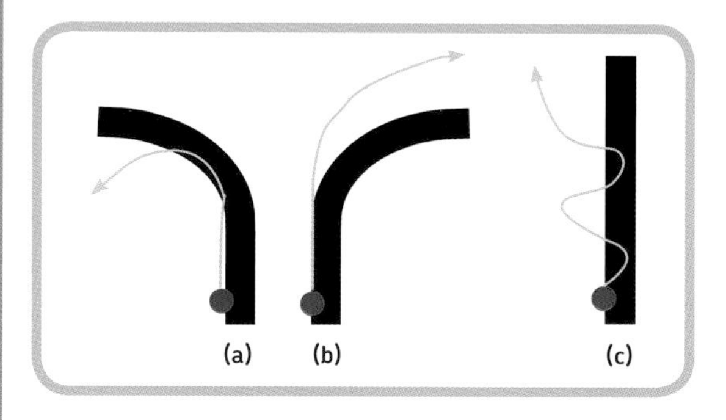

Figure 19-6: Moving into the Light Side Bad Zone

A proportional controller that works well in the Good Zone will likely use Steering values that are very large when the robot is in this Bad Zone, often causing the robot to spin in a circle. If the robot can get close to the line again (in the Good Zone), it can recover and continue following the line. On the other hand, if it gets too far away—for example, if the line makes a sharp U-turn to the right—then the robot will spin in an endless circle.

the dark side bad zone

To the right of the Good Zone is the Dark Side Bad Zone, where the robot is too far over the edge of the line. There are two parts to this area. Between 7 and 12 cm from the midpoint position, the sensor reading is essentially constant. This gives us the same problems as when the robot is in the Light Side Bad Zone: We can't really tell how far the robot is from the edge of the line, and the Steering value the program uses is likely to make the robot spin.

The situation is actually worse when the distance increases beyond 12 cm because the sensor reading starts to increase. This makes the program behave as if the robot is approaching the edge, when it's actually getting farther away.

Figure 19-7 shows how the robot can find itself in this area. The program can overcompensate when the line turns to the right (Figure 19-7a) or not adjust enough when the line turns to the left (Figure 19-7b). Large oscillations in the robot's motion along a straight line can also put the robot in this area (Figure 19-7c).

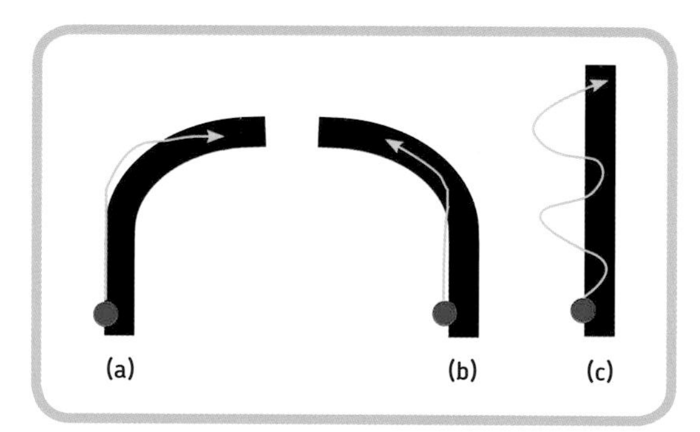

Figure 19-7: Moving into the Dark Side Bad Zone

When the robot is in this area, the program tries to steer it back toward the edge, although it may not steer it quickly enough. If the program can push the robot back to the left, into the Good Zone, then it can recover and continue following the line. If the robot goes too far to the right, it enters the Catastrophe Zone.

the catastrophe zone

The Catastrophe Zone starts at the position where the sensor reading is greater than the midpoint value on the wrong side of the line. When the robot reaches this point, the program begins to steer the robot in the wrong direction, moving to the right in an effort to make the sensor reading go down. At this point, the robot cannot recover: It will spin in a circle, or find the wrong side of the line and start following it in the opposite direction (which can be fun to see!).

These four zones, and how they affect the program, are somewhat specific to the line-follower program. Any program that uses a PID controller will have a set of sensor readings that resemble the Good Zone, where the reading is proportional to the value being controlled. However, what happens outside this zone depends on the specific program. For example, a program that makes the robot follow you can easily be programmed to stop when the distance is 0, or go full speed if it gets too far behind. So this program will only have a Good Zone and a Bad Zone (when it is far away but can't tell how far). On the other hand, a balancing robot will have a Good Zone and two Catastrophe Zones—one on either side—because the robot will fall over if it tips too far forward or too far backward.

selecting the target

Now that you know how the controller works in detail, how does this affect our choice of the Target value? The midpoint between the maximum and minimum sensor readings is reasonable because with this setting the program will try to keep the robot right in the middle of the Good Zone, where the program is well behaved, but perhaps we can do better.

We can make the program more robust (meaning it will fail less often) by increasing the Target value slightly. For example, if we use 40 instead of 35, the robot will travel to the left of the center of the Good Zone (about 2 cm, from Figure 19-5). This change makes it a more likely that the robot will move too far to the left and less likely that it will move too far to the right. This is an improvement because the robot has a better chance of recovering in the Light Side Bad Zone than it does in the Dark Side Bad Zone or Catastrophe Zone. Moving too far to the left is bad, but moving too far to the right is worse, so it makes sense to bias the program a little to the left.

The best Target value depends on your test path. When the robot has to make both left and right turns, using a target that puts the robot slightly to the left of the Good Zone's center works well. However, if your path is an oval and the robot only has to turn in one direction, you can adjust the target to make turning in that direction more reliable.

If you have the robot traveling on the inside of the oval, as I do in my testing, then the robot always makes left turns and when the program fails it's because the robot moves across the line. In this case, setting the target that puts the robot a little more toward the left might work better.

On the other hand, if the robot is traveling on the outside of the oval, then the most common failure happens when the robot doesn't turn quickly enough and moves too far off the line. In this case, setting the target to the center or slightly to the right of the Good Zone might work best.

collecting the min and max sensor readings

Because the Target value will probably need to change for the program to work with different lines, sensors, robot designs, and lighting, we'll create a calibration program called *LineFollowerCal* that collects the Color Sensor's minimum and maximum reading and saves the two values to a file. Then we'll change the *LineFollower* program to read the values from the file and use them to calculate the target.

Like the *LightTest* program, the *LineFollowerCal* program moves the TriBot across the line and monitors the reflected light reading from the Color Sensor. Instead of logging all of the readings, this program just keeps track of the highest and lowest readings. When the robot stops moving, the program displays the two values and gives you a chance to accept or reject them. This both lets you know the limits that will be used by the *LineFollower* program and lets you avoid any problems caused by a bad calibration run (for example, if you had the sensor plugged into the wrong port, or started the robot too close or too far from the line).

The program, shown in Figure 19-8, is long but not very complicated. It starts by initializing the Min and Max variables to 100 and 0, respectively. Next, the wheels start moving slowly forward, and the program enters a loop and starts reading the Color Sensor using Measure - Reflected Light Intensity mode. If the new reading is greater than the current Max value or less than the current Min value, the appropriate variable is updated. Note that the first time through the loop, the reading will (almost always) be less than 100 and greater than 0, so both variables will be updated.

After the TriBot travels forward for one rotation, the loop exits and the motors stop. Two DisplayNumber blocks are used to show the new Min and Max values and then the program waits for you to press a button. If the numbers look reasonable, press the Center button to delete the old *LightFollowerCal* file and write the two values to a new file. If the numbers look wrong—for example, if you didn't place the robot close enough to the line—then press the Left button, and the program will end without saving the values. (There are no blocks on the Left button case of the Switch block.)

Run the program. If everything works, you should see it display the minimum and maximum Color Sensor readings. Press the Center button to save the values to the file. You can use the Memory Browser to make sure the file exists, and run the *FileReader* program from Chapter 16 to make sure the two values you saw displayed really got written to the file.

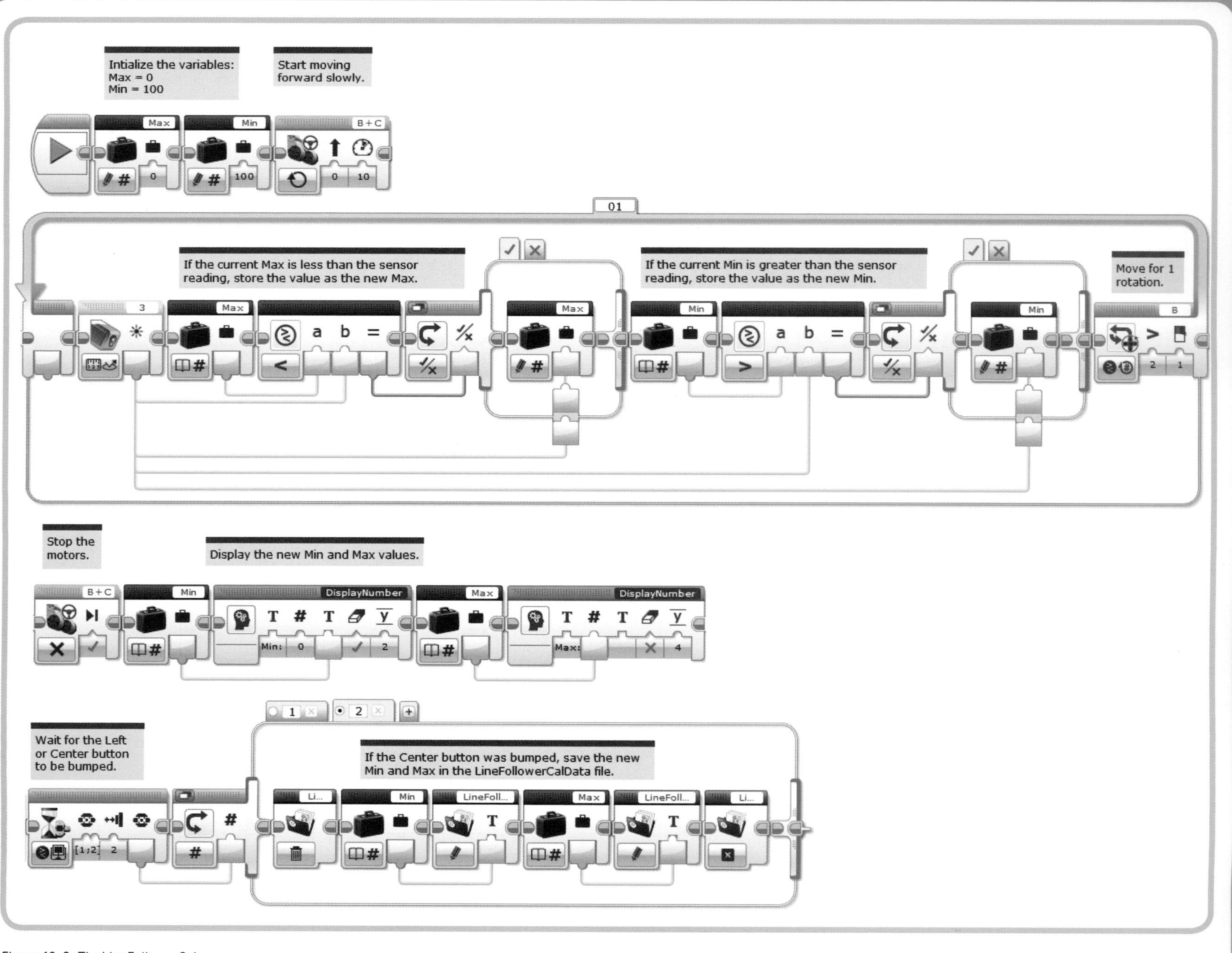

Figure 19-8: The LineFollowerCal *program*

normalizing the sensor reading and target values

The *LineFollowerCal* program collects the minimum and maximum Color Sensor readings so we can calculate the Target value. This helps make the *LineFollower* program adapt to different test lines, but there's one more step we can take to make the program even more adaptable.

Here's the problem: Let's say I run the *LineFollowerCal* program on my test line and get Min and Max values of 5 and 65, and you run the program on your line and get values of 15 and 55. The midpoint of both sets of values is 35, but the program really should react differently for each line because the range of values is different. A sensor reading of 55 on my line means the robot is to the left of the line but still near it, while the same reading on your line means that the robot is completely off the line. The Steering value that the program uses to get the robot back to where it belongs should be different for your line versus my line.

To address this issue, we'll use a process known as *normalizing* the data, which simply means converting the data so that it always uses the same range. Instead of looking at the raw sensor reading, which will be between 5 and 65 for my line and 15 and 55 for your line, we can convert each reading into a percentage of the expected range of values (in other words, we convert each sensor reading to fall in the range 0 to 100). The formula to calculate the normalized sensor reading from the raw value and the Min and Max values is this:

$$\text{Normalized reading} = \frac{100 \times (\text{Sensor reading} - \text{Min})}{(\text{Max} - \text{Min})}$$

Figure 19-9 shows how a Color Sensor block and Math block can be used to calculate the normalized sensor reading. The result from the Math block will be between 0 and 100 and reflects how much light is reflected relative to the range of expected values. A sensor reading of 35 will be normalized to 50 using the ranges from either of our test lines. But a sensor reading of 55 using my test line will be normalized to 83, whereas using your test line will yield 100. This allows the *LineFollower* program to respond appropriately based on the range of expected values, turning more sharply on your test line than it would on mine.

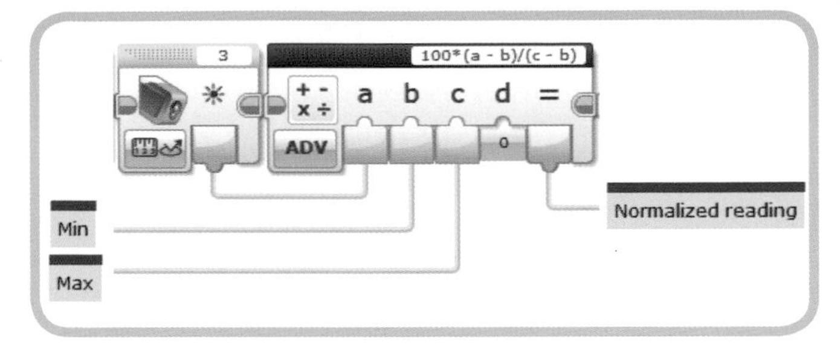

Figure 19-9: Normalizing the Color Sensor reading

When we write the *LineFollower* program, we use the blocks in Figure 19-9 to normalize the sensor reading, which means that we also need to specify the target using the same 0-to-100 range. A target of 50 corresponds to the midpoint of the Min and Max values. As discussed in "Selecting the Target" on page 233, we may want to use a slightly larger value. A raw sensor value of 40, using my Min and Max values of 5 and 65, corresponds to a normalized value of about 60. So in the *LineFollower* program, I'll use 60 for the target.

enhancing the proportional control LineFollower

With all the preliminaries out of the way, we can now add the code to our proportional control line-follower program (building on the version from Chapter 13) so that it uses the values from the *LineFollowerCalData* file and normalizes the sensor and Target values.

I've made a few other smaller changes to make the program easier to understand. Four variables are set at the beginning of the program so that making adjustments and tuning the controller is simpler.

* The Power variable is used to control the robot's speed.
* The Target variable holds the normalized Target value.
* *Kp* is the proportional gain.
* Direction is set to either 1 or -1 depending on whether the sign of the calculated Steering value needs be changed. To follow the left side of the line, this value should be set to -1. Pulling this value out into a variable makes the control part of the code more reusable.

In the Chapter 13 program, the variable holding the proportional gain was named *Gain*. I've changed the name in this program to *Kp* because we will eventually have three gains, one

COLOR SENSOR CALIBRATE MODES

The Color Sensor has built-in support for sensor reading normalization, using the Calibrate modes of the Color Sensor block (Figure 19-10). Set the minimum value you expect to see using Calibrate - Reflected Light Intensity - Minimum mode and the largest value you expect to see using Calibrate - Reflected Light Intensity - Maximum mode, and the sensor will normalize the readings to the given range. The Calibrate - Reflected Light Intensity - Reset mode will return the Color Sensor back to its normal operating condition.

When you use the Calibrate modes to set the sensor's Min and Max, the EV3 remembers these values and continues to use them, even for other programs. The limits are even remembered if you turn off the EV3 and turn it back on again. To clear these calibration values, you must run a Color Sensor block using Calibrate - Reflected Light Intensity - Reset mode. So if you use this feature in any of your programs and it later seems like your Color Sensor has stopped working, try using Reset mode and see if that solves your problem.

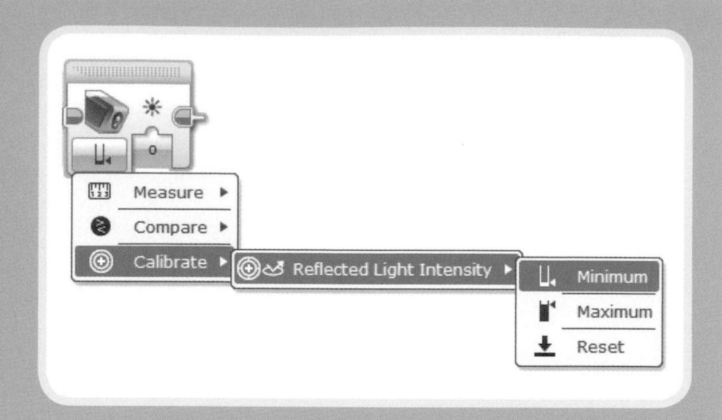

Figure 19-10: The Color Sensor block's Calibrate modes

I chose to perform the normalization step in the *LineFollower* program rather than using the Calibrate modes because I prefer to have the sensor always behave the same way without being affected by sensor calibrations from other programs, even if it means adding an extra Math block in the program. Plus, you'll need to explicitly perform this step in any program that uses a different sensor (the other sensors don't have these Calibrate modes).

for each term (proportional, integral, and derivative). The letter *K* is commonly used in formulas for a constant value, such as we have here, and the *p* is used here because this is the proportional gain.

The first part of the program, which initializes the variables, reads the calibration file, and starts the robot moving forward, is shown in Figure 19-11.

Figure 19-12 shows the main loop of the program, where the sensor is read and the Steering value adjusted. Note that the Error value is stored in a variable instead of being passed straight to the Math block—this will make things easier as the program gets bigger. The Direction value is applied using a separate Math block for the same reason.

NOTE The Error value is stored in a variable and then retrieved solely for clarity; the value could be passed directly to the Math block where it's used. This will make more of a difference in the following sections as the program gets bigger. The Direction value is applied using a separate Math block for the same reason.

Test the program, and tune it to find the right Gain value and Power parameter, just like you did for the version in Chapter 13. I found reasonable performance with a gain of 0.7 at a Power parameter of 40. The new program, combined with the *LineFollowerCal* program, is more easily adaptable to different lines and lighting conditions. However, the program should behave the same as the previous version because we didn't change control algorithm.

implementing PID control

To make the program even more reliable, we're going to turn it from a proportional controller to a full-on PID controller by adding two more terms to the formula we use to calculate the new Steering value. First we'll add the derivative term, which will help the *LineFollower* handle sudden changes in the direction of the line. Then we'll add the integral term, which will account for any constant errors. When you're done, you'll have a robust controller that works great for line following and is easily adaptable for other programs that use a sensor to control the motors.

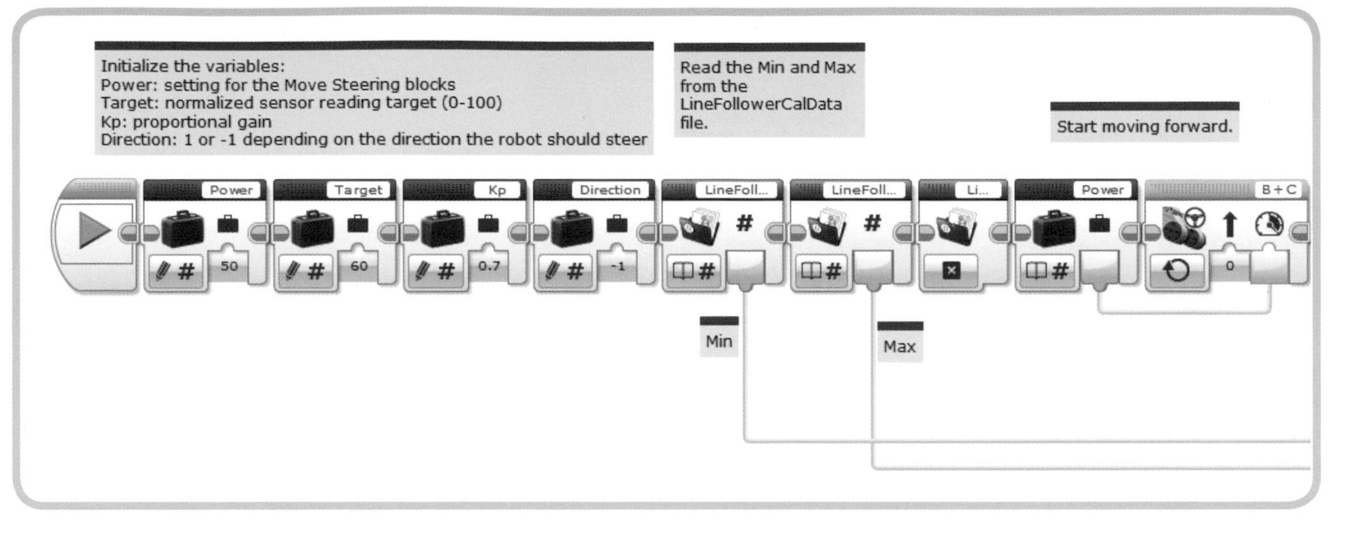

Figure 19-11: The proportional control LineFollower *program, part 1*

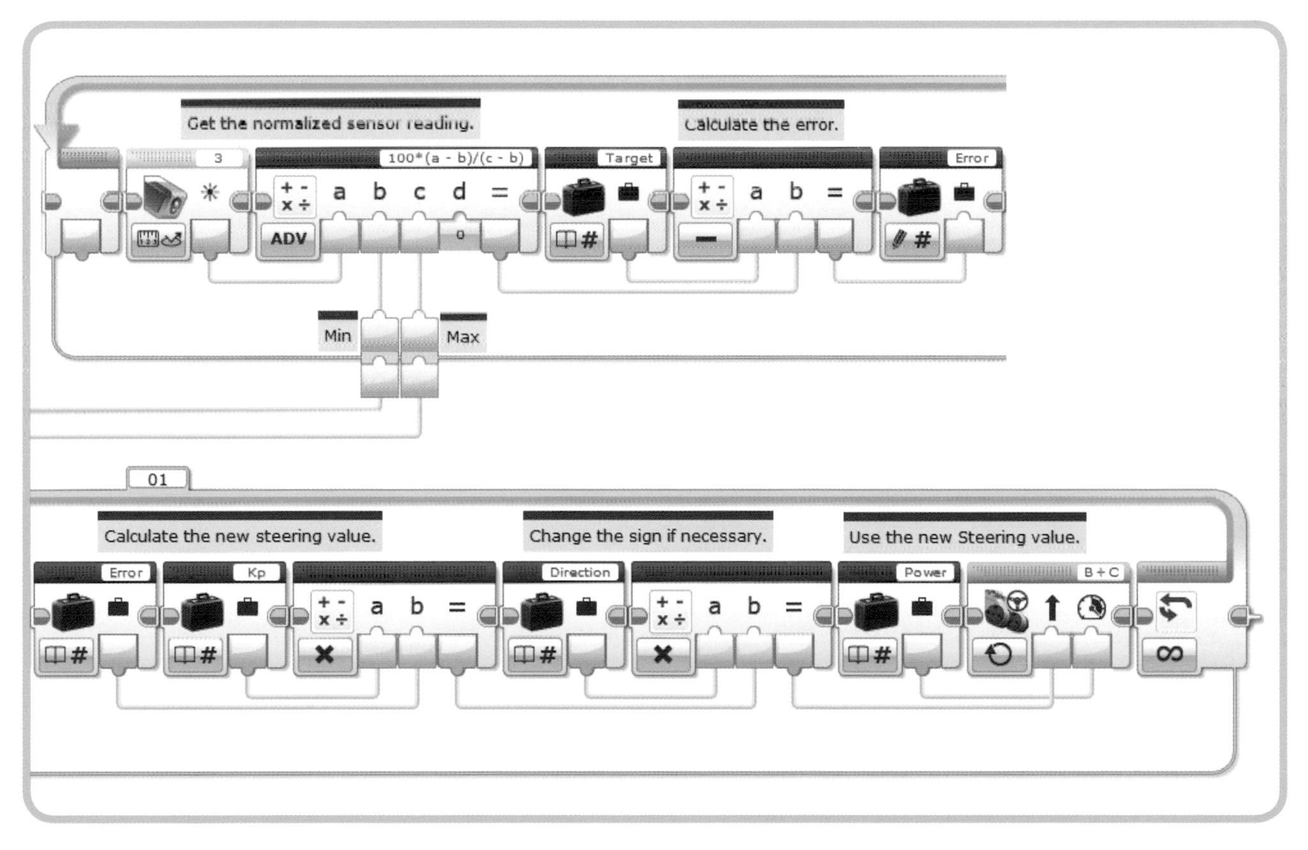

Figure 19-12: The proportional control LineFollower *program, part 2*

adding the derivative term

The proportional controller works very well if the line is straight, but it can struggle with turns if the robot is moving fast or the turn is too sharp. The proportional gain used by the controller determines how much the Steering value changes based on the Error value. The problem is that a gain that works well for a straight line is too small to handle sharp turns, and one that's big enough to handle the turns causes large oscillations when the line is straight.

Conceptually, we want the program to make only small changes when the line is mostly straight and make big changes when the line turns. To accomplish this, we need to add another term to our expression for calculating the new Steering value: the *derivative term*. The derivative measures the change in the Error value. When the line is straight, the derivative will be small because the sensor reading doesn't change much. When the robot gets to a turn, the reading will suddenly start changing by a large amount as the robot moves either away from or over the line.

A simple yet effective way of approximating the derivative of the error is to subtract the previous Error value from the current Error value. To calculate the new Steering value, we'll keep the proportional term (the *Kp* gain times the error) and add the derivative multiplied by another gain factor, called the derivative gain and usually denoted as *Kd*. The formula for the Steering value now becomes this:

Derivative = Error − LastError
Steering = *Kp* × Error + *Kd* × Derivative

When the robot follows a straight line, there won't be much difference between successive Error values, and the derivative will be small or 0, so this term won't affect the robot's motion. When the robot reaches a turn, the difference between successive error readings will grow and the derivative term will add a significant change to the Steering value. This has the effect of giving the robot a larger push in the right direction when it reaches a turn. How big the push is depends on the derivative gain.

To incorporate the derivative into our program, we need two new variables: *Kd* holds the derivative gain, and LastError will hold the previous Error value. *Kd* needs to be set at the beginning of the program to a value you arrive at after some experimentation, and LastError will be initialized to 0.

Figure 19-13 shows the code we'll use to calculate the derivative and save the LastError value. Figure 19-14 shows the derivative term added to the calculation of the new Steering value. We'll use these snippets of code in the final program, but before we can do that, I need to introduce the third term in a PID controller: the integral term.

adding the integral term

The formula we've developed so far has one flaw: It assumes that if the error is 0, then the Steering value is also 0. An Error value of 0 means that the TriBot is in the right spot, so it makes sense to tell the Move Steering block to move straight. But there are several factors that might make it so that a Steering value of 0 doesn't actually make the robot move perfectly straight (for example, if the robot is unbalanced, if there are

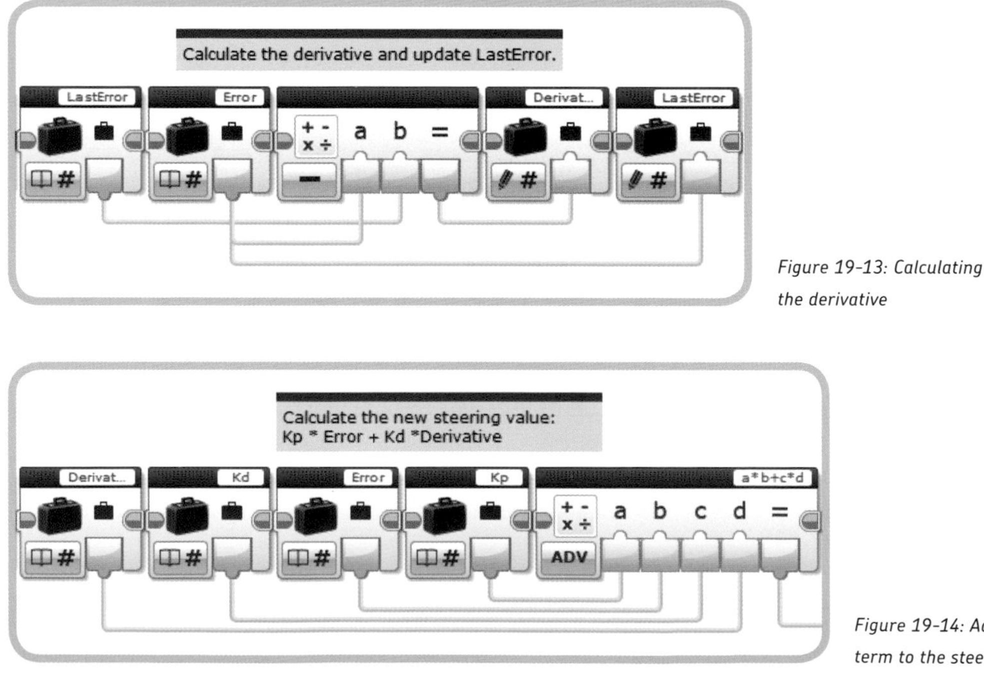

Figure 19-13: Calculating the derivative

Figure 19-14: Adding the derivative term to the steering calculation

small differences in wheel diameters, or if slippage occurs), and if that's the case, there will be a constant error that needs to be adjusted for.

For example, let's say the robot is driving on a slanted surface that makes it pull a little to the left so that it really needs the steering to be set at 2 to go straight. As the robot follows a straight line, it will slowly drift to the left and the error will increase. The change in the error will be small, so the derivative term won't correct for it. The proportional term will change the steering and move the robot gently back toward to the edge of the line. But then the robot will again drift to the left and then be brought gently back. This will continue forever, resulting in small oscillations to one side of the line, as shown in Figure 19-15.

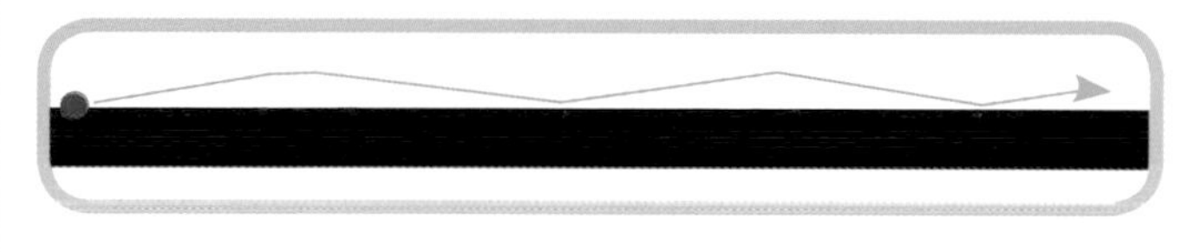

Figure 19-15: Small oscillations caused by a constant error

To fix this, we need to add a third term to our expression: the *integral term*. The integral adds up all the errors the program has seen so far. If everything is balanced, then the sum of all the Error values will add up to 0 because some will be positive and some negative, and they should cancel each other out. If the robot is drifting to one side, as in the example just described, then the integral will be a measure of how much the robot is drifting, and we can use the integral to fix the error.

One way to calculate the integral is to simply add up all the Error values. The problem with this approach is that it treats errors that happened a long time ago the same as recent errors. For example, the integral may become large when the robot moves around a tight corner because it adds up many large errors. Unless the program sees the same total of errors in the other direction, the integral may stay large even if the program starts moving along a straight part of the line where the Error value stays consistently near 0.

The solution to this problem is to shrink the integral each time through the loop before adding the new Error value. This has the effect of removing old Error values over time. In the *LineFollower* program, we'll use this expression to calculate the new integral value from the previous integral value and the error:

New integral = 0.5 × Integral + Error

When the Error values are large, the integral will still add up quickly, and when the Error value is consistently near 0, the integral will eventually become very small. We'll use the code shown in Figure 19-16 to calculate and store the new integral value.

We'll add a third Gain value to our program—*Ki*, the integral gain—and change the steering formula to this:

Steering = *Kp* × Error + *Kd* × Derivative + *Ki* × Integral

Figures 19-17 and 19-18 show the EV3 program. The arrangement of the Math blocks and data wires closely follows the description of the formulas in the text. If you're using a USB connection to the EV3, you can monitor the values on the data wires while the robot is following the line to see how the values change or to find errors you might have made copying the program. When you have the program working, you can shorten it by consolidating some of the Math blocks.

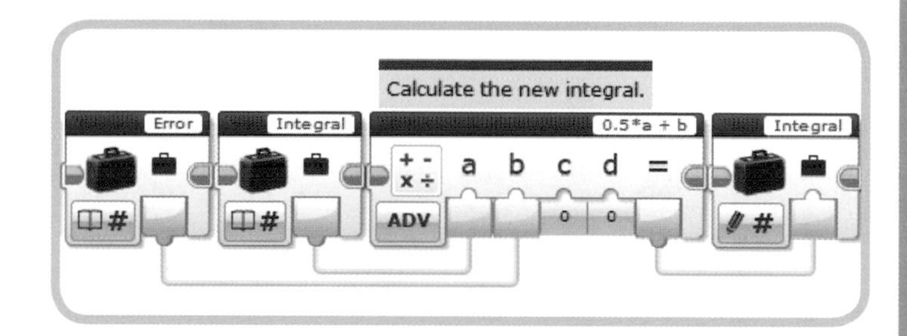

Figure 19-16: Calculating the integral

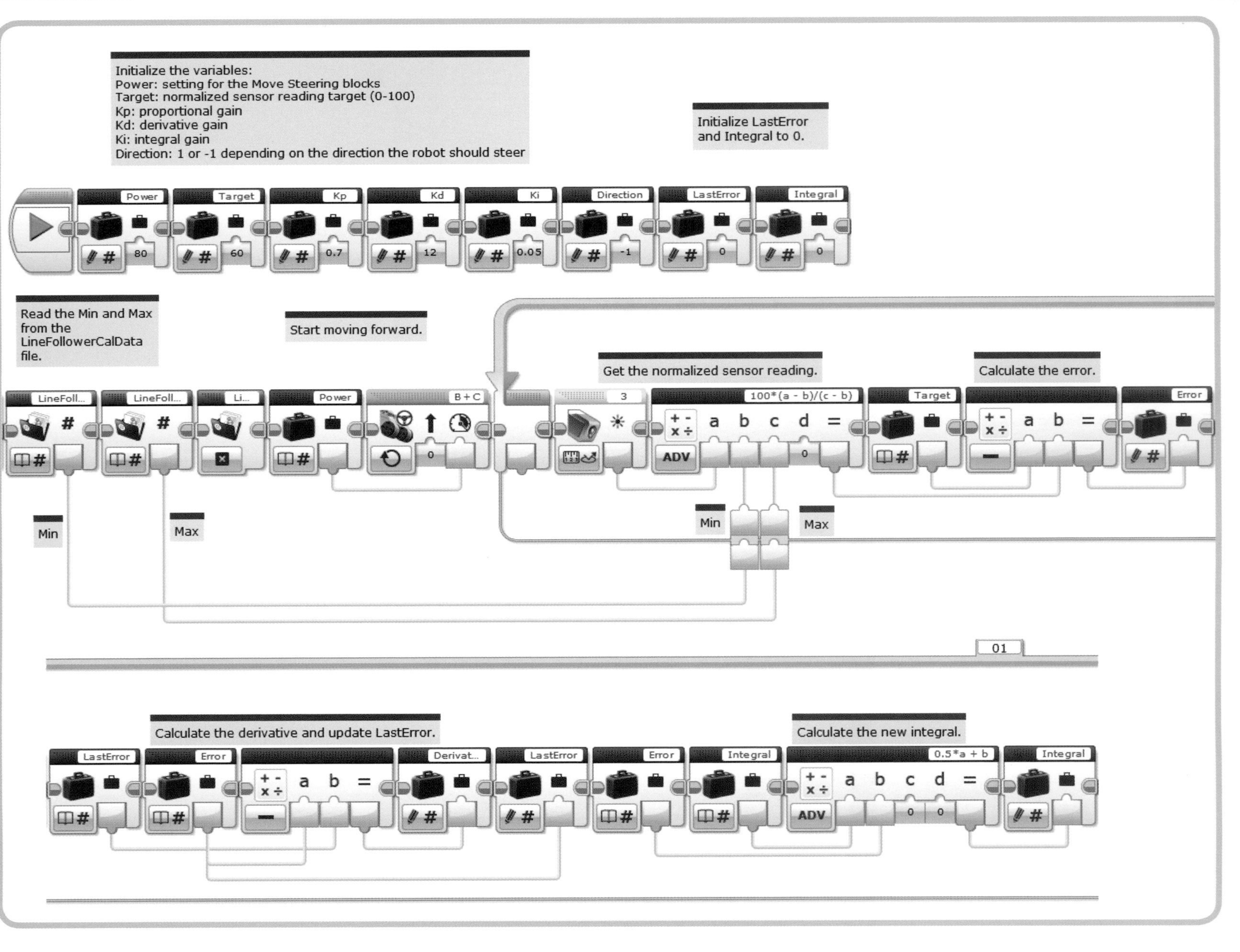

Figure 19-17: The LineFollower *program with PID controller, part 1*

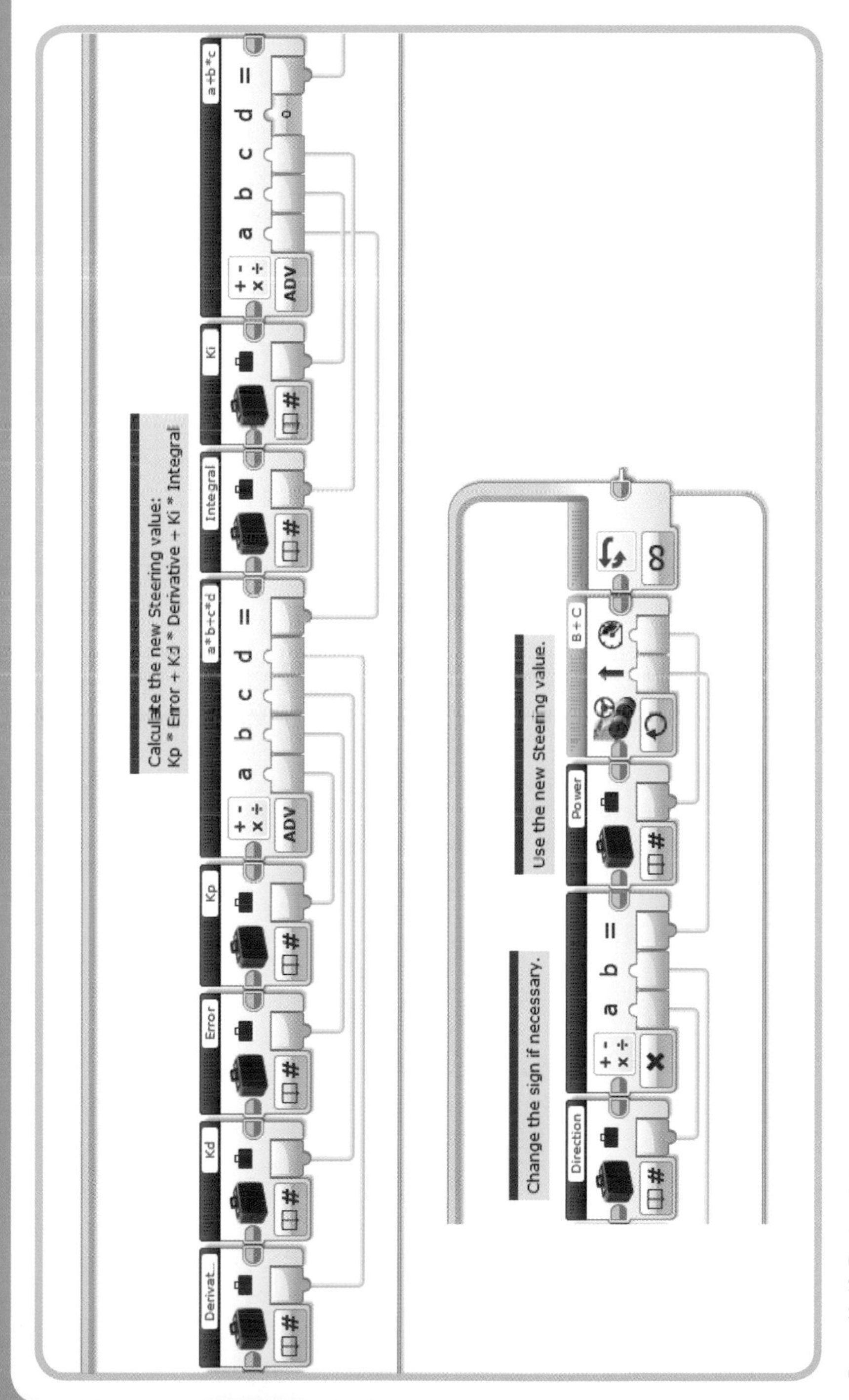

Figure 19-18: The LineFollower program with PID controller, part 2

tuning the controller

One feature that makes a PID controller an attractive solution to many problems is that the algorithm, and consequently the code, doesn't have to change for different conditions or even different applications. Adjusting the code to work for your particular problem often simply requires *tuning* the controller, which means selecting the values for the three gains: *Kp*, *Kd*, and *Ki*. The goal is to set the gains so that the program stays in the Good Zone. The derivative and integral terms won't necessarily help the program cope with being outside the Good Zone, but if these terms are tuned correctly, they can help keep the program from leaving that zone.

Here are some steps you can use to tune your PID-controlled line follower:

1. Set the Power to 50.
2. Start with *Kd* and *Ki* at 0 and *Kp* at 1. With our target at 60, this will make the Steering value change between -60 and 40 as the normalized sensor reading goes between 0 and 100.
3. Start by testing with just a straight line. A *Kp* of 1 is likely too large and will cause noticeable oscillation. Progressively reduce *Kp* by 0.05 until the robot follows a line with no side-to-side movement or only small movement to one side of the edge.
4. Progressively increase *Ki* by 0.01 until the robot follows the edge of a straight line with no oscillation. If the robot does not constantly drift to one side, you may be able to leave *Ki* at 0. Be aware that setting *Ki* too high (above 0.05) will cause the oscillations to grow bigger.
5. Now test the program on a line with curves. Increase the Power variable until the robot is unable to make the turn.
6. Progressively increase *Kd* by 1 until the robot can traverse the entire path.

I was able to set the Power to 80 on my test line with *Kp* = 0.7, *Kd* = 12, and *Ki* = 0.05. For your test setup, you might need to use slightly different values.

For other programs, the gains might be quite different, as they are very dependent on the relationship between the sensor reading and the parameter being controlled (usually either Steering or Power). The relationship between the three gains depends on how the Error values are likely to change. For our line follower, we expect a small error while following a straight line, which gives us a small *Kp*. With a robot that doesn't drift much to one side, there is little or no steady state error, so *Ki* is small or 0. Turns cause large Error values that change quickly, so a relatively large *Kd* is needed to keep the program on track.

further exploration

Try these activities to learn more about line following and the PID controller:

1. Tune the *LineFollower* program for different paths and for going in different directions on the same line (for example, clockwise and counterclockwise around an oval). See if you can find settings that work well in general, and tweak those settings for a little improvement in each situation.
2. Create a My Block from the blocks that make up the PID controller. Use Input parameters for the target, normalized sensor reading, *Kp*, *Kd*, *Ki*, and Direction value, and use an Output parameter for the result.
3. Use the PID Controller My Block in the *WallFollower* program in place of the two-state controller.
4. Create a *RemoteFollower* program that makes the TriBot follow the Infrared Remote. Use the heading to control the steering and the proximity to control the speed. Hint: This requires two PID controllers.

conclusion

Writing a line-following program is a classic robotics exercise that can challenge you to use all your EV3 skills. The *LineFollowerCal* program and the accompanying changes to the *LineFollower* program from Chapter 13 demonstrate how to use a file to store a setting for a program. This lets you avoid hard-coding values, making the program more flexible. You can use this technique for any program where the sensor Target values or other settings may need to be changed.

The final version of the *LineFollower* program uses a PID control algorithm to improve the TriBot's responsiveness to changes in the direction of the line. Using the Color Senor reading and some complex concepts to determine how much the robot should turn allows the robot to both move more quickly and stay closer to the line. Using the readings from sensors to control a robot's motors is a basic part of many programs, and experimenting with different control algorithms is a great way to expand your knowledge of robotics while honing your programming abilities.

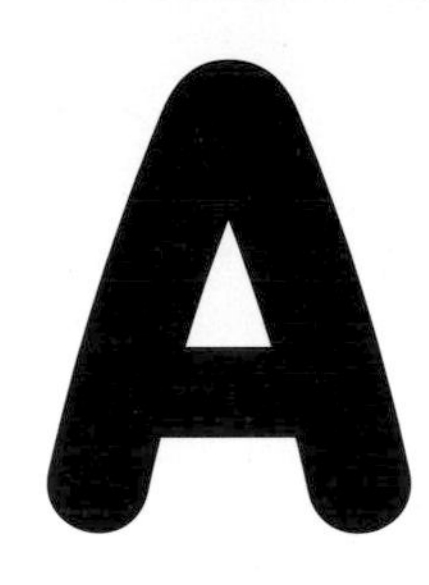

NXT and EV3 compatibility

LEGO has done an admirable job of ensuring that the EV3 Brick and software work well with the existing NXT hardware. In fact, the NXT motors and sensors all work just fine with the EV3 set. So if you have an NXT set, you can use the parts with your EV3 set to expand the range of possible robots you can build. For schools and FLL teams, this backwards compatibility means that the investment made in NXT products isn't lost if you move to the EV3 products.

Using the NXT motors and sensors with the EV3 Brick and software is easy; everything will work as you would expect. However, for the most part you can't go in the other direction; the EV3 sensors won't work with the NXT Brick (although the EV3 motors will), and the NXT software can't be used to program the EV3 Brick. It is possible to use the EV3 software to program the NXT Brick, with some limitations.

motors

The NXT motor and large EV3 motor are very similar, and you can use any motor with either the EV3 or NXT Brick and software. The EV3 medium motor will also work with the NXT Brick and software.

sensors

The NXT sensors all work with the EV3 Brick and software. The EV3 programming blocks are able to work with the sensors from either system; for example, the Color Sensor block works with either the NXT or EV3 Color Sensor.

The EV3 Home Edition software doesn't support the Ultrasonic or Sound Sensor by default. If you want to use either of these sensors, you can download the blocks for them from *http://www.lego.com/en-us/mindstorms/downloads/*. After downloading the block, select the **Tools ▸ My Block Import** menu item to open the Block Import and Export window (Figure A-1). Use the **Browse** button to select the folder to which you downloaded the blocks. Then select the blocks from the list and click the **Import** button. New sensor blocks will appear on the Sensor palette and the options for the Wait, Loop, and Switch blocks will include options for the sensors.

Figure A-1: The Block Import and Export window

The NXT Light Sensor will work with the Color Sensor block (and associated Wait, Loop, and Switch modes) using either Reflected or Ambient Light mode. However, if you select one of the Color modes for the Light Sensor, the ambient light reading will be used and your program will not work as desired.

The only problem with using NXT sensors that I know of is that some older NXT Touch Sensors from the NXT 1.0 set are wired differently and won't work with the EV3 Brick. You can use Port View to determine if your Touch Sensor will work. Make sure the EV3 is connected to the software and attach the sensor. If the sensor shows up in Port View, it will work. If the sensor doesn't appear in Port View, it won't work with the EV3 Brick.

As mentioned earlier, you cannot use the EV3 sensors with the NXT Brick.

software

You cannot use the NXT software to program the EV3 Brick, but you can use the EV3 software to program the NXT Brick, with the following caveats. Although I think that in most cases it's simpler to use the NXT software with the NXT Brick rather than using the EV3 software, one exception is a classroom environment where both NXT and EV3 sets are being used. In this case, it might be easier to have everyone use the same version of the software. Here's how to use the EV3 software with the NXT Brick:

* You must use the USB cable to connect the NXT Brick to your computer. The Bluetooth connection won't work.
* The NXT screen is smaller than the EV3 screen (100x64 versus 178x128) so many images won't display correctly; the bottom and right side may be cut off. If the NXT Brick is connected to the software, the Display block's preview will show the clipped image (Figure A-2).

Figure A-2: The Big smile image file is clipped using an NXT Brick.

* With the EV3 Brick, the Display block starts numbering pixels from the top left corner, as shown in Figure A-3. With the NXT Brick, the numbering starts in the bottom left corner, as shown in Figure A-4. This means that any drawing code needs to be adjusted for the NXT Brick.

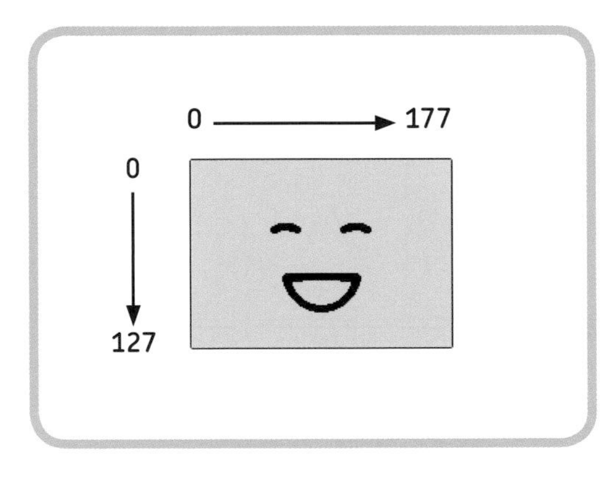

Figure A-3: EV3 pixel numbering

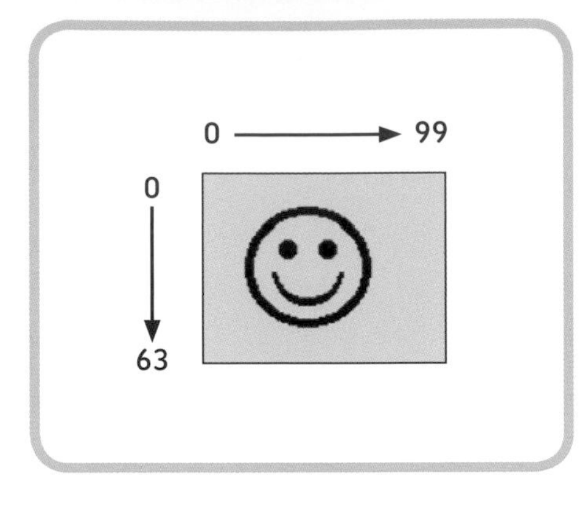

Figure A-4: NXT pixel numbering

* The Display block doesn't support the Font or Color parameters.
* The Brick Status Light, Array Operation, Invert Motor, and Medium Motor blocks are not supported. If you have the NXT Brick connected to the software while using one of these blocks, you'll get a clear warning as shown in Figure A-5.

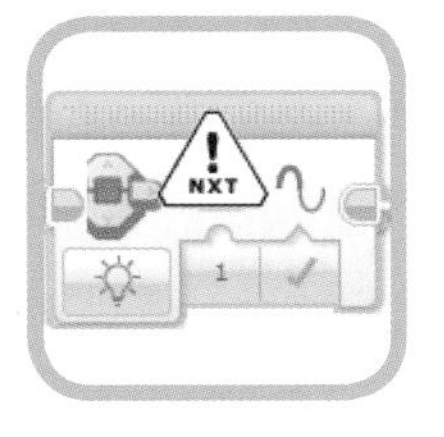

Figure A-5: The Brick Status Light is not supported with the NXT Brick.

* The Math block's Advanced and Exponent modes are not supported.
* The Bluetooth Connection and Messaging blocks are not supported.
* The NXT Light Sensor doesn't work using the Color Sensor block or modes. There is a workaround for this; you need to use the Sound Sensor block and select dB mode to measure reflected light and dBa mode to measure ambient light.

B

EV3 websites

This is a list of websites that contain useful information about EV3 programming. Many of these sites also contain links to other resources that you may be interested in, such as building instructions for EV3 robots and more general robotic topics.

http://mindstorms.lego.com/

The official LEGO MINDSTORMS site contains the latest official EV3 news and support information. This site also hosts a large collection of user-supplied projects.

http://www.thenxtstep.com/

The NXT STEP blog covers news and events of all things related to EV3 robotics.

http://www.mindboards.net/

This site's message forum provides lots of useful information, including the answers to most of the common problems encountered by new EV3 users. This site is frequented by many knowledgeable users who are very generous with their time and is a great place to get your questions answered.

http://forums.usfirst.org/

This site hosts forums for participants in the FIRST LEGO League competition. The programming forums contain a wealth of great information that is useful even if you're not (yet) involved in competitive robotics.

http://www.legoengineering.com/

This site is supported through a partnership between the Tufts University Center for Engineering Education and Outreach (CEEO) and LEGO Education. The focus of this site is the use of LEGO MINDSTORMS to engage students in Science, Technology, Engineering, and Math (STEM).

https://groups.google.com/forum/#!forum/legoengineering

This is a Google group devoted to using robotics, including the EV3 sets, in the classroom.

http://bricks.stackexchange.com/

This is a question-and-answer site for LEGO enthusiasts and is another good option for finding answers to your programming problems and questions.

index

Page numbers shown in italic indicate figures.

A

B

C

D

E

F

G

H

I

L

M

N

O

P

R

S

T

U

V

W

X

The Art of LEGO MINDSTORMS EV3 Programming is set in Chevin. The book was printed and bound by Lake Book Manufacturing in Melrose Park, Illinois. The paper is 70# Orion Satin.

updates and downloads

Visit *http://www.nostarch.com/artofev3programming* for updates, errata, and downloadable versions of the programs.